Annals of Information Systems

D0711788

Series Editors
Ramesh Sharda
Oklahoma State University
Stillwater, OK, USA

Stefan Voß
University of Hamburg
Hamburg, Germany

For further volumes:
http://www.springer.com/series/7573

Robert Stahlbock · Sven F. Crone · Stefan Lessmann

Editors

Data Mining

Special Issue in Annals of Information Systems

 Springer

Editors

Robert Stahlbock
Department of Business Administration
University of Hamburg
Institute of Information Systems
Von-Melle-Park 5
20146 Hamburg
Germany
stahlbock@econ.uni-hamburg.de

Sven F. Crone
Department of Management Science
Lancaster University
Management School
Lancaster
United Kingdom LA1 4YX
sven.f.crone@crone.de

Stefan Lessmann
Department of Business Administration
University of Hamburg
Institute of Information Systems
Von-Melle-Park 5
20146 Hamburg
Germany
lessmann@econ.uni-hamburg.de

ISSN 1934-3221   e-ISSN 1934-3213
ISBN 978-1-4419-1279-4   e-ISBN 978-1-4419-1280-0
DOI 10.1007/978-1-4419-1280-0
Springer New York Dordrecht Heidelberg London

Library of Congress Control Number: 2009910538

Printed on acid-free paper

Springer is part of Springer Science+Business Media (www.springer.com)

Preface

Data mining has experienced an explosion of interest over the last two decades. It has been established as a sound paradigm to derive knowledge from large, heterogeneous streams of data, often using computationally intensive methods. It continues to attract researchers from multiple disciplines, including computer sciences, statistics, operations research, information systems, and management science. Successful applications include domains as diverse as corporate planning, medical decision making, bioinformatics, web mining, text recognition, speech recognition, and image recognition, as well as various corporate planning problems such as customer churn prediction, target selection for direct marketing, and credit scoring. Research in information systems equally reflects this inter- and multidisciplinary approach. Information systems research exceeds the software and hardware systems that support data-intensive applications, analyzing the systems of individuals, data, and all manual or automated activities that process the data and information in a given organization.

The *Annals of Information Systems* devotes a special issue to topics at the intersection of information systems and data mining in order to explore the synergies between information systems and data mining. This issue serves as a follow-up to the *International Conference on Data Mining (DMIN)* which is annually held in conjunction within WORLDCOMP, the largest annual gathering of researchers in computer science, computer engineering, and applied computing. The special issue includes significantly extended versions of prior DMIN submissions as well as contributions without DMIN context.

We would like to thank the members of the DMIN program committee. Their support was essential for the quality of the conferences and for attracting interesting contributions. We wish to express our sincere gratitude and respect toward Hamid R. Arabnia, general chair of all WORLDCOMP conferences, for his excellent and tireless support, organization, and coordination of all WORLDCOMP conferences. Moreover, we would like to thank the two series editors, Ramesh Sharda and Stefan Voß, for their valuable advice, support, and encouragement. We are grateful for the pleasant cooperation with Neil Levine, Carolyn Ford, and Matthew Amboy from Springer and their professional support in publishing this volume. In addition, we would like to thank the reviewers for their time and their thoughtful reviews. Finally, we would like to thank all authors who submitted their work for consideration to this focused issue. Their contributions made this special issue possible.

Hamburg, Germany   Robert Stahlbock
Hamburg, Germany   Stefan Lessmann
Lancaster, UK   Sven F. Crone

Contents

1 Data Mining and Information Systems: Quo Vadis? 1
Robert Stahlbock, Stefan Lessmann, and Sven F. Crone
  1.1   Introduction 1
  1.2   Special Issues in Data Mining 3
        1.2.1   Confirmatory Data Analysis 3
        1.2.2   Knowledge Discovery from Supervised Learning 4
        1.2.3   Classification Analysis 6
        1.2.4   Hybrid Data Mining Procedures 8
        1.2.5   Web Mining 10
        1.2.6   Privacy-Preserving Data Mining 11
  1.3   Conclusion and Outlook 12
  References 13

Part I Confirmatory Data Analysis

**2 Response-Based Segmentation Using Finite Mixture Partial Least
   Squares** 19
Christian M. Ringle, Marko Sarstedt, and Erik A. Mooi
  2.1   Introduction 20
        2.1.1   On the Use of PLS Path Modeling 20
        2.1.2   Problem Statement 22
        2.1.3   Objectives and Organization 23
  2.2   Partial Least Squares Path Modeling 24
  2.3   Finite Mixture Partial Least Squares Segmentation 26
        2.3.1   Foundations 26
        2.3.2   Methodology 28
        2.3.3   Systematic Application of FIMIX-PLS 31
  2.4   Application of FIMIX-PLS 34
        2.4.1   On Measuring Customer Satisfaction 34
        2.4.2   Data and Measures 34
        2.4.3   Data Analysis and Results 36

2.5   Summary and Conclusion 44
References 45

Part II Knowledge Discovery from Supervised Learning

3 Building Acceptable Classification Models 53
David Martens and Bart Baesens
3.1   Introduction 54
3.2   Comprehensibility of Classification Models 55
        3.2.1   Measuring Comprehensibility 57
        3.2.2   Obtaining Comprehensible Classification Models 58
3.3   Justifiability of Classification Models 59
        3.3.1   Taxonomy of Constraints 60
        3.3.2   Monotonicity Constraint 62
        3.3.3   Measuring Justifiability 63
        3.3.4   Obtaining Justifiable Classification Models 68
3.4   Conclusion 70
References 71

**4 Mining Interesting Rules Without Support Requirement: A
   General Universal Existential Upward Closure Property** 75
Yannick Le Bras, Philippe Lenca, and Stéphane Lallich
4.1   Introduction 76
4.2   State of the Art 77
4.3   An Algorithmic Property of Confidence 80
        4.3.1   On UEUC Framework 80
        4.3.2   The UEUC Property 80
        4.3.3   An Efficient Pruning Algorithm 81
        4.3.4   Generalizing the UEUC Property 82
4.4   A Framework for the Study of Measures 84
        4.4.1   Adapted Functions of Measure 84
        4.4.2   Expression of a Set of Measures of $\mathcal{D}_{d_{conf}}$ 87
4.5   Conditions for GUEUC 90
        4.5.1   A Sufficient Condition 90
        4.5.2   A Necessary Condition 91
        4.5.3   Classification of the Measures 92
4.6   Conclusion 94
References 95

5 Classification Techniques and Error Control in Logic Mining 99
Giovanni Felici, Bruno Simeone, and Vincenzo Spinelli
5.1   Introduction 100
5.2   Brief Introduction to Box Clustering 102
5.3   *BC*-Based Classifier 104
5.4   Best Choice of a Box System 108
5.5   Bi-criterion Procedure for *BC*-Based Classifier 111

   5.6   Examples 112
       5.6.1   The Data Sets 112
       5.6.2   Experimental Results with *BC* 113
       5.6.3   Comparison with Decision Trees 115
   5.7   Conclusions 117
   References 117

Part III Classification Analysis

6 An Extended Study of the Discriminant Random Forest 123
Tracy D. Lemmond, Barry Y. Chen, Andrew O. Hatch,
and William G. Hanley

   6.1   Introduction 123
   6.2   Random Forests 124
   6.3   Discriminant Random Forests 125
       6.3.1   Linear Discriminant Analysis 126
       6.3.2   The Discriminant Random Forest Methodology 127
   6.4   DRF and RF: An Empirical Study 128
       6.4.1   Hidden Signal Detection 129
       6.4.2   Radiation Detection 132
       6.4.3   Significance of Empirical Results 136
       6.4.4   Small Samples and Early Stopping 137
       6.4.5   Expected Cost 143
   6.5   Conclusions 143
   References 145

7 Prediction with the SVM Using Test Point Margins 147
Süreyya Özöğür-Akyüz, Zakria Hussain, and John Shawe-Taylor

   7.1   Introduction 147
   7.2   Methods 151
   7.3   Data Set Description 154
   7.4   Results 154
   7.5   Discussion and Future Work 155
   References 157

**8 Effects of Oversampling Versus Cost-Sensitive Learning for
Bayesian and SVM Classifiers** 159
Alexander Liu, Cheryl Martin, Brian La Cour, and Joydeep Ghosh

   8.1   Introduction 159
   8.2   Resampling 161
       8.2.1   Random Oversampling 161
       8.2.2   Generative Oversampling 161
   8.3   Cost-Sensitive Learning 162
   8.4   Related Work 163
   8.5   A Theoretical Analysis of Oversampling Versus Cost-Sensitive
       Learning 164

8.5.1 Bayesian Classification 164
8.5.2 Resampling Versus Cost-Sensitive Learning in
        Bayesian Classifiers 165
8.5.3 Effect of Oversampling on Gaussian Naive Bayes 166
8.5.4 Effects of Oversampling for Multinomial Naive Bayes .. 168
8.6 Empirical Comparison of Resampling and Cost-Sensitive
      Learning.. 170
8.6.1 Explaining Empirical Differences Between Resampling
        and Cost-Sensitive Learning 170
8.6.2 Naive Bayes Comparisons on Low-Dimensional
        Gaussian Data 171
8.6.3 Multinomial Naive Bayes........................... 176
8.6.4 SVMs 178
8.6.5 Discussion 181
8.7 Conclusion... 182
Appendix 183
References 190

9   The Impact of Small Disjuncts on Classifier Learning 193
    Gary M. Weiss
    9.1   Introduction 193
    9.2   An Example: The Vote Data Set 195
    9.3   Description of Experiments 197
    9.4   The Problem with Small Disjuncts 198
    9.5   The Effect of Pruning on Small Disjuncts 202
    9.6   The Effect of Training Set Size on Small Disjuncts 210
    9.7   The Effect of Noise on Small Disjuncts 213
    9.8   The Effect of Class Imbalance on Small Disjuncts 217
    9.9   Related Work.. 220
    9.10 Conclusion... 223
    References 225

Part IV Hybrid Data Mining Procedures

10 Predicting Customer Loyalty Labels in a Large Retail Database: A
    Case Study in Chile... 229
    Cristián J. Figueroa
    10.1 Introduction 229
    10.2 Related Work.. 231
    10.3 Objectives of the Study 233
    10.3.1 Supervised and Unsupervised Learning 234
    10.3.2 Unsupervised Algorithms........................... 234
    10.3.3 Variables for Segmentation 238
    10.3.4 Exploratory Data Analysis 239
    10.3.5 Results of the Segmentation 240
    10.4 Results of the Classifier 241

10.5   Business Validation 244
         10.5.1   In-Store Minutes Charges for Prepaid Cell Phones 245
         10.5.2   Distribution of Products in the Store 246
10.6   Conclusions and Discussion 248
Appendix 250
References 252

11   PCA-Based Time Series Similarity Search 255
Leonidas Karamitopoulos, Georgios Evangelidis, and Dimitris Dervos
11.1   Introduction 256
11.2   Background 258
         11.2.1   Review of PCA 258
         11.2.2   Implications of PCA in Similarity Search 259
         11.2.3   Related Work 261
11.3   Proposed Approach 263
11.4   Experimental Methodology 265
         11.4.1   Data Sets 265
         11.4.2   Evaluation Methods 266
         11.4.3   Rival Measures 267
11.5   Results 268
         11.5.1   1-NN Classification 268
         11.5.2   k-NN Similarity Search 271
         11.5.3   Speeding Up the Calculation of APEdist 272
11.6   Conclusion 274
References 274

**12   Evolutionary Optimization of Least-Squares Support Vector
     Machines** 277
Arjan Gijsberts, Giorgio Metta, and Léon Rothkrantz
12.1   Introduction 278
12.2   Kernel Machines 278
         12.2.1   Least-Squares Support Vector Machines 279
         12.2.2   Kernel Functions 280
12.3   Evolutionary Computation 281
         12.3.1   Genetic Algorithms 281
         12.3.2   Evolution Strategies 282
         12.3.3   Genetic Programming 283
12.4   Related Work 283
         12.4.1   Hyperparameter Optimization 284
         12.4.2   Combined Kernel Functions 284
12.5   Evolutionary Optimization of Kernel Machines 286
         12.5.1   Hyperparameter Optimization 286
         12.5.2   Kernel Construction 287
         12.5.3   Objective Function 288
12.6   Results 289
         12.6.1   Data Sets 289

        12.6.2   Results for Hyperparameter Optimization 290
        12.6.3   Results for EvoKM^{GP} 293
   12.7   Conclusions and Future Work 294
   References 295

13 Genetically Evolved kNN Ensembles 299
   Ulf Johansson, Rikard König, and Lars Niklasson
   13.1   Introduction 299
   13.2   Background and Related Work 301
   13.3   Method 302
        13.3.1   Data sets 305
   13.4   Results 307
   13.5   Conclusions 312
   References 313

Part V Web-Mining

14 Behaviorally Founded Recommendation Algorithm for Browsing
   Assistance Systems 317
   Peter Géczy, Noriaki Izumi, Shotaro Akaho, and Kôiti Hasida
   14.1   Introduction 317
        14.1.1   Related Works 318
        14.1.2   Our Contribution and Approach 319
   14.2   Concept Formalization 319
   14.3   System Design 323
        14.3.1   A Priori Knowledge of Human–System Interactions 323
        14.3.2   Strategic Design Factors 323
        14.3.3   Recommendation Algorithm Derivation 325
   14.4   Practical Evaluation 327
        14.4.1   Intranet Portal 328
        14.4.2   System Evaluation 330
        14.4.3   Practical Implications and Limitations 331
   14.5   Conclusions and Future Work 332
   References 333

15 Using Web Text Mining to Predict Future Events: A Test
   of the Wisdom of Crowds Hypothesis 335
   Scott Ryan and Lutz Hamel
   15.1   Introduction 335
   15.2   Method 337
        15.2.1   Hypotheses and Goals 337
        15.2.2   General Methodology 339
        15.2.3   The 2006 Congressional and Gubernatorial Elections.... 339
        15.2.4   Sporting Events and Reality Television Programs 340
        15.2.5   Movie Box Office Receipts and Music Sales 341
        15.2.6   Replication 342

15.3   Results and Discussion 343

        15.3.1   The 2006 Congressional and Gubernatorial Elections. . . . 343

        15.3.2   Sporting Events and Reality Television Programs 345

        15.3.3   Movie and Music Album Results 347

15.4   Conclusion 348

References 349

Part VI Privacy-Preserving Data Mining

16   Avoiding Attribute Disclosure with the (Extended) p-Sensitive k-Anonymity Model 353

Traian Marius Truta and Alina Campan

16.1   Introduction 353

16.2   Privacy Models and Algorithms 354

        16.2.1   The p-Sensitive k-Anonymity Model and Its Extension . . 354

        16.2.2   Algorithms for the p-Sensitive k-Anonymity Model 357

16.3   Experimental Results 360

        16.3.1   Experiments for p-Sensitive k-Anonymity 360

        16.3.2   Experiments for Extended p-Sensitive k-Anonymity 362

16.4   New Enhanced Models Based on p-Sensitive k-Anonymity 366

        16.4.1   Constrained p-Sensitive k-Anonymity 366

        16.4.2 p-Sensitive k-Anonymity in Social Networks 370

16.5   Conclusions and Future Work 372

References 372

17   Privacy-Preserving Random Kernel Classification of Checkerboard Partitioned Data 375

Olvi L. Mangasarian and Edward W. Wild

17.1   Introduction 375

17.2   Privacy-Preserving Linear Classifier for Checkerboard Partitioned Data 379

17.3   Privacy-Preserving Nonlinear Classifier for Checkerboard Partitioned Data 381

17.4   Computational Results 382

17.5   Conclusion and Outlook 384

References 386

Chapter 1
Data Mining and Information Systems: Quo Vadis?

Robert Stahlbock, Stefan Lessmann, and Sven F. Crone

1.1 Introduction

Information and communication technology has been a steady source of innovations which have considerably impacted the way companies conduct business in the digital as well as the physical world. Today, information systems (IS) holistically support virtually all aspects of corporations and nonprofit institutions, along internal processes from purchasing and operations management toward sales, marketing, and eventually the customer (horizontally along the supply chain), from these operational functions toward finance, accounting, and upper management activities (vertically across the hierarchy) and externally to collaborate with external partners, suppliers, or customers. The holistic support of internal business processes and external relationships by means of IS has, in turn, led to the vast growth of internal and external data being stored and processed within corporate environments.

The progressive gathering of very large and heterogeneous data sets, accompanied by the increasing computational power and evolving database technology, summoned an increasing interest in data mining (DM) as a (novel) tool for discovering knowledge in data. In addition to technological advances, the success of DM – at least in corporate environments – can also be attributed to changes in the business environment. For example, increasing competition through the advent of electronic commerce and the removal barriers for new market entrants, more informed and thus

Robert Stahlbock
Institute of Information Systems, University of Hamburg, Von-Melle-Park 5, D-20146 Hamburg, Germany; Lecturer at the FOM University of Applied Sciences, Essen/Hamburg, Germany, e-mail: `stahlbock@econ.uni-hamburg.de`

Stefan Lessmann
Institute of Information Systems, University of Hamburg, Von-Melle-Park 5, D-20146 Hamburg, Germany, e-mail: `lessmann@econ.uni-hamburg.de`

Sven F. Crone
Centre for Forecasting, Lancaster University Management School, Lancaster LA1 4YX, UK, e-mail: `sven.f.crone@crone.de`

R. Stahlbock et al. (eds.), *Data Mining,* Annals of Information Systems 8,
DOI 10.1007/978-1-4419-1280-0_1, © Springer Science+Business Media, LLC 2010

1

demanding customers as well as increasing saturation in many markets created a need for enhanced insight, understanding, and actionable plans that allow companies to systematically manage and deepen customer relationships (e.g., insurance companies identifying those individuals most likely to purchase additional policies, retailers seeking those customers most likely to respond to marketing activities, or banks determining the creditworthiness of new customers). The corresponding developments in the areas of corporate data warehousing, computer-aided planning, and decision support systems constitute some of the major topics in the discipline of IS.

As deriving knowledge from data has historically been a statistical endeavor [22], it is not surprising that size of data sets is emphasized as a constituting factor in many definitions of DM (see, e.g., [3, 7, 20, 24]). In particular, traditional tools for data analysis had not been designed to cope with vast amounts of data. Therefore, the size and structure of the data sets naturally determined the topics that emerged first, and early activities in DM research concentrated mainly on the development and advancement of highly scalable algorithms. Given this emphasis on methodological issues, many contributions to the advancement of DM were made by statistics, computer science, and machine learning, as well as database technologies. Examples include the well-known Apriori algorithm for mining associations and identifying frequent itemsets [1] and its many successors, procedures for solving clustering, regression, and time series problems, as well as paradigms like ensemble learning and kernel machines (see [52] for a recent survey regarding the top-10 DM methods). It is important to note that data set size does refer not only to the number of examples in a sample but also to the number of attributes being measured per case. Particularly applications in the medical sciences and the field of information retrieval naturally produce an extremely large number of measurements per case, and thus very high-dimensional data sets. Consequently, algorithms and induction principles were needed which overcome the curse of dimensionality (see, e.g., [25]) and facilitate processing data sets with many thousands of attributes, as well as data sets with a large number of instances at the same time. As an example, without the advancements in statistical learning [45–47], many applications like the analysis of gene expression data (see, e.g., [19]) or text classification (see, e.g., [27, 28]) would not have been possible. The particular impact of related disciplines – and efforts to develop DM as a discipline in its own right – may also be seen in the development of a distinct vocabulary within similar taxonomies; DM techniques are routinely categorized according to their primary objective into predictive and descriptive approaches (see, e.g., [10]), which mirror the established distinction of supervised and unsupervised methods in machine learning. We are not in a position to argue whether DM has become a discipline in its own right (see, e.g., the contributions by Hand [22, 21]). At least, DM is an interdisciplinary field with a vast and nonexclusive list of contributors (although many contributors to the field may not consider themselves "data miners" at all, and perceive their developments solely within the frame of their own established discipline).

The discipline of IS however, it seems, has failed to leave its mark and make substantial contributions to DM, despite its apparent relevance in the analytical support of corporate decisions. In accordance with the continuing growth of data, we are

able to observe an ever-increasing interest in corporate DM as an approach to analyze large and heterogeneous data sets for identifying hidden patterns and relationships, and eventually discerning actionable knowledge. Today, DM is ubiquitous
and has even captured the attention of mainstream literature through best sellers
(e.g., [2]) that thrive as much on the popularity of DM as on the potential knowledge one can obtain from conventional statistical data analysis. However, DM has
remained focused on methodological topics that have captured the attention of the
technical disciplines contributing to it and selected applications, routinely neglect-
ing the decision context of the application or areas of potential research, such as
the use of company internal data for DM activities. It appears that the DM community has primarily developed independently without any significant contributions
from IS. The discipline of IS continues to serve as a mediator between management
and computer science, driving the management of information at the interface of
technological aspects and business decision making. While original contributions
on methods, algorithms, and underlying database structure may rightfully develop
elsewhere, IS can make substantial contributions in bringing together the managerial
decision context and the technology at hand, bridging the gap between real-world
applications and algorithmic theory.

Based on the framework provided in this brief review, this special issue seeks
to explore the opportunities for innovative contributions at the interface of IS with
DM. The chapters contained in this special issue embrace many of the facets of
DM as well as challenging real-world applications, which, in turn, may motivate
and facilitate the development of novel algorithms – or enhancements to established
ones – in order to effectively address task-specific requirements. The special issue
is organized into six sections in order to position the original research contributions
within the field of DM it aims to contribute to: confirmatory data analysis (one chapter), knowledge discovery from supervised learning (three chapters), classification
analysis (four chapters), hybrid DM procedures (four chapters), web mining (two
chapters), and privacy-preserving DM (two chapters). We hope that the academic
community as well as practitioners in the industry will find the 16 chapters of this
volume interesting, informative, and useful.

1.2 Special Issues in Data Mining

1.2.1 Confirmatory Data Analysis

In their seminal paper, Fayyad et al. [10] made a clear and concise distinction between DM and the encompassing process of knowledge discovery in data (KDD),
whereas these terms are mainly used interchangeably in contemporary work. Still,
the general objective of identifying novel, relevant, and actionable patterns in data
(i.e., knowledge discovery) is emphasized in many, if not all, formal definitions of

DM. In contrast, techniques for confirmatory data analysis (that emphasize the reliable confirmation of preconceived ideas rather than the discovery of new ones) have received much less attention in DM and are rarely considered within the adjacent communities of machine learning and computer science. However, techniques such as structural equation modeling (SEM) that are employed to verify a theoretical model of cause and effect enjoy ongoing popularity not only in statistics and econometrics but also in marketing and information systems (with the most popular models being LISREL and AMOS). The most renowned example in this context is possibly the application of partial least squares (PLS) path modeling in Davis' famous technology acceptance model [9]. However, earlier applications of causal modeling predominantly employed relatively small data sets which were often collected from surveys.

Recently, the rapid and continuing growth of data storage paired with internet-based technologies to easily collect user information online facilitates the use of significantly larger volumes of data for SEM purposes. Since the underlying principles for induction and estimation of SEM are similar to those encountered in other DM applications, it is desirable to investigate the potential of DM techniques to aid SEM in more detail. In this sense, the work of Ringle et al. [41] serves as a first step to increase the awareness of SEM within the DM community. Ringle et al. introduce finite-mixture PLS as a state-of-the-art approach toward SEM and demonstrate its potential to overcome many of the limitations of ordinary PLS. The particular merit of their approach originates from the fact that the possible existence of subgroups within a data set is automatically taken into account by means of a latent class segmentation approach. Data clusters are formed, which are subsequently examined independently in order to avoid an estimation bias because of heterogeneity. This approach differs from conventional clustering techniques and exploits the hypothesized relationships within the causal model instead of finding segments by optimizing some distance measure of, e.g., intercluster heterogeneity. The possibility to incorporate ideas from neural networks or fuzzy clustering into this segmentation step has so far been largely unexplored and therefore represents a promising route toward future research at the interface of DM and confirmatory data analysis.

1.2.2 Knowledge Discovery from Supervised Learning

The preeminent objective of DM – discovering novel and useful knowledge from data – is most naturally embodied in the unsupervised DM techniques and their corresponding algorithms for identifying frequent itemsets and clusters. In contrast, contributions in the field of supervised learning commonly emphasize principles and algorithms for constructing predictive models, e.g., for classification or regression, where the quality of a model is assessed in terms of predictive accuracy. However, a predictive model may also fulfill objectives concerned with "knowledge discovery" in a wider sense, if the model's underlying rules (i.e., the relationships discerned from data) are made interpretable and understandable to human decision makers.

Whereas a vast assortment of valid and reliable statistical indicators has been developed for assessing the accuracy of regression and classification models, an objective measurement of model comprehensibility remains elusive, and its justification a nontrivial undertaking. Martens and Baesens [36] review research activities to conceptualize comprehensibility and further extend these ideas by proposing a general framework for acceptable prediction models. Acceptability requires a third constraint besides accuracy and comprehensibility to be met. That is, a model must also be in line with domain knowledge, i.e., the user's belief. Martens and Baesens refer to such accordance as justifiability and propose techniques to measure this concept.

The interpretability of DM procedures, and classification models in particular, is also taken up by Le Bras et al. [31]. They focus on rule-based classifiers, which are commonly credited for being (relatively easily) comprehensible. However, their analysis emphasizes yet another important property that a prediction model has to fulfill in the context of knowledge discovery: its results (i.e., rules) have to be interesting. In this sense, the concept of interestingness complements Martens and Baesens [36] considerations on adequate and acceptable models. And although issues of measuring interestingness have enjoyed more attention in the past (see, e.g., Freitas [14], Liu et al. [34], and the recent survey by Geng and Hamilton [17]), designing respective measures remains as challenging as in the case of comprehensibility and justifiability. Drawing on the wealth of well-developed approaches in the field of association rule mining, Le Bras et al. consider so-called associative classifiers which consist of association rules whose consequent part is a class label. Two key statistics in association rule mining are support and confidence, which measure the number of cases (i.e., the database transactions in association rule mining) that contain a rule's antecedent and consequent parts and the number of cases that contain the consequent part among those containing the antecedent part, respectively. In that sense, support and confidence may be interpreted as measures of a rule's interestingness. In addition, these figures are of pivotal importance for the task of developing efficient rule induction algorithms. For the case of associative classification, it has been shown that the confidence measure possesses the so-called universal existential upward closure property, which facilitates a fast top-down derivation of classification rules. Le Bras et al. generalize this measure and provide necessary and sufficient conditions for the existence of this property. Furthermore, they demonstrate that several alternative measures of rule interestingness also exhibit general universal existential upward closure. This is important because the suitability of interestingness measures depends upon the specific requirements of an application domain. Therefore, the contribution of Le Bras et al. will allow users to select from a broad range of measures of a rule's interestingness, and to develop tailor-made ones, while maintaining the efficiency and feasibility of a rule mining algorithm.

The field of logic mining represents a special form of classification rule mining in the sense that the resulting models are expressed as logic formulas. As this type of model representation may again be seen as particularly easy to interpret, logic mining techniques represent an interesting candidate for knowledge discovery in general, and for resolving classification problems that require comprehensible models in particular. A respective approach, namely the box-clustering technique, is considered by Felici et al. [11]. Box clustering offers the advantage that preprocessing activities to transform a data set into a logical form, as required by any logic mining technique, are performed implicitly. Although logic mining in general and box clustering in particular are appealing due to their inherent model comprehensibility, they also suffer from an important limitation: algorithms to construct a model from empirical data are less developed than for alternative classifiers. In particular, methodologies and best practices for avoiding the well-known problem of overfitting are very mature in the case of, e.g., support vector machines (SVMs) or artificial neural networks (ANNs). On the contrary, overfitting remains a key challenge in box clustering. To overcome this problem, Felici et al. propose a bi-criterion procedure to select the best box-clustering solution for a given classification problem and balance the two goals of having a predictive and at the same time simple model. Therefore, these procedures can be seen as an approach to implement the principles of statistical learning theory [46] in logic mining, providing potential advancements both in accuracy and in robustness for logic mining.

1.2.3 Classification Analysis

In predictive DM, the area of classification analysis has received unrivalled attention – both within literature and in practice. Classification has proven its effectiveness to support decision making and to solve complex planning tasks in various real-world application domains, including credit scoring (see, e.g., Crook et al. [8]) and direct marketing (see, e.g., Bose and Xi [4]). The predominant popularity of developing novel classification algorithms in the DM community seems to be only surpassed by the (often marginal) extension of existing algorithms in fine-tuning them to a particular data set or problem at hand. Consequently, Hand reflects that much of the claimed progress in DM research may turn out to be only illusive [23]. This leads to his reasonable expectation that advances will be based rather upon progress in computer hardware with more powerful data storage and processing ability than on building fine-tuned models of ever-increasing complexity. However, Friedman argues in a recent paper [15] that the development of kernel methods (e.g., SVMs) and ensemble classifiers, which form predictions by aggregating multiple basic models, both within the field of machine learning and DM, has further "revitalized" research within this field. Those methods may be seen as promising approaches toward future research in classification.

A novel ensemble classifier is introduced by Lemmond et al. [32] who draw inspiration from Breiman's random forest algorithm [6] and construct a random forest of linear discriminant models. Compared to classification trees used in the original algorithm, the base classifiers of linear discriminant analysis perform multivariate splits and are capable of exhibiting a higher diversity, which constitute novel and promising properties. It is theorized that these features may allow the resulting ensemble to achieve an even higher accuracy than the original random forest.

Lemmond et al. consider examples of the field of signal detection and conduct several empirical experiments to confirm the validity of this hypothesis.

SVM classifiers are employed in the work of Özöğür-Akyüz et al. [40], who propose a new paradigm for using this popular algorithm more effectively and efficiently in practical applications. Contrary to ensemble classifiers, standard practice in using SVMs stipulates the use of a single suitable model selected from a candidate pool determined by the algorithm's parameters. Regardless of potential disadvantages of this explicit "model selection" with respect to the reliability and robustness of the results, this principle is particularly counterintuitive because, prior to selecting this single model, a large number of SVM classifiers have to be built in order to determine suitable parameter settings in the absence of a robust methodology in specifying SVMs for data sets with distinct properties. In other words, the prevailing approach to employ SVMs is to first construct a (large) number of models, then to discard all but one of them and use this one to generate predictions. The approach by Özöğür-Akyüz et al. proposes to keep all classifier candidates and select either a single "most suitable" SVM or a collection of suitable classifiers for each individual case that is to be predicted. This procedure achieves appealing results in terms of forecasting accuracy and also computational efficiency, and it serves to integrate the established solutions of ensembles (an aggregate model selection) and individual model selection. Moreover, the general idea of reusing classifiers constructed within model selection and integrating them to produce ensemble forecasts can be directly transferred to other algorithms such as ANNs and other wrapper-based approaches, and thus contributes considerably to the general understanding of how such procedures can/should be used effectively.

Irrespective of substantial methodological and algorithmic advancements, the task of specifying classification models capable of dealing with imbalanced class distributions remains a particular challenge. In many empirical classification problems (where the target variable to be predicted takes on a nominal scale) one target class in the database is heavily underrepresented. Whereas such minority groups are usually of key importance for the respective application (e.g., detecting anomalous behavior of credit card use or predicting the probability of a customer defaulting on a loan), algorithms that strive to maximize the number of correct classifications will always be biased toward the majority class and impair their predictive accuracy on the minority group (see, e.g., [26, 50]). This problem is also considered by Liu et al. [33] in the context of classification with naive Bayes and SVM classifiers. Two popular approaches to increase a classifier's sensitivity for examples of the minority class involve either resampling schemes to elevate their frequency, e.g., through duplication of instances or the creation of artificial examples, or cost sensitive learning, essentially making misclassification of minority examples more costly. Whereas both techniques have been used successfully in previous work, a clear understanding as to how and under what conditions an approach works is yet lacking. To overcome this shortcoming, Liu et al. examine the formal relationship between cost-sensitive learning and different forms of resampling, most notably both from a theoretical and from an empirical perspective.

Learning in the presence of class and/or cost imbalance is one example where classification on empirical data sets proves difficult. Markedly, it has been observed that some applications do not enable a high classification accuracy to be obtained per se. The study of Weiss [49] aims at shedding light on the origins of this artifact. In particular, small disjuncts are identified as one influential source of high error rates, providing the motivation to examine their influence on classifier learning in detail. The term disjunct refers to a part of a classification model, e.g., a single rule within a rule-based classifier or one leaf within a decision tree, whereby the size of a disjunct is defined as the number of training examples that it correctly classifies. Previous research suggests that small disjuncts are collectively responsible for many individual classification errors across algorithms. Weiss develops a novel metric, error concentration, that captures the degree to which this pattern occurs in a data set and provides a single number measurement. Using this measure, an exhaustive empirical study is conducted that investigates several factors relevant to classifier learning (e.g., training set size, noise, and imbalance) with respect to their impact on small disjuncts and error concentration in particular.

1.2.4 Hybrid Data Mining Procedures

As a natural result to the predominant attention of classification algorithms in DM a myriad of hybrid algorithms have been explored for specific classification tasks, combining neuro, fuzzy genetic, and evolutionary approaches. But there are also promising innovations beyond the mere hybridization of an algorithm tailored to a specific task. In practical applications, DM techniques for classification, regression, or clustering are rarely used in isolation but in conjunction with other methods, e.g., to integrate the respective merits of complementary procedures while avoiding their demerits and, thereby, best meet the requirements of a specific application. This is particularly evident from the perception of DM within the process of knowledge discovery from databases [10], which exemplifies an iterative and modular combination of different algorithms. Although a purposive combination of different techniques may be particularly valuable beyond the singular optimization within each step of the KDD process, this has often been neglected in research. This special issue includes four examples of such hybrid approaches.

A joint use of supervised and unsupervised methods within the process of KDD is considered by Figueroa [12] and Karamitopoulos et al. [30]. Figueroa conducts a case study within the field of customer relationship management and develops an approach to estimate customer loyalty in a retailing setting. Loyalty has been identified as one of the key drivers of customer value, and the concepts of customer lifetime value have been firmly established beyond DM. Therefore, it may prove sensible to devote particular attention to the loyal customers and, e.g., target marketing campaigns for cross-/up-selling specifically to this subgroup. However, defining the concept of loyalty is, in itself, a nontrivial undertaking, especially in noncontractual settings where changes in customer behavior are difficult to identify. The task is further complicated by the fact that a regular and frequent update of respective information is essential. Figueroa proposes a possible solution to address these challenges: supervised and unsupervised learning methods are integrated to first identify customer subgroups and loyalty labels. This facilitates a subsequent application of ANNs to score novel customers according to their (estimated) loyalty.

Unsupervised methods are commonly employed as a means of reducing the size of data sets prior to building a prediction model using supervised algorithms. A respective approach is discussed by Karamitopoulos et al. who consider the case of multivariate time series analysis for similarity detection. Large volumes of time series data are routinely collected by, e.g., motion capturing or video surveillance systems that record multiple measurements for a single object at the same time interval. This generates a matrix of observations (i.e., measurements × discrete time periods) for each object, whereas standard DM routines such as clustering or classification would require objects being represented by row vectors. As a simple data transformation would produce extremely high-dimensional data sets, it would thereby further complicate analysis of such time series data. To alleviate this difficulty, Karamitopoulos et al. suggest reducing data set size and dimensionality by means of principal component analysis (PCA). This well-explored statistical approach will generate a novel representation of the data, which consists of a vector of the m largest eigenvalues (with m being a user-defined parameter) and a matrix of respective eigenvectors of the original data set's covariance matrix. As Karamitopoulos et al. point out, if two multivariate time series are similar, their PCA representations will be similar as well. That is, the produced matrices will be close in some sense. Consequently, Karamitopoulos et al. design a novel similarity measure based upon a time series' PCA signature. The concept of measuring similarity is at the core of many time series DM tasks, including clustering, classification, novelty detection, motif, or rule discovery as well as segmentation or indexing. Thus, it ensures broad applicability of the proposed approach. The main difference from other methods is that the novel similarity measure does not require applying a computer-intensive PCA to a query object: resource-intensive computations are conducted only once to build up a database of PCA signatures, which allows the identification of a query object's most similar correspondent in the database quickly. The potential of this novel similarity measure is supported by evidence from empirical experimentation using nearest neighbor classification.

Another branch of hybridization by integrating different DM techniques is explored by Johansson et al. [29] and Gijsberts et al. [18], who employ algorithms from the field of meta-heuristics to construct predictive classification and regression models. Meta-heuristics can be characterized as general search procedures to solve complex optimization problems (see, e.g., Voß [48]). Within DM, they are routinely employed to select a subset of attributes for a predictive model (i.e., feature selection), to construct a model from empirical data (e.g., as in the case of rule-based classification) or to tune the (hyper-)parameters of a specific model to adapt it to a given data set. The latter case is considered by Gijsberts et al. who evaluate evolutionary strategies (ES) to parameterize a least-square support vector regression (SVR) model. Whereas this task is commonly approached by means of genetic algorithms,

ES may be seen as a more natural choice because they avoid a transformation of the continuous SVR parameters into a binary representation. In addition, Gijsberts et al. examine the potential of genetic programming (GP) for SVR model building. SVR belongs to the category of kernel methods that employ a kernel function to perform an implicit mapping of input data into a higher dimensional feature space in order to account for nonlinear patterns within data. Exploiting the mathematical properties of such kernel functions, Gijsberts et al. develop a second approach that utilizes GP to "learn" an appropriate kernel function in a data-driven manner.

A related approach is designed by Johansson et al. for classification, where they employ GP to optimize the parameters of a k-nearest neighbor (kNN) classifier, most importantly the number of neighbors (i.e., k) and the weight individual features receive within distance calculations. In their study, Johansson et al. encompass classifier ensembles, whereby a collection of base kNN models is produced utilizing the stochasticity of GP to ensure diversity among ensemble members. As the general robustness of kNN with respect to resampling (i.e., the prevailing approach to construct diverse base classifiers) has hindered an application of kNN within an ensemble context, the approach of employing GP is particularly appealing to overcome this obstacle. Furthermore, Johansson et al. show that the predictive performance of the GP–kNN hybrid can be further increased by partitioning the input space into subregions and optimizing k and the feature weights locally within these regions. A large-scale empirical comparison across different 27 UCI data sets provides valid and reliable evidence of the efficacy of the proposed model.

1.2.5 Web Mining

The preceding papers concentrate mainly on the methodological aspects of DM. Clearly, the relevance of sophisticated data analysis tools in general, and their advancements in particular, is given by their broad range of challenging applications in various domains well beyond that of business and management. One domain of particular importance for corporate decision making, information systems and DM alike, is the World Wide Web, which has provided a new set of challenges through novel applications and data to many disciplines. In the context of DM, the term web mining has been coined to refer to the three branches of website structure, website content, and website usage mining.

A novel approach to improve website usability is proposed by Geczy et al. [16]. They focus on knowledge portals in corporate intranets and develop a recommendation algorithm to assist a user's navigation by predicting which resource the user is ultimately interested in and provide direct access to this resource by making the respective link available. This concept improves upon traditional techniques that usually aim only at estimating the next page within a navigation path. Consequently, providing the opportunity to access a potentially desired resource in a more direct manner would help to save a user's time, computational resources of servers, and bandwidth of networks.

A second branch of web mining is concerned with analyzing website content, e.g., to automatically categorize websites into predefined groups or to judge a page's relevance for a given user query in information retrieval. As techniques for natural language processing have reached a mature stage, unstructured data (such as web pages) can be transformed into a machine-readable format to facilitate DM with relative ease. The opportunities arising thereof is exploited by Ryan and Hamel [42]. The internet is considered as a pool of opinions, where current topics are discussed and shared among users, whose aggregation may facilitate the generation of accurate forecasts. Their research aims at constructing a forecasting model on the basis of search engine query results in order to predict future events. The proposed techniques allow the internet to be used as one large prediction market and, as such, represent an innovative approach toward forecasting. Current and future developments within the scope of Web 2.0 (e.g., social networking, blogging) as well as the Semantic Web can be expected to further increase the potential of this idea. This idea, in turn, will require the development of supporting IS (e.g., for gathering query results, transforming text data into machine-readable formats, as well as aggregating and possibly weighting resulting information) for a successful development in the long run.

1.2.6 Privacy-Preserving Data Mining

The availability of very large data sets of detailed customer-centric information, e.g., on the purchasing behavior of an individual consumer or detailed information on a surfer's web usage behavior, not only offers opportunities from a DM perspective but also summons serious concerns regarding data privacy. As a consequence, both the relevance of privacy issues in DM and the awareness thereof continuously increase. This is mirrored by the increasing research activities within the field of privacy-preserving DM. In particular, substantial work has been conducted to conceptualize different models of privacy and develop privacy-preserving data analysis procedures. Privacy models like k-anonymity require that, after deleting identifiers from a data set, tuples of attributes which may serve as so-called quasi-identifiers (e.g., age, zip code.) show identical values across at least k data records. This prohibits a reidentification of instances and hence insures privacy. Achieving k-anonymity or extended variants may thus necessitate some transformation of the original attributes, whereby inherent information has to be sustained to the largest degree possible in order to not impede subsequent DM activities. Truta and Campan [44] review alternative privacy models and propose two novel algorithms for achieving privacy levels of extended p-sensitive k-anonymity. Both techniques compare favorably to the established *Incognito* algorithm in terms of three different performance metrics (i.e., discernibility, normalized average cluster size, and running time) within an empirical comparison. Furthermore, Truta and Campan propose new privacy models that allow decision makers to constrain the degree to which quasi-identifier attributes are generalized within data anonymization. These models are more aligned with the needs of real-world application by enabling a user to control the trade-off between privacy on the one hand and specific DM objectives (e.g., forecasting accuracy and between-cluster heterogeneity.) on the other explicitly. One of these models is tailored to the specific requirements of privacy in social networks, which have experienced a rapid growth within the last years. Up to now, their proliferation has not been accompanied by sufficient efforts to maintain privacy of users as well as their network relationships. In this context, the novel model for *p*-sensitive *k*-anonymity social networks may be seen as a particularly important and timely contribution.

Employing the techniques described by Truta and Campan allows anonymization of a single data set, so that an identification of individual data records through quasi-identifier attributes becomes impossible. However, such precautions can be circumvented if multiple data sets are linked and related to each other. For example, a respective case has been reported within the scope of the *Netflix* competition. A large data set of movie ratings from anonymous users has been published within this challenge to develop and assess recommendation algorithms. However, it was shown that users could be reidentified by linking the anonymous rating data with some other sources [37], which indicates the risk of severely violating privacy through linking data sets. On the other hand, a strong desire exists to share data sets with collaborators and engage in joint DM activities, e.g., within the scope of supply chain management or medical diagnosis to support and improve decision making. To enable this, Mangasarian and Wild [35] develop an approach that facilitates a distributed use of data for a DM, but avoids actually sharing it between participating entities. Mangasarian and Wild exploit the particular characteristics of kernel methods and develop a privacy-preserving SVM classifier, which is shown to effectively overcome the alleged trade-off between privacy and accuracy. Short of a true trade-off between accuracy and privacy, the proposed technique not only preserves privacy but also achieves equivalent accuracy as a classifier that has access to all data.

1.3 Conclusion and Outlook

Quo vadis, IS and DM? IS have been a key originator of corporate data growth and remain to have a core interest in the advancement of sophisticated approaches to analytical decision support in management. Processes, systems, and techniques in this field are commonly referred to as business intelligence (BI) within the IS community, and DM is acknowledged as part of corporate BI. However, in comparison to other analytical approaches such as OLAP (online analytical processing) or data warehouses, it has received only limited attention. On the contrary, disciplines like statistics, computer sciences, machine learning, and, more recently, operational research (see, e.g., [39, 38, 13]) have been most influential, which explains the emphasis on methodological aspects in the DM domain. This focus is well justified when considering the ever-growing number of novel applications and respective requirements DM methods have to fulfill. Continuously sustaining such compliance with application needs requires that research activities do not only focus on established direction like procedures for predictive and descriptive data analysis but are also geared toward concrete decision contexts. Very recently, this understanding gave rise to two novel streams in DM research, namely utility-based DM (UBDM) (see, e.g., [51]) and domain-driven DM (see, e.g., [53]). Both acknowledge the importance of novel algorithms, but stress that their development should be guided by real-world decision contexts and constraints. This is precisely the approach toward decision support that has always been prevalent within the IS community. Consequently, more research along this line is highly desirable and needed to systematically exploit the core competencies found in IS and DM, respectively, and further improve the support of managerial decision making. Noteworthy examples of how this may be achieved have recently appeared in leading IS journals [43, 5] and reemphasize the potential of research at the interface between these two fields.

To the understanding of the reviewers and editors, the chapters in this special issue have captured those essential aspects in a convincing and clear manner and provide interesting, original, and significant contributions to the advancement of both DM and IS in the context of decision making. Therefore, in some sense, they can be considered as building blocks of the road that shows at least one possible direction for the further development of DM and IS. Of course, it is far beyond our goals and means to suggest one beatific direction. However, for DM and IS may be a fruitful answer to "Where are you going?" could be "Wherever we will go, we should accompany each other."

Acknowledgments We would like to thank all authors who submitted their work for consideration to this focused issue. Their contributions made this special issue possible. We would like to thank especially the reviewers for their time and their thoughtful reviews. Finally, we would like to thank the two series editors, Ramesh Sharda and Stefan Voß, for their valuable advice and encouragement, and the editorial staff for their support in the production of this special issue (Hamburg, June 2009).

References

1. Agrawal, R. and Srikant, R. Fast algorithms for mining association rules in large databases. In: Bocca, J. B., Jarke, M., and Zaniolo, C. (eds.), *Proc. of the 20th Intern. Conf. on Very Large Databases (VLDB'94)*, pp. 487–499, Santiago de Chile, Chile, 1994. Morgan Kaufmann.
2. Ayres, I. *Super Crunchers: Why Thinking-By-Numbers Is the New Way to Be Smart*. Bantam Dell, New York, 2007.
3. Berry, M. J. A. and Linoff, G. *Data Mining Techniques: For Marketing, Sales and Customer Relationship Management*. Wiley, New York, 2nd ed., 2004.
4. Bose, I. and Xi, C. Quantitative models for direct marketing: A review from systems perspective. *European Journal of Operational Research*, 195(1):1–16, 2009.
5. Boylu, F., Aytug, H., and Köhler, G. J. Induction over strategic agents. *Information Systems Research*, forthcoming.
6. Breiman, L. Random forests. *Machine Learning*, 45(1):5–32, 2001.
7. Cabena, P., Hadjnian, P., Stadler, R., Verhees, J., and Zanasi, A. *Discovering Data Mining: From Concept to Implementation*. Prentice Hall, London, 1997.

8. Crook, J. N., Edelman, D. B., and Thomas, L. C. Recent developments in consumer credit risk assessment. *European Journal of Operational Research*, 183(3):1447–1465, 2007.
9. Davis, F. D. Perceived usefulness, perceived ease of use, and user acceptance of information technology. *MIS Quarterly*, 13(3):319–340, 1989.
10. Fayyad, U., Piatetsky-Shapiro, G., and Smyth, P. From data mining to knowledge discovery in databases: An overview. *AI Magazine*, 17(3):37–54, 1996.
11. Felici, G., Simeone, B., and Spinelli, V. Classification techniques and error control in logic mining. *Annals of Information Systems*, in this issue.
12. Figueroa, C. J. Predicting customer loyalty labels in a large retail database: A case study in Chile. *Annals of Information Systems*, in this issue.
13. Fildes, R., Nikolopoulos, K., Crone, S. F., and Syntetos, A. A. Forecasting and operational research: A review. *Journal of the Operational Research Society*, 59:1150–1172, 2006.
14. Freitas, A. On rule interestingness measures. *Knowledge-Based Systems*, 12(5–6):309–315, October 1999. URL http://www.cs.kent.ac.uk/pubs/1999/1407.
15. Friedman, J. H. Recent advances in predictive (machine) learning. *Journal of Classification*, 23(2):175–197, 2006.
16. Geczy, P., Izumi, N., Akaho, S., and Hasida, K. Behaviorally founded recommendation algorithm for browsing assistance systems. *Annals of Information Systems*, in this issue.
17. Geng, L. and Hamilton, H. J. Interestingness measures for data mining: A survey. *ACM Computing Surveys*, 38(3):Article No. 9, 2006.
18. Gijsberts, A., Metta, G., and Rothkrantz, L. Evolutionary optimization of least-squares support vector machines. *Annals of Information Systems*, in this issue.
19. Guyon, I., Weston, J., Barnhill, S., and Vapnik, V. Gene selection for cancer classification using support vector machines. *Machine Learning*, 46(1-3):389–422, 2002.
20. Han, J. and Kamber, M. *Data mining: Concepts and Techniques*. The Morgan Kaufmann series in data management systems. Morgan Kaufmann, San Francisco, 7th ed., 2004.
21. Hand, D. J. Data mining: Statistics and more? *American Statistician*, 52(2):112–118, 1998.
22. Hand, D. J. Statistics and data mining: Intersecting disciplines. *ACM SIGKDD Explorations Newsletter*, 1(1):16–19, 1999.
23. Hand, D. J. Classifier technology and the illusion of progress. *Statistical Science*, 21(1):1–14, 2006.
24. Hand, D. J., Mannila, H., and Smyth, P. *Principles of Data Mining*. Adaptive computation and machine learning. MIT Press, Cambridge, London, 2001.
25. Hastie, T., Tibshirani, R., and Friedman, J. *The Elements of Statistical Learning: Data Mining, Inference, and Prediction*. Springer, New York, 2002.
26. Japkowicz, N. and Stephen, S. The class imbalance problem: A systematic study. *Intelligent Data Analysis*, 6(5):429–450, 2002.
27. Joachims, T. Text categorization with support vector machines: Learning with many relevant features. In: Nedellec, C. and Rouveirol, C. (eds.), *Proc. of the 10th European Conf. on Machine Learning*, vol. 1398 of *Lecture Notes in Computer Science*, pp. 137–142, Chemnitz, Germany, 1998. Springer.
28. Joachims, T. Making large-scale SVM learning practical. In: Schölkopf, B., Burges, C. J. C., and Smola, A. J. (eds.), *Advances in Kernel Methods: Support Vector Learning*, pp. 169–184. MIT Press, Cambridge, 1999.
29. Johansson, U., König, R., and Niklasson, L. Genetically evolved kNN ensembles. *Annals of Information Systems*, in this issue.
30. Karamitopoulos, L., Evangelidis, G., and Dervos, D. PCA-based time series similarity search. *Annals of Information Systems*, in this issue.
31. Le Bras, Y., Lenca, P., and Lallich, S. Mining interesting rules without support requirement: A general universal existential upward closure property. *Annals of Information Systems*, in this issue.
32. Lemmond, T. D., Chen, B. Y., Hatch, A. O., and Hanley, W. G. An extended study of the discriminant random forest. *Annals of Information Systems*, in this issue.

33. Liu, A., Martin, C., La Cour, B., and Ghosh, J. Effects of oversampling versus cost-sensitive learning for Bayesian and SVM classifiers. *Annals of Information Systems*, in this issue.
34. Liu, B., Hsu, W., Chen, S., and Ma, Y. Analyzing the subjective interestingness of association rules. *IEEE Intelligent Systems*, 15(5):47–55, 2000.
35. Mangasarian, O. L. and Wild, E. W. Privacy-preserving random kernel classification of checkerboard partitioned data. *Annals of Information Systems*, in this issue.
36. Martens, D. and Baesens, B. Building acceptable classification models. *Annals of Information Systems*, in this issue.
37. Narayanan, A. and Shmatikov, V. How to break anonymity of the Netflix prize dataset, 2006. URL http://www.citebase.org/abstract?id=oai:arXiv.org:cs/0610105.
38. Olafsson, S. Introduction to operations research and data mining. *Computers and Operations Research*, 33(11):3067–3069, 2006.
39. Olafsson, S., Li, X., and Wu, S. Operations research and data mining. *European Journal of Operational Research*, 187(3):1429–1448, 2008.
40. Özöğür-Akyüz, S., Hussain, Z., and Shawe-Taylor, J. Prediction with the SVM using test point margins. *Annals of Information Systems*, in this issue.
41. Ringle, C. M., Sarstedt, M., and Mooi, E. A. Repose-based segmentation using finite mixture partial least squares. *Annals of Information Systems*, in this issue.
42. Ryan, S. and Hamel, L. Using web text mining to predict future events: A test of the wisdom of crowds hypothesis. *Annals of Information Systems*, in this issue.
43. Saar-Tsechansky, M. and Provost, F. Decision-centric active learning of binary-outcome models. *Information Systems Research*, 18(1):4–22, 2007.
44. Truta, T. M. and Campan, A. Avoiding attribute disclosure with the (extended) p-sensitive k-anonymity model. *Annals of Information Systems*, in this issue.
45. Vapnik, V. N. *Estimation of Dependences Based on Empirical Data*. Springer, New York, 1982.
46. Vapnik, V. N. *The Nature of Statistical Learning Theory*. Springer, New York, 1995.
47. Vapnik, V. N. *Statistical Learning Theory*. Wiley, New York, 1998.
48. Voß, S. Meta-heuristics: The state of the art. In: Nareyek, A. (ed.), *Local Search for Planning and Scheduling*, vol. 2148 of *Lecture Notes in Artificial Intelligence*, pp. 1–23. Springer, Berlin, 2001.
49. Weiss, G. M. The impact of small disjuncts on classifier learning. *Annals of Information Systems*, in this issue.
50. Weiss, G. M. Mining with rarity: A unifying framework. *ACM SIGKDD Explorations Newsletter*, 6(1):7–19, 2004.
51. Weiss, G. M., Zadrozny, B., and Saar-Tsechansky, M. Guest editorial: special issue on utility-based data mining. *Data Mining and Knowledge Discovery*, 17(2):129–135, 2008.
52. Wu, X., Kumar, V., Ross Quinlan, J., Ghosh, J., Yang, Q., Motoda, H., McLachlan, G., Ng, A., Liu, B., Yu, P., Zhou, Z.-H., Steinbach, M., Hand, D., and Steinberg, D. Top 10 algorithms in data mining. *Knowledge and Information Systems*, 14(1):1–37, 2008.
53. Yu, P. (ed.). *Proc. of the 2007 Intern. Workshop on Domain Driven Data Mining*. ACM, New York, 2007.

Part I
Confirmatory Data Analysis

Chapter 2

Response-Based Segmentation Using Finite Mixture Partial Least Squares

Theoretical Foundations and an Application to American Customer Satisfaction Index Data

Christian M. Ringle, Marko Sarstedt, and Erik A. Mooi

Abstract When applying multivariate analysis techniques in information systems and social science disciplines, such as management information systems (MIS) and marketing, the assumption that the empirical data originate from a single homogeneous population is often unrealistic. When applying a causal modeling approach, such as partial least squares (PLS) path modeling, segmentation is a key issue in coping with the problem of heterogeneity in estimated cause-and-effect relationships. This chapter presents a new PLS path modeling approach which classifies units on the basis of the heterogeneity of the estimates in the inner model. If unobserved heterogeneity significantly affects the estimated path model relationships on the aggregate data level, the methodology will allow homogenous groups of observations to be created that exhibit distinctive path model estimates. The approach will, thus, provide differentiated analytical outcomes that permit more precise interpretations of each segment formed. An application on a large data set in an example of the American customer satisfaction index (ACSI) substantiates the methodology's effectiveness in evaluating PLS path modeling results.

Christian M. Ringle

Institute for Industrial Management and Organizations, University of Hamburg, Von-Melle-Park 5, 20146 Hamburg, Germany, e-mail: cringle@econ.uni-hamburg.de, and Centre for Management and Organisation Studies (CMOS), University of Technology Sydney (UTS), 1-59 Quay Street, Haymarket, NSW 2001, Australia, e-mail: christian.ringle@uts.edu.au

Marko Sarstedt

Institute for Market-based Management, University of Munich, Kaulbachstr. 45, 80539 Munich, Germany, e-mail: sarstedt@bwl.lmu.de

Erik A. Mooi

Aston Business School, Aston University, Room NB233 Aston Triangle, Birmingham B47ET, UK, e-mail: e.a.mooi@aston.ac.uk

R. Stahlbock et al. (eds.), *Data Mining,* Annals of Information Systems 8, DOI 10.1007/978-1-4419-1280-0_2, © Springer Science+Business Media, LLC 2010

2.1 Introduction

2.1.1 On the Use of PLS Path Modeling

Since the 1980s, applications of structural equation models (SEMs) and path modeling have increasingly found their way into academic journals and business practice. Currently, SEMs represent a quasi-standard in management research when it comes to analyzing the cause–effect relationships between latent variables. Covariance-based structural equation modeling [CBSEM; 38, 59] and partial least squares analysis [PLS; 43, 80] constitute the two matching statistical techniques for estimating causal models.

Whereas CBSEM has long been the predominant approach for estimating SEMs, PLS path modeling has recently gained increasing dissemination, especially in the field of consumer and service research. PLS path modeling has several advantages over CBSEM, for example, when sample sizes are small, the data are non-normally distributed, or non-convergent results are likely because complex models with many variables and parameters are estimated [e.g., 20, 4]. However, PLS path modeling should not simply be viewed as a less stringent alternative to CBSEM, but rather as a complementary modeling approach [43]. CBSEM, which was introduced as a confirmatory model, differs from PLS path modeling, which is prediction-oriented.

PLS path modeling is well established in the academic literature, which appreciates this methodology's advantages in specific research situations [20]. Important applications of PLS path modeling in the management sciences discipline are provided by [23, 24, 27, 76, 18]. The use of PLS path modeling can be predominantly found in the fields of marketing, strategic management, and management information systems (MIS). The employment of PLS path modeling in MIS draws mainly on Davis's [10] technology acceptance model [TAM; e.g., 1, 25, 36]. In marketing, the various customer satisfaction index models – such as the European customer satisfaction index [ECSI; e.g., 15, 30, 41] and Festge and Schwaiger's [18] driver analysis of customer satisfaction with industrial goods – represent key areas of PLS use. Moreover, in strategic management, Hulland [35] provides a review of PLS path modeling applications. More recent studies focus specifically on strategic success factor analyses [e.g., 62].

Figure 2.1 shows a typical path modeling application of the American customer satisfaction index model [ACSI; 21], which also serves as an example for our study. The squares in this figure illustrate the manifest variables (indicators) derived from a survey and represent customers' answers to questions while the circles illustrate latent, not directly observable, variables. The PLS path analysis predominantly focuses on estimating and analyzing the relationships between the latent variables in the inner model. However, latent variables are measured by means of a block of manifest variables, with each of these indicators associated with a particular latent variable. Two basic types of outer relationships are relevant to PLS path modeling: formative and reflective models [e.g., 29]. While a formative measurement model has cause–effect relationships between the manifest variables and the latent index (independent causes), a reflective measurement model involves paths from the latent construct to the manifest variables (dependent effects).

The selection of either the formative or the reflective outer mode with respect to the relationships between a latent variable and its block of manifest variables builds on theoretical assumptions [e.g., 44] and requires an evaluation by means of empirical data [e.g., 29]. The differences between formative and reflective measurement models and the choice of the correct approach have been intensively discussed in the literature [3, 7, 11, 12, 19, 33, 34, 68, 69]. An appropriate choice of measurement model is a fundamental issue if the negative effects of measurement model misspecification are to be avoided [44].

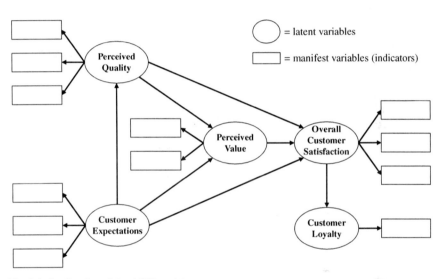

Fig. 2.1 Application of the ACSI model

While the outer model determines each latent variable, the inner path model involves the causal links between the latent variables, which are usually a hypothesized theoretical model. In Fig. 2.1, for example, the latent construct "Overall Customer Satisfaction" is hypothesized to explain the latent construct "Customer Loyalty." The goal of prediction-oriented PLS path modeling method is to minimize the residual variance of the endogenous latent variables in the inner model and, thus, to maximize their R^2 values (i.e., for the key endogenous latent variables such as customer satisfaction and customer loyalty in an ACSI application). This goal underlines the prediction-oriented character of PLS path modeling.

2.1.2 Problem Statement

While the use of PLS path modeling is becoming more common in management disciplines such as MIS, marketing management, and strategic management, there are at least two critical issues that have received little attention in prior work. First, unobserved heterogeneity and measurement errors are endemic in social sciences. However, PLS path modeling applications are usually based on the assumption that the analyzed data originate from a single population. This assumption of homogeneity is often unrealistic, as individuals are likely to be heterogeneous in their perceptions and evaluations of latent constructs. For example, in customer satisfaction studies, users may form different segments, each with different drivers of satisfaction. This heterogeneity can affect both the measurement part (e.g., different latent variable means in each segment) and the structural part (e.g., different relationships between the latent variables in each segment) of a causal model [79]. In their customer satisfaction studies, Jedidi et al. [37] Hahn et al. [31] as well as Sarstedt, Ringle and Schwaiger [72] show that an aggregate analysis can be seriously misleading when there are significant differences between segment-specific parameter estimates. Muthén [54] too describes several examples, showing that if heterogeneity is not handled properly, SEM analysis can be seriously distorted. Further evidence of this can be found in [16, 66, 73]. Consequently, the identification of different groups of consumers in connection with estimates in the inner path model is a serious issue when applying the path modeling methodology to arrive at decisive interpretations [61]. Analyses in a path modeling framework usually do not address the problem of heterogeneity, and this failure may lead to inappropriate interpretations of PLS estimations and, therefore, to incomplete and ineffective conclusions that may need to be revised.

Second, there are no well-developed statistical instruments with which to extend and complement the PLS path modeling approach. Progress toward uncovering unobserved heterogeneity and analytical methods for clustering data have specifically lagged behind their need in PLS path modeling applications. Traditionally, heterogeneity in causal models is taken into account by assuming that observations can be assigned to segments a priori on the basis of, for example, geographic or demographic variables. In the case of a customer satisfaction analysis, this may be achieved by identifying high and low-income user segments and carrying out multigroup structural equation modeling. However, forming segments based on a priori information has serious limitations. In many instances there is no or only incomplete substantive theory regarding the variables that cause heterogeneity. Furthermore, observable characteristics such as gender, age, or usage frequency are often insufficient to capture heterogeneity adequately [77]. Sequential clustering procedures have been proposed as an alternative. A researcher can partition the sample into segments by applying a clustering algorithm, such as k-means or k-medoids, with respect to the indicator variables and then use multigroup structural equation modeling for each segment. However, this approach has conceptual shortcomings: "Whereas researchers typically develop specific hypotheses about the relationships between the variables of interest, which is mirrored in the structural equation model tested in the second step, traditional cluster analysis assumes independence among these variables" [79, p. 2]. Thus, classical segmentation strategies cannot account for heterogeneity in the relationships between latent variables and are often inappropriate for forming groups of data with distinctive inner model estimates [37, 61, 73, 71].

2.1.3 Objectives and Organization

A result of these limitations is that PLS path modeling requires complementary techniques for model-based segmentation, which allows treating heterogeneity in the inner path model relationships. Unlike basic clustering algorithms that identify clusters by optimizing a distance criterion between objects or pairs of objects, model-based clustering approaches in SEMs postulate a statistical model for the data. These are also often referred to as latent class segmentation approaches. Sarstedt [74] provides a taxonomy (Fig. 2.2) and a review of recent latent class segmentation approaches to PLS path modeling such as PATHMOX [70], FIMIX-PLS [31, 61, 64, 66], PLS genetic algorithm segmentation [63, 67], Fuzzy PLS Path Modeling [57], or REBUS-PLS [16, 17]. While most of these methodologies are in an early or experimental stage of development, Sarstedt [74] concludes that the finite mixture partial least squares approach (FIMIX-PLS) can currently be viewed as the most comprehensive and commonly used approach to capture heterogeneity in PLS path modeling. Hahn et al. [31] pioneered this approach in that they also transferred Jedidi et al.'s [37] finite mixture SEM methodology to the field of PLS path modeling. However, knowledge about the capabilities of FIMIX-PLS is limited.

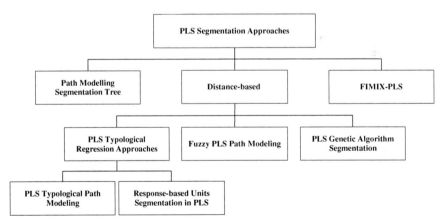

Fig. 2.2 Methodological taxonomy of latent class approaches to capture unobserved heterogeneity in PLS path models [74]

This chapter's main contribution to the body of knowledge on clustering data in PLS path modeling is twofold. First, we present FIMIX-PLS as recently implemented in the statistical software application SmartPLS [65] and, thereby, made broadly available for empirical research in the various social sciences disciplines. We thus present a systematic approach to applying FIMIX-PLS as an appropriate and necessary means to evaluate PLS path modeling results on an aggregate data level. PLS path modeling applications can exploit this approach to response-based market segmentation by identifying certain groups of customers in cases where unobserved moderating factors cause consumer heterogeneity within inner model relationships. Second, an application of the methodology to a well-established marketing example substantiates the requirement and applicability of FIMIX-PLS as an analytical extension of and standard test procedure for PLS path modeling.

This study is particularly important for researchers and practitioners who can exploit the capabilities of FIMIX-PLS to ensure that the results on the aggregate data level are not affected by unobserved heterogeneity in the inner path model estimates. Furthermore, FIMIX-PLS indicates that this problem can be handled by forming groups of data. A multigroup comparison [13, 32] of the resulting segments indicates whether segment-specific PLS path estimates are significantly different. This allows researchers to further differentiate their analysis results. The availability of FIMIX-PLS capabilities (i.e., in the software application SmartPLS) paves the way to a systematic analytical approach, which we present in this chapter as a standard procedure to evaluate PLS path modeling results.

We organize the remainder of this chapter as follows: First, we introduce the PLS algorithm – an important issue associated with its application. Next, we present a systematic application of the FIMIX-PLS methodology to uncover unobserved heterogeneity and form groups of data. Thereafter, this approach's application to a well-substantiated and broadly acknowledged path modeling application in marketing research illustrates its effectiveness and the need to use it in the evaluation process of PLS estimations. The final section concludes with implications for PLS path modeling and directions regarding future research.

2.2 Partial Least Squares Path Modeling

The PLS path modeling approach is a general method for estimating causal relationships in path models that involve latent constructs which are indirectly measured by various indicators. Prior publications [80, 43, 8, 75, 32] provide the methodological foundations, techniques for evaluating the results [8, 32, 43, 75, 80], and some examples of this methodology. The estimation of a path model, such as the ACSI example in Fig. 2.1, builds on two sets of outer and inner model linear equations. The basic PLS algorithm, as proposed by Lohmöller [43], allows the linear relationships' parameters to be estimated and includes two stages, as presented in Table 2.1.

Table 2.1 The basic PLS algorithm [43]

Stage 1: Iterative estimation of latent variable scores

#1   Inner weights

$$v_{ji} = \begin{cases} \text{sign cov}(Y_j; Y_i) & \text{if } Y_j \text{ and } Y_i \text{ are adjacent} \\ 0 & \text{otherwise} \end{cases}$$

#2   Inside approximation

$$\tilde{Y}_j := \sum_i v_{ji} Y_i$$

#3   Outer weights; solve for

$$y_{k_j n} = \tilde{w}_{k_j} \tilde{Y}_{jn} + e_{k_{jn}} \qquad \text{Mode A}$$
$$\tilde{Y}_{jn} = \sum_{k_j} \tilde{w}_{k_j} y_{k_{jn}} + d_{jn} \qquad \text{Mode B}$$

#4   Outside approximation

$$Y_{jn} := \sum_{k_j} \tilde{w}_{k_j} y_{k_j n}$$

Variables: *Parameters:*

y = manifest variables (data)   v = inner weights

Y = latent variables   w = weight coefficients

d = validity residuals

e = outer residuals

Indices:

$i = 1, \ldots, I$ for blocks of manifest variables

$j = 1, \ldots, J$ for latent variables

$k_j = 1, \ldots, K$ for manifest variables counted within block j

$n = 1, \ldots, N$ for observational units (cases)

Stage 2: Estimation of outer weights, outer loadings, and inner path model coefficients

---

In the measurement model, manifest variables' data – on a metric or quasi-metric scale (e.g., a seven-point Likert scale) – are the input for the PLS algorithm that starts in step 4 and uses initial values for the weight coefficients (e.g., "+1" for all weight coefficients). Step 1 provides values for the inner relationships and Step 3 for the outer relationships, while Steps 2 and 4 compute standardized latent variable scores. Consequently, the basic PLS algorithm distinguishes between reflective (Mode A) and formative (Mode B) relationships in step 3, which affects the generation of the final latent variable scores. In step 3, the algorithm uses Mode A to obtain the outer weights of reflective measurement models (single regressions for the relationships between the latent variable and each of its indicators) and Mode B for formative measurement models (multiple regressions through which the latent variable is the dependent variable). In practical applications, the analysis of reflective measurement models focuses on the loading, whereas the weights are used to analyze formative relationships. Steps 1 to 4 in the first stage are repeated until convergence is obtained (e.g., the sum of changes of the outer weight coefficients in step 4 is below a threshold value of 0.001). The first stage provides estimates for the latent variable scores. The second stage uses these latent variable scores for ordinary least squares (OLS) regressions to generate the final (standardized) path coefficients for the relationships between the latent variables in the inner model as well as the final (standardized) outer weights and loadings for the relationships between a latent variable and its block of manifest variables [32].

A key issue in PLS path modeling is the evaluation of results. Since the PLS algorithm does not optimize any global scalar function, fit measures that are well known from CBSEM are not available for the nonparametric PLS path modeling approach. Chin [8] therefore presents a catalog of nonparametric criteria to separately assess the different model structures' results. A systematic application of these criteria is a two-step process [32]. The evaluation of PLS estimates begins with the measurement models and employs decisive criteria that are specifically associated with the formative outer mode (e.g., significance, multicollinearity) or reflective outer mode (e.g., indicator reliability, construct reliability, discriminant validity). Only if the latent variable scores show evidence of sufficient reliability and validity is it worth pursuing the evaluation of inner path model estimates (e.g., significance of path coefficients, effect sizes, R^2 values of latent endogenous variables). This assessment also includes an analysis of the PLS path model estimates regarding their capabilities to predict the observed data (i.e., the predictive relevance). The estimated values of the inner path coefficients allow the relative importance of each exogenous latent variable to be decided in order to explain an endogenous latent variable in the model (i.e., R^2 value). The higher the (standardized) path coefficients – for example, in the relationship between "Overall Customer Satisfaction" and "Customer Loyalty" in Fig. 2.1 – the higher the relevance of the latent predecessor variable in explaining the latent successor variable. The ACSI model assumes significant inner path model relationships between the key constructs "Overall Customer Satisfaction" and "Customer Loyalty" as well as substantial R^2 values for these latent variables.

2.3 Finite Mixture Partial Least Squares Segmentation

2.3.1 Foundations

Since its formal introduction in the 1950s, market segmentation has been one of the primary marketing concepts for product development, marketing strategy, and understanding customers. To segment data in a SEM context, researches frequently use sequential procedures in which homogenous subgroups are formed by means of a priori information to explain heterogeneity, or they revert to the application of cluster analysis techniques, followed by multigroup structural equation modeling. However, none of these approaches is considered satisfactory, as observable characteristics often gloss over the true sources of heterogeneity [77]. Conversely, the application of traditional cluster analysis techniques suffers from conceptual shortcomings and cannot account for heterogeneity in the relationships between latent variables. This weakness is broadly recognized in the literature and, consequently, there has been a call for model-based clustering methods.

In data mining, model-based clustering algorithms have recently gained increasing attention, mainly because they allow researchers to identify clusters based on their shape and structure rather than on proximity between data points [50]. Several approaches, which form a statistical model based on large data sets, have been proposed. For example, Wehrens et al. [78] propose methods that use one or several samples of data to construct a statistical model which serves as a basis for a subsequent application on the entire data set. Other authors [c.g., 45] developed procedures to identify a set of data points which can be reasonably classified into clusters and iterate the procedure on the remainder. Different procedures do not derive a statistical model from a sample but apply strategies to scale down massive data sets [14] or use reweighted data to fit a new cluster to the mixture model [49]. Whereas these approaches to model-based clustering have been developed within a data mining context and are thus exploratory in nature, SEMs rely on a confirmatory concept as researchers need to specify a hypothesized path model in the first step of the analysis. This path model serves as the basis for subsequent cluster analyses but is supposed to remain constant across all segments.

In CBSEM, Jedidi et al. [37] pioneered this field of research and proposed the finite mixture SEM approach, i.e., a procedure that blends finite mixture models and the expectation-maximization (EM) algorithm [46, 47, 77]. Although the original technique extends CBSEM and is implemented in software packages for statistical computations [e.g., Mplus; 55], the method is inappropriate for PLS path modeling due to unlike methodological assumptions. Consequently, Hahn et al. [31] introduced the finite FIMIX-PLS method that combines the strengths of the PLS path modeling method with the maximum likelihood estimation's advantages when deriving market segments with the help of finite mixture models. A finite mixture approach to model-based clustering assumes that the data originate from several subpopulations or segments [48]. Each segment is modeled separately and the overall population is a mixture of segment-specific density functions. Consequently, homogeneity is no longer defined in terms of a set of common scores, but at a distributional level. Thus, finite mixture modeling enables marketers to cope with heterogeneity in data by clustering observations and estimating parameters simultaneously, thus avoiding well-known biases that occur when models are estimated separately [37]. Moreover, there are many versatile or parsimonious models, as well as clustering algorithms available that can be customized with respect to a wide range of substantial problems [48].

Based on this concept, the FIMIX-PLS approach simultaneously estimates the model parameters and ascertains the heterogeneity of the data structure within a PLS path modeling framework. FIMIX-PLS is based on the assumption that heterogeneity is concentrated in the inner model relationships. The approach captures this heterogeneity by assuming that each endogenous latent variable η_i is distributed as a finite mixture of conditional multivariate normal densities. According to Hahn et al. [31, p. 249], since "the endogenous variables of the inner model are a function of the exogenous variables, the assumption of the conditional multivariate normal distribution of the η_i is sufficient." From a strictly theoretical viewpoint, the imposition of a distributional assumption on the endogenous latent variable may prove to be problematic. This criticism gains force when one considers that PLS path modeling is generally preferred to covariance structure analysis in circumstances where assumptions of multivariate normality cannot be made [4, 20]. However, recent simulation evidence shows the algorithm to be robust, even in the face of distributional misspecification [18]. By differentiating between dependent (i.e., endogenous latent) and explanatory (i.e., exogenous latent) variables in the inner model, the approach follows a mixture regression concept [77] that allows the estimation of separate linear regression functions and the corresponding object memberships of several segments.

2.3.2 Methodology

Drawing on a modified presentation of the relationships in the inner model (Table 2.2 provides a description of all the symbols used in the equations presented in this chapter.),

$$B\eta_i + \Gamma\xi_i = \zeta_i, \tag{2.1}$$

it is assumed that η_i is distributed as a finite mixture of densities $f_{i|k}(\cdot)$ with K ($K < \infty$) segments

$$\eta_i \sim \sum_{k=1}^{K} \rho_k f_{i|k}(\eta_i|\xi_i, B_k, \Gamma_k, \Psi_k), \tag{2.2}$$

whereby $\rho_k > 0 \; \forall k$, $\sum_{k=1}^{K} \rho_k = 1$ and ξ_i, B_k, Γ_k, Ψ_k depict the segment-specific vector of unknown parameters for each segment k. The set of mixing proportions ρ determines the relative mixing of the K segments in the mixture. Substituting $f_{i|k}(\eta_i|\xi_i, B_k, \Gamma_k, \Psi_k)$ results in the following equation:[1]

$$\eta_i \sim \sum_{k=1}^{K} \rho_k \left[\frac{1}{(2\pi)^{M/2}\sqrt{|\Psi_k|}} \right] e^{-\frac{1}{2}\left((\bar{I}-B_k)\eta_i+(-\Gamma_k)\xi_i\right)'\Psi_k^{-1}\left((\bar{I}-B_k)\eta_i+(-\Gamma_k)\xi_i\right)}. \tag{2.3}$$

Equation 2.4 represents an EM formulation of the complete log-likelihood (lnL_c) as the objective function for maximization:

$$\text{Ln}L_C = \sum_{i=1}^{I} \sum_{k=1}^{K} z_{ik} \ln(f(\eta_i|\xi_i, B_k, \Gamma_k, \Psi_k)) + \sum_{i=1}^{I} \sum_{k=1}^{K} z_{ik} \ln(\rho_k) \tag{2.4}$$

An EM formulation of the FIMIX-PLS algorithm (Table 2.3) is used for statistical computations to maximize the likelihood and to ensure convergence in this model. The expectation of Equation 2.4 is calculated in the E-step, where z_{ik} is 1

---

[1] Note that the following presentations slightly differ from Hahn et al.'s [31] original paper.

Table 2.2 Explanation of symbols

A_m	Number of exogenous variables as regressors in regression m
a_m	exogenous variable a_m with $a_m = 1, \ldots, A_m$
B_m	number of endogenous variables as regressors in regression m
b_m	endogenous variable b_m with $b_m = 1, \ldots, B_m$
$\gamma_{a_m mk}$	regression coefficient of a_m in regression m for class k
$\beta_{b_m mk}$	regression coefficient of b_m in regression m for class k
τ_{mk}	$((\gamma_{a_m mk}), (\beta_{b_m mk}))'$ vector of the regression coefficients
ω_{mk}	cell(m × m) of Ψ_k
c	constant factor
$f_{i\|k}(\cdot)$	probability for case i given a class k and parameters $(\cdot)$
I	number of cases or observations
i	case or observation i with $i = 1, \ldots, I$
J	number of exogenous variables
j	exogenous variable j with $j = 1, \ldots, J$
K	number of classes
k	class or segment k with $k = 1, \ldots, K$
M	number of endogenous variables
m	endogenous variable m with $m = 1, \ldots, M$
N_k	number of free parameters defined as $(K-1) + KR + KM$
P_{ik}	probability of membership of case i to class k
R	number of predictor variables of all regressions in the inner model
S	stop or convergence criterion
V	large negative number
X_{mi}	case values of the regressors for regression m of individual i
Y_{mi}	case values of the regressant for regression m of individual i
z_{ik}	$z_{ik} = 1$, if the case i belongs to class k; $z_{ik} = 0$ otherwise
ζ_i	random vector of residuals in the inner model for case i
η_i	vector of endogenous variables in the inner model for case i
ξ_i	vector of exogenous variables in the inner model for case i
B	$M \times M$ path coefficient matrix of the inner model for the relationships between endogenous latent variables
Γ	$M \times J$ path coefficient matrix of the inner model for the relationships between exogenous and endogenous latent variables
$\tilde{I}$	$M \times M$ identity matrix
Δ	difference of $current_{lnL_c}$ and $last_{lnL_c}$
B_k	$M \times M$ path coefficient matrix of the inner model for latent class k for the relationships between endogenous latent variables
Γ_k	$M \times J$ path coefficient matrix of the inner model for latent class k for the relationships between exogenous and endogenous latent variables
Ψ_k	$M \times M$ matrix for latent class k containing the regression variances
ρ	$(\rho_1, \ldots, \rho_K)$, vector of the K mixing proportions of the finite mixture
ρ_k	mixing proportion of latent class k

iff subject i belongs to class k (or 0 otherwise). The mixing proportion ρ_k (i.e., the relative segment size) and the parameters ξ_i, B_k, Γ_k, and Ψ_k of the conditional probability function are given (as results of the M-step), and provisional estimates (expected values) $E(z_{ik}) = P_{ik}$, for z_{ik} are computed according to Bayes's [5] theorem (Table 2.3).

Table 2.3 The FIMIX-PLS algorithm

set random starting values for P_{ik} ; set $last_{lnL_C} = V$; set $0 < S < 1$
// *run initial M-step*

// *run EM-algorithm until convergence*
repeat do
// *the E-step starts here*
if $\Delta \geq S$ then
$P_{ik} = \frac{\rho_k f_{i|k}(\eta_i|\xi_i,B_k,\Gamma_k,\Psi_k)}{\sum_{k=1}^{K} \rho_k f_{i|k}(\eta_i|\xi_i,B_k,\Gamma_k,\Psi_k)} \forall i,k$
$last_{lnL_C} = current_{lnL_C}$

// *the M-step starts here*
$\rho_k = \frac{\sum_{i=1}^{l} P_{ik}}{I} \forall k$
determine B_k, Γ_k, Ψ_k, $\forall k$
calculate $current_{lnL_C}$
$\Delta = current_{lnL_C} - last_{lnL_C}$
until $\Delta < S$

Equation 2.4 is maximized in the M-step (Table 2.3). This part of the FIMIX-PLS algorithm accounts for the most important changes in order to fit the finite mixture approach to PLS path modeling, compared to the original finite mixture structural equation modeling technique [37]. Initially, we calculate new mixing proportions ρ_k through the average of the adjusted expected values P_{ik} that result from the previous E-step. Thereafter, optimal parameters are determined for B_k, Γ_k, and Ψ_k through independent OLS regressions (one for each relationship between the latent variables in the inner model). The ML estimators of coefficients and variances are assumed to be identical to OLS predictions. We subsequently apply the following equations to obtain the regression parameters for endogenous latent variables:

$$Y_{mi} = \eta_{mi} \quad and \quad X_{mi} = (E_{mi}, N_{mi})' \tag{2.5}$$

$$E_{mi} = \begin{cases} \{\xi_1, ..., \xi_{A_m}\}, A_m \geq 1, a_m = 1, ..., A_m \wedge \xi_{a_m} \text{ is regressor of } m \\ \emptyset \text{ else} \end{cases} \tag{2.6}$$

$$N_{mi} = \begin{cases} \{\eta_1, ..., \eta_{B_m}\} B_m \geq 1, b_m = 1, ..., B_m \wedge \eta_{b_m} \text{ is regressor of } m \\ \emptyset \text{ else} \end{cases} \tag{2.7}$$

The closed-form OLS analytic formula for τ_{mk} and ω_{mk} is expressed as follows:

$$\tau_{mk} = [X'_m P_k X_m]^{-1} [X'_m P_k Y_m] \tag{2.8}$$

$$\omega_{mk} = \left[(Y_m - X_m \tau_{mk})' \left((Y_m - X_m \tau_{mk}) P_k \right) \right] / \widetilde{I} \rho_k \qquad (2.9)$$

As a result, the M-step determines the new mixing proportions ρ_k, and the independent OLS regressions are used in the next E-step iteration to improve the outcomes of P_{ik}. The EM algorithm stops whenever lnL_C no longer improves noticeably, and an a priori-specified convergence criterion is reached.

2.3.3 Systematic Application of FIMIX-PLS

To fully exploit the capabilities of the approach, we propose the systematic approach to FIMIX-PLS clustering as depicted in Fig. 2.3. In FIMIX-PLS step 1, the basic PLS algorithm provides path modeling results, using the aggregate set of data. Step 2 uses the resulting latent variable scores in the inner path model to run the FIMIX-PLS algorithm as described above. The most important computational results of this step are the probabilities P_{ik}, the mixing proportions ρ_k, class-specific estimates B_k and Γ_k for the inner relationships of the path model, and Ψ_k for the (unexplained) regression variances.

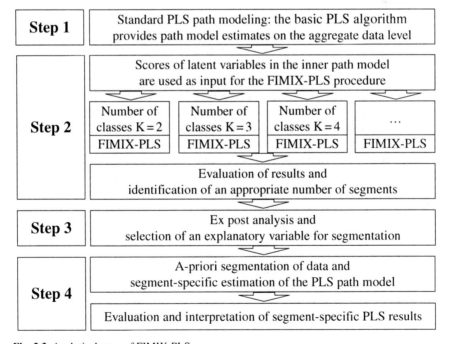

Fig. 2.3 Analytical steps of FIMIX-PLS

The methodology fits each observation with the finite mixture's probabilities P_{ik} into each of the predetermined number of classes. However, on the basis of the FIMIX-PLS results, it must be specifically decided whether the approach detects and treats heterogeneity in the inner PLS path model estimates by (unobservable) discrete moderating factors. This objective is explored in step 2 by analyzing the results of different numbers of K classes (approaches to guide this decision are presented in the next section).

When applying FIMIX-PLS, the number of segments is usually unknown. The process of identifying an appropriate number of classes is not straightforward. For various reasons, there is no statistically satisfactory solution for this analytical procedure [77]. One such reason is that the mixture models are not asymptotically chi-square distributed and do not allow the calculation of the likelihood ratio statistic with respect to obtaining a clear-cut decision criterion. Another reason is that the EM algorithm converges for any given number of K classes. One never knows if FIMIX-PLS stops at a local optimum solution. The algorithm should be started several times (e.g., 10 times) for each number of segments for different starting partitions [47]. Thereafter, the analysis should draw on the maximum log-likelihood outcome of each alternative number of classes. Moreover, the FIMIX-PLS model may result in the computation of non-interpretable segments for endogenous latent variables with respect to the class-specific estimates B_k and Γ_k of the inner path model relationships and with respect to the regression variances Ψ_k when the number of segments is increased. Consequently, segment size is a useful indicator to stop the analysis of additional numbers of latent classes to avoid incomprehensible FIMIX-PLS results. At a certain point, an additional segment is just very small, which explains the marginal heterogeneity in the overall data set.

In practical applications, researchers can compare estimates of different segment solutions by means of heuristic measures such as Akaike's information criterion (AIC), consistent AIC (CAIC), or Bayesian information criterion (BIC). These information criteria are based on a penalized form of the likelihood, as they simultaneously take a model's goodness-of-fit (likelihood) and the number of parameters used to achieve that fit into account. Information criteria generally favor models with a large log-likelihood and few parameters and are scaled so that a lower value represents a better fit. Operationally, researchers examine several competing models with varying numbers of segments and pick the model which minimizes the value of the information criterion. Researchers usually use a combination of criteria and simultaneously revert to logical considerations to guide the decision.

Although the preceding heuristics explain over-parameterization through the integration of a penalty term, they do not ensure that the segments are sufficiently separated in the selected solution. As the targeting of markets requires segments to be differentiable, i.e., the segments are conceptually distinguishable and respond differently to certain marketing mix elements and programs [40], this point is of great practical interest. Classification criteria that are based on an entropy statistic, which indicates the degree of separation between segments, can help to assess whether the analysis produces well-separated clusters [77]. Within this context, the normed entropy statistic [EN; 58] is a critical criterion for analyzing segment-specific FIMIX-PLS results. This criterion indicates the degree of all observations' classification and their estimated segment membership probabilities P_{ik} on a case-by-case basis and subsequently reveals the most appropriate number of latent segments for a clear-cut segmentation:

$$\text{EN}_K = 1 - \frac{[\sum_i \sum_k - P_{ik} ln(P_{ik})]}{I ln(K)} \qquad (2.10)$$

The EN ranges between 0 and 1 and the quality of the classification commensurates with the increase in EN_K. The more the observations exhibit high membership probabilities (e.g., higher than 0.7), the better they uniquely belong to a specific class and can thus be properly classified in accordance with high EN values. Hence, the entropy criterion is especially relevant for assessing whether a FIMIX-PLS solution is interpretable or not. Applications of FIMIX-PLS provide evidence that EN values above 0.5 result in estimates of P_{ik} that permit unambiguous segmentation [66, 71, 72].

An explanatory variable must be uncovered in the ex post analysis (step 3) in situations where FIMIX-PLS results indicate that heterogeneity in the overall data set can be reduced through segmentation by using the best fitting number of K classes. In this step, data are classified by means of an explanatory variable, which serves as input for segment-specific computations with PLS path modeling. An explanatory variable must include both the similar grouping of data, as indicated by the FIMIX-PLS results, and the interpretability of the distinctive clusters. However, the ex post analysis is a very challenging FIMIX-PLS analytical step. Ramaswamy et al. [58] propose a statistical procedure to conduct an ex post analysis of the estimated FIMIX-PLS probabilities. Logistic regressions, or in the case of large data sets, CHAID analyses, and classification and regression trees [9] may likewise be applied to identify variables that can be used to classify additional observations in one of the designed segments. While these systematic searches uncover explanatory variables that fit the FIMIX-PLS results well in terms of data structure, a logical search, in contrast, mostly focuses on the interpretation of results. In this case, certain variables with high relevance with respect to explaining the expected differences in segment-specific PLS path model computations are examined regarding their ability to form groups of observations that match FIMIX-PLS results.

The process of identifying an explanatory variable is essential for exploiting FIMIX-PLS results. The findings are also valuable to researchers to confirm that unobserved heterogeneity in the path model estimates is not an issue, or they allow this problem to be dealt with by means of segmentation and, thereby, facilitate multigroup PLS path modeling analyses [13, 32] in step 4. Significantly different group-specific path model estimations impart further differentiated interpretations of PLS modeling results and may foster the origination of more effective strategies.

2.4 Application of FIMIX-PLS

2.4.1 On Measuring Customer Satisfaction

When researchers work with empirical data and do not have a priori segmentation assumptions to capture heterogeneity in the inner PLS path model relationships, FIMIX-PLS is often not as clear-cut as in the simulation studies presented by Ringle [61] as well as Esposito Vinzi et al. [16]. To date, research efforts to apply FIMIX-PLS and assess its usefulness with respect to expanding the methodological toolbox were restricted by the lack of statistical software programs for this kind of analysis. Since such functionalities have recently been provided as a module in the SmartPLS software, FIMIX-PLS can be applied more easily to empirical data, thereby increasing our knowledge of the approach and its applicability. As a means of presenting the benefits of the method for PLS path modeling in marketing research, we focus on customer satisfaction to identify and treat heterogeneity in consumers through segmentation. However, the general approach of this analysis can be applied to any PLS application such as the various TAM model estimations in MIS.

Customer satisfaction has become a fundamental and well-documented construct in marketing that is critical with respect to demand and for any business's success given its importance and established relation with customer retention and corporate profitability [2, 52, 53]. Although it is often acknowledged that there are no truly homogeneous segments of consumers, recent studies report that there is indeed substantial unobserved customer heterogeneity within a given product or service class [81]. Dealing with this unobserved heterogeneity in the overall sample is critical for forming groups of consumers that are homogeneous in terms of the benefits that they seek or their response to marketing programs (e.g., product offering, price discounts). Segmentation is therefore a key element for marketers in developing and improving their targeted marketing strategies.

2.4.2 Data and Measures

We applied FIMIX-PLS to the ACSI model to measure customer satisfaction as presented by Fornell et al. [21] in the *Journal of Marketing* but used empirical data from their subsequent survey in 1999.[2] These data are collected quarterly to assess customers' overall satisfaction with the services and products that they buy from a number of organizations. The ACSI study has been conducted since 1994 for consumers of 200 publicly traded Fortune 500 firms as well as several US public administration and government departments. These firms and departments comprise more than 40% of the US gross domestic product. The sample selection mechanism ensures that all

---

[2] The data were provided by Fornell, Claes. AMERICAN CUSTOMER SATISFACTION INDEX, 1999 [Computer file]. ICPSR04436-v1. Ann Arbor, MI: University of Michigan. Ross School of Business, National Quality Research Center/Reston, VA: Wirthlin Worldwide [producers], 1999. Ann Arbor, MI: Inter-University Consortium for Political and Social Research [distributor], 2006-06-09. We would like to thank Claes Fornell and the ICPSR for making the data available.

types of organizations are included across all economic sectors considered. For the 1999 survey, about 250 consumers of each organization's products/services were selected via telephone. Each call identified the person in the household (for household sizes >1) whose birthday was closest, after which this person (if older than 18 years) was asked about the durables he or she had purchased during the last 3 years and about the nondurables purchased during the last month. If the products or services mentioned originated from one of the 200 organizations, a short questionnaire was administered that contained the measures described in Table 2.4.

The data-gathering process was carried out in such a manner that the final data were comparable across industries [21]. The ACSI data set has frequently been used in diverse areas in the marketing field, using substantially different methodologies such as event history modeling or simultaneous equations modeling. However, past research has not yet accounted for unobserved heterogeneity.

Table 2.4 Measurement scales, items, and descriptive statistics

Construct	Items
Overall Customer Satisfaction	Overall satisfaction
	Expectancy disconfirmation (performance falls short of or exceeds expectations)
	Performance versus the customer's ideal product or service in the category
Customer Expectations of Quality	Overall expectations of quality (prior to purchase)
	Expectation regarding customization, or how well the product fits the customer's personal requirements (prior to purchase)
	Expectation regarding reliability, or how often things would go wrong (prior to purchase)
Perceived Quality	Overall evaluation of quality experience (after purchase)
	Evaluation of customization experience, or how well the product fits the customer's personal requirements (after purchase)
	Evaluation of reliability experience, or how often things have gone wrong (after purchase)
Perceived Value	Rating of quality given price
	Rating of price given quality
Customer Complaints	Has the customer complained either formally or informally about the product or service?
Customer Loyalty	Likelihood rating prior to purchase
Covariates	
Age	Average = 43, Standard deviation = 15, minimum = 18, maximum = 84
Gender	42% male, 58% female
Education	Less than high school = 4.8%, high school graduate = 21.9%, some college = 34.6%, college graduate = 23.1%, post graduate = 15.5%
Race	White = 82.4%, Black/African American = 7.2%, American Indian = 1.1%, Asian or Pacific Islander = 1.8%, other race = 3.7%
Total Annual Family Income	Under $20.000 = 13.5%, $20.000–$30.000 = 13.9%, $30.000– $40.000 = 14.9%, $40.000–$60.000 = 22.3%, $60.000–$80.000 = 15.1%, $80.000–$100.000 = 8.4%, Over $100.000 = 11.9%

To illustrate the capabilities of FIMIX-PLS, we used data from the first quarter of 1999 ($N = 17,265$). To ensure the validity of our analysis, we adjusted the data set by carrying out a missing value analysis. In standard PLS estimations, researchers frequently revert to mean replacement algorithms. However, when replacing relatively high numbers by missing values per variable and case by mean values, FIMIX-PLS will most likely form its own segment of these observations. Consequently, we applied case-wise replacement. As this procedure would have led to the exclusion of a vast number of observations, we decided to reduce the original ACSI model as presented by Fornell et al. [21]. Consequently, we excluded two items from the "Customer Loyalty" construct, as they had a high number of missing values. Furthermore, we omitted the construct "Customer Complaints," measured by a binary single item, because we wanted to use this variable as an explanatory variable in the ex post analysis (step 3 in Fig. 2.3).

As our goal is to demonstrate the applicability of FIMIX-PLS regarding empirical data and to illustrate a cause–effect relationship model with respect to customer satisfaction, we do not regard the slight change in the model setup as a debilitating factor. Consequently, the final sample comprised $N = 10,417$ observations. Figure 2.1 illustrates the path model under consideration.

Fornell et al. [21] identified the three driver constructs "Perceived Quality," "Customer Expectations of Quality," and "Perceived Value," which are measured by three and two reflective indicators, with respect to "Overall Customer Satisfaction." The ACSI construct itself directly relates to the "Customer Loyalty" construct. Both latent variables also employ a reflective measurement operationalization. Table 2.4 provides the measurement scales and the items used in our study plus various descriptive statistics of the full sample.

2.4.3 Data Analysis and Results

Methodological considerations that are relevant to the analysis include the assessment of the measures' reliability, their discriminant validity. As the primary concern of the FIMIX-PLS algorithm is to capture heterogeneity in the inner model, the focus of the comparison lies on the evaluation of the overall goodness-of-fit of the models. Nevertheless, as the existence of reliable and valid measures is a prerequisite for deriving meaningful solutions, we also deal with these aspects.

As depicted in Fig. 2.3, the basic PLS algorithm [43] is applied to estimate the overall model by using the SmartPLS 2.0 [64] in step 1. To evaluate the PLS estimates, we follow the suggestions by Chin [8] and Henseler et al. [32]. On assessing the empirical results, almost all factor loadings exhibit very high values of above 0.8. The smallest loading of 0.629 still ranges well above the commonly suggested threshold value of 0.5 [35], thus supporting item reliability. Composite reliability is assessed by means of composite reliability ρ_c and Cronbach's α. Both measures' values are uniformly high around 0.8, thus meeting the stipulated thresholds [56]. To assess the discriminant validity of the reflective measures, two approaches are applied. First, the indicators' cross loadings are examined, which reveals that no indicator loads higher on the opposing endogenous constructs. Second, the Fornell and Larcker [22] criterion is applied, in which the square root of each endogenous construct's average variance extracted (AVE) is compared with its bivariate correlations with all opposing endogenous constructs [cp. 28, 32]. The results show that in all cases, the square root of AVE is greater than the variance shared by each construct and its opposing constructs. Consequently, we can also presume a high degree of discriminant validity with respect to all constructs in this study.

The central criterion for the evaluation of the inner model is the R^2. Whereas ACSI exhibits a highly satisfactory R^2 value of 0.777, all other constructs show only moderate values of below 0.5 (Table 2.8).

In addition to the evaluation of R^2 values, researchers frequently revert to the cross-validated redundancy measure Q^2 (Stone–Geisser test), which has been developed to assess the predictive validity of the exogenous latent variables and can be computed using the blindfolding procedure. Values greater than zero imply that the exogenous constructs have predictive relevance for the endogenous construct under consideration, whereas values below zero reveal a lack of predictive relevance [8]. All Q^2 values range significantly above zero, thus indicating the exogenous constructs' high predictive power. Another important analysis concerns the significance of hypothized relationships between the latent constructs. For example, "Perceived Quality" as well as "Perceived Value" exert a strong positive influence on the endogenous variable "Overall Customer Satisfaction," whereas the effect of "Customer Expectations of Quality" is close to zero. To test whether path coefficients differ significantly from zero, t values were calculated using bootstrapping with 10,417 cases and 5000 subsamples [32]. The analysis reveals that all relationships in the inner path model exhibit statistically significant estimates (Table 2.8).

In the next analytical step, the FIMIX-PLS module of SmartPLS was applied to segment observations based on the estimated latent variable scores (step 2 in Fig. 2.3). Initially, FIMIX-PLS results are computed for two segments (see settings in Fig. 2.4). Thereafter, the number of segments is increased sequentially. A comparison of the segment-specific information and classification criteria, as presented in Table 2.5, reveals that the choice of two groups is appropriate for customer segmentation purposes. All relevant evaluation criteria increase considerably in the ensuing numbers of classes.

The choice of two segments is additionally supported by the EN value of 0.504. As illustrated in Table 2.6, more than 80% of all our observations are assigned to one of the two segments with a probability P_{ik} of more than 0.7. These probabilities decline considerably with respect to higher numbers of K classes, which indicates an increased segmentation fuzziness that is also depicted by the lower EN. An EN of 0.5 or higher for a certain number of segments allows the unambiguous segmentation of data.

Next, observations are assigned to each segment according to their segment membership's maximum probability. Table 2.7 shows the segment sizes with respect to the different segment solutions, which allows the heterogeneity that affects the analysis to be specified: (a) As the number of segments increases, the smaller seg-

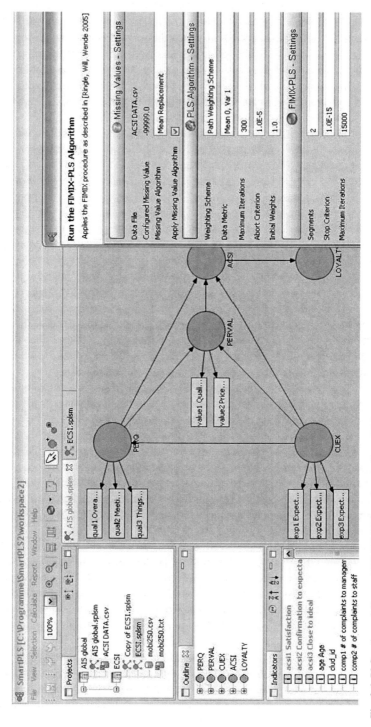

Fig. 2.4 PLS path modeling and FIMIX-PLS settings in SmartPLS

Table 2.5 Information and classification criteria for varying K

K	lnL	AIC	BIC	CAIC	EN
2	−44,116.354	88,278.708	88,445.486	88,468.486	0.504
3	−46,735.906	93,541.811	93,795.563	93,830.563	0.431
4	−47,276.720	94,647.440	94,988.246	95,035.246	0.494
5	−49,061.353	98,240.706	98,668.527	98,727.527	0.447
6	−50,058.503	100,259.006	100,773.840	100,844.840	0.443

Table 2.6 Overview of observations' highest probability of segment membership

P_{ik}	K = 2	K = 3	K = 4	K = 5	K = 6
[0.9, 1.0]	0.510	0.158	0.134	0.054	0.046
[0.8, 0.9)	0.211	0.279	0.237	0.093	0.061
[0.7, 0.8)	0.118	0.182	0.195	0.253	0.171
[0.6, 0.7)	0.090	0.153	0.173	0.225	0.198
[0.5, 0.6)	0.071	0.142	0.151	0.198	0.236
[0.4, 0.5)		0.076	0.087	0.147	0.225
[0.3, 0.4)		0.009	0.022	0.030	0.061
[0.2, 0.3)			0.001		0.002
[0.1, 0.2)					
[0, 0.1)					
Sum	1.000	1.000	1.000	1.000	1.000

Table 2.7 Segment sizes for different numbers of segments

K	ρ_1	ρ_2	ρ_3	ρ_4	ρ_5	ρ_6	$\sum_k \rho_k$
2	0.673	0.327					1.000
3	0.179	0.219	0.602				1.000
4	0.592	0.075	0.075	0.258			1.000
5	0.534	0.036	0.245	0.096	0.089		1.000
6	0.079	0.313	0.449	0.037	0.081	0.041	1.000

ment is gradually split up to create additional segments, while the size of the larger segment remains relatively stable (about 0.6 for $K \in \{2, 3, 4\}$ and 0.5 for $K \in \{5, 6\}$). (b) The decline in the outcomes of additional numbers of classes based on the EN criterion allows us to conclude that the overall set of observations regarding this particular analysis of the ACSI consists of a large, stable segment and a small fuzzy one. (c) FIMIX-PLS cannot further reduce the fuzziness of the smaller segment.

In the process of increasing the number of segments, FIMIX-PLS can still identify the larger segment with comparably high probabilities of membership but is ambivalent when processing the small group with heterogeneous observations. Consequently, the probability of membership P_{ik} declines, resulting in decreasing

EN values. This indicates methodological complexity in the process of assigning the observations in this data set to additional segments. FIMIX-PLS computation forces observations to fit within a given number of K classes. As a result, FIMIX-PLS generates outcomes that are statistically problematic for the segment-specific estimates B_k and for Γ_k, i.e., regarding the inner relationships of the path model, and for Ψ_k, i.e., regarding the regression variances of endogenous latent variables. In this example, results exhibiting inner path model relationships and/or regression variances above one are obtained with respect to $K = 7$ classes. Consequently, the analysis of additional numbers of classes can stop at this juncture in accordance with the development of segment sizes in Table 2.7.

Table 2.8 presents the global model and FIMIX-PLS results of two latent segments. Before evaluating goodness-of-fit measures and inner model relationships, all outcomes with respect to segment-specific path model estimations were tested with regard to reliability and discriminant validity. The analysis showed that all measures satisfy the relevant criteria for model evaluation [32]. As in the global model, all paths are significant at a level of 0.01.

When comparing the global model with the results derived from FIMIX-PLS, one finds that the relative importance of the driver constructs "Overall Customer Satisfaction" differs quite substantially within the two segments. For example, the global model suggests that the perceived quality is the most important driver construct with

Table 2.8 Global model and FIMIX-PLS results of two latent segments

	Global	FIMIX-PLS		
		k = 1	k = 2	t[mgp]
Customer Expectations of Quality → Perceived Quality	0.556*** (56.755)	0.807*** (168.463)	0.258*** (13.643)	26.790***
Customer Expectations of Quality → Perceived Value	0.072*** (7.101)	0.218*** (16.619)	−0.107*** (6.982)	15.571***
Customer Expectations of Quality → Overall Customer Satisfaction	0.021*** (3.294)	0.117*** (14.974)	−0.068*** (6.726)	14.088***
Perceived Quality → Overall Customer Satisfaction	0.557*** (63.433)	0.425*** (50.307)	0.633*** (49.038)	10.667***
Perceived Quality → Perceived Value	0.619*** (62.943)	0.582*** (46.793)	0.544*** (42.394)	1.899**
Perceived Value → Overall Customer Satisfaction	0.394*** (44.846)	0.455*** (62.425)	0.308*** (21.495)	7.922***
Overall Customer Satisfaction → Customer Loyalty	0.687*** (93.895)	0.839*** (208.649)	0.481*** (31.794)	19.834***
ρ_k	1.000	0.673	0.327	
$R^2_{\text{Perceived Quality}}$	0.309	0.651	0.067	
$R^2_{\text{Perceived Value}}$	0.439	0.591	0.277	
$R^2_{\text{Overall Customer Satisfaction}}$	0.777	0.848	0.679	
$R^2_{\text{Customer Loyalty}}$	0.471	0.704	0.231	

t[mgp] = t-value for multi-group comparison test
*** sig. at 0.01, **sig. at 0.05, *sig. at 0.1

respect to customer overall satisfaction. As "Perceived Quality" describes an ex post evaluation of quality, companies should emphasize product and service quality and their fit with use, which can be achieved through informative advertising. However, in the first segment of FIMIX-PLS, the most important driver construct with respect to customer satisfaction is "Perceived Value." In addition, also "Customer Expectations of Quality" exerts an increased positive influence on customer satisfaction. Likewise, both segments differ considerably with regard to the relationships between the three driver constructs "Perceived Quality," "Customer Expectations of Quality," and "Perceived Value."

However, only significant differences between the segments offer valuable interpretations for marketing practice. Consequently, we performed a multigroup comparison to assess whether segment-specific path coefficients differ significantly. The PLS path modeling multigroup analysis (PLS-MGA) applies the permutation test (5000 permutations) as described by [13] and which has recently been implemented as an experimental module in the SmartPLS software.

Multigroup comparison results show that all paths differ significantly between $k = 1$ and $k = 2$. Thus, consumers in each segment exhibit significantly different drivers with respect to their overall satisfaction, which allows differentiated marketing activities to satisfy customers' varying wants better. At the same time, all endogenous constructs have increased R^2 values, ranging between 2% ("Overall Customer Satisfaction") and 49% ("Perceived Quality") higher than in the global model. These were calculated as the sum of each endogenous construct's R^2 values across the two segments, weighted by the relative segment size.

The next step involves the identification of explanatory variables that best characterize the two uncovered customer segments. We consequently applied the QUEST [42] and Exhaustive CHAID [6] algorithm, using SPSS Answer Tree 3.1 on the covariates to assess if splitting the sample according to the sociodemographic variables' modalities leads to a statistically significant discrimination in the dependent measure. In the latter, continuous covariates were first transformed into ordinal predictors. In both approaches, "age" and "total annual family income" showed the greatest potential for meaningful a priori segmentation, with Exhaustive CHAID producing more accurate results. The result is shown in Fig. 2.5. The percentages in the nodes denote the share of total observations (as described in the root node) with respect to each segment. These mark the basis of the a priori segmentation of observations based on the maximum percentages for each node.

Segment one ($n_{k1} = 6,314$) comprises middle-aged customers ($age \in (28, 44]$) with a total annual family income between $40,000 and less than $100,000. Furthermore, customers aged 44 and above belong to this segment. Segment two ($n_{k2} = 4,103$) consists of young customers (age ≤ 28) as well as middle-aged customers with a total annual family income of less than $40,000 or more than $100,000. The resulting classification corresponds to 56.878% of the FIMIX-PLS classification. In addition to this clustering according to sociodemographic variables, we used the behavioral variable "Customer Complaints" (Table 2.4) to segment the data. Segment one ($n_{k1} = 7,393$) represents customers that have not yet complained about a product or service, whereas segment two ($n_{k2} = 3,023$) contains customers who

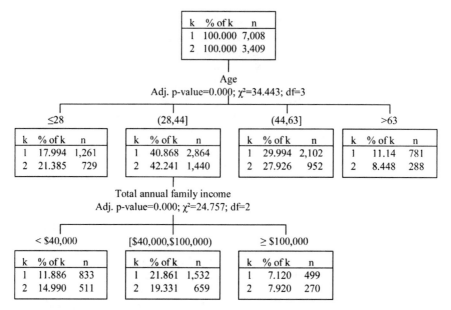

Fig. 2.5 Segmentation tree results of the exhaustive CHAID analysis

have complained in the past (consistency with FIMIX-PLS classification: 62.811%). Table 2.9 documents the results of the ex post analysis. The evaluation of the PLS path modeling estimates [8] with respect to these four a priori segmented data sets confirms that the results are satisfactory.

Similar results as those with the FIMIX-PLS analysis were obtained with regard to the ex post analysis using the Exhaustive CHAID algorithm. Again, the goodness-of-fit measures of the first segment exhibit increased values. Furthermore, the path coefficients differ significantly between the two segments. For example, the large segment exhibits a substantial relationship between "Customer Expectations of Quality" and "Overall Customer Satisfaction," which is highly relevant from a marketing perspective. With respect to this group of mostly older consumers, satisfaction is also explained by expected quality, which can potentially be controlled by marketing activities. For example, non-informative advertising (e.g., sponsorship programs) can primarily be used as a signal of expected product quality [39, 51]. However, it must be noted that with respect to the global model, the differences are less pronounced than those in the FIMIX-PLS analysis. Even though there are several differences observable, the path coefficient estimates are more balanced across the two segments, thus diluting response-based segmentation results. Similar figures result with respect to the ex post analysis based on the variable "customer complaints."

Despite the encouraging results of the ex post analysis, the analysis showed that the covariates available in the ACSI data set only offer a limited potential for meaningful a priori segmentation. Even though one segment's results improved, the dif-

Table 2.9 Inner model path coefficients with *t* values and goodness-of-fit measures

	Global	Ex Post CHAID			Ex Post Cust. Compl.		
		k = 1	k = 2	t[mgp]	k = 1	K = 2	t[mgp]
Customer Expectations of Quality → Perceived Quality	0.556*** (56.755)	0.575*** (45.184)	0.526*** (29.956)	2.599***	0.589*** (50.844)	0.511*** (29.224)	3.457***
Customer Expectations of Quality → Perceived Value	0.072*** (7.101)	0.072*** (5.511)	0.067*** (3.581)	0.246	0.089*** (6.169)	0.071*** (4.025)	0.825
Customer Expectations of Quality → Overall Customer Satisfaction	0.021*** (3.294)	0.036*** (4.334)	−0.002 (0.252)	2.761***	0.047*** (5.336)	−0.001 (0.094)	3.252***
Perceived Quality → Overall Customer Satisfaction	0.557*** (63.433)	0.548*** (45.653)	0.572*** (35.861)	1.283*	0.517*** (45.217)	0.578*** (35.806)	3.179***
Perceived Quality → Perceived Value	0.619*** (62.943)	0.635*** (50.417)	0.599*** (36.269)	1.716**	0.519*** (35.082)	0.659*** (43.397)	6.518***
Perceived Value → Overall Customer Satisfaction	0.394*** (44.846)	0.400*** (34.377)	0.384*** (24.571)	0.819	0.402*** (35.728)	0.390*** (23.599)	0.608
Overall Customer Satisfaction → Customer Loyalty	0.687*** (93.895)	0.677*** (68.975)	0.698*** (57.207)	1.440*	0.616*** (58.929)	0.705*** (57.805)	6.237***
ρ_k	1	0.606	0.394	0.710	0.290		
$R^2_{\text{Perceived Quality}}$	0.309	0.331	0.277	0.347	0.261		
$R^2_{\text{Perceived Value}}$	0.439	0.461	0.406	0.332	0.488		
$R^2_{\text{Overall Customer Satisfaction}}$	0.777	0.793	0.752	0.713	0.798		
$R^2_{\text{Customer Loyalty}}$	0.471	0.459	0.488	0.380	0.497		

t[mgp] = t-value for multi-group comparison test
*** sig. at 0.01, ** sig. at 0.05, * sig. at 0.1

ferences between the segments were considerably smaller when compared to those
of the FIMIX-PLS results.

2.5 Summary and Conclusion

Unobserved heterogeneity and measurement errors are common problems in social
sciences. Jedidi et al. [37] have addressed these problems with respect to CBSEM.
Hahn et al. [31] have further developed their finite mixture SEM methodology for
PLS path modeling, which is an important alternative to CBSEM for researchers
and practitioners. This chapter introduced and discussed the FIMIX-PLS approach,
as it has recently been implemented in the software application SmartPLS. Consequently, researchers from marketing and other disciplines can exploit this approach
to response-based segmentation by identifying certain groups of customers. We
demonstrate the potentials of FIMIX-PLS by applying the procedure on data from
the ACSI model. We thus extend prior research work on this important model by
explaining unobserved heterogeneity in the inner model path estimates. Moreover,
we show that, contrary to existing work on the same data set, there are different
segments, which has significant implications.

Our example application demonstrates how FIMIX-PLS reliably identifies an appropriate number of customer segments, provided that unobserved moderating factors account for consumer heterogeneity within inner model path relationships. In
this kind of very likely situation, FIMIX-PLS enables us to identify two segments
with distinct inner model path estimates that differ substantially from the aggregate-
level analysis. For example, unlike in the global model, "Customer Expectations of
Quality" exerts a pronounced influence on the customers' perceived value. Furthermore, the FIMIX-PLS analysis achieved a considerably increased model fit in the
larger segment.

In the course of an ex post analysis, two explanatory variables ("Age" and "Total Annual Family Income") were uncovered. An a priori segmentation based on
the exhaustive CHAID analysis results, followed by segment-specific path analyses yielded similar findings as the FIMIX-PLS procedure. The same holds for segmenting along the modalities of the behavioral variable "Customer Complaints."
These findings allow marketers to formulate differentiated, segment-specific marketing activities to better satisfy customers' varying wants. Researchers can exploit
these additional analytic potentials where theory essentially supports path modeling
in situations with heterogeneous data. We expect that these conditions will hold true
in many marketing-related path modeling applications.

Future research will require the extensive use of FIMIX-PLS on marketing examples with heterogeneous data to illustrate the applicability and the problematic
aspects of the approach from a practical point of view. Researchers will also need
to test the FIMIX-PLS methodology by means of simulated data with a wide range
of statistical distributions and a large variety of path model setups to gain additional
implications. Finally, theoretical research should provide satisfactory improvements

of problematic areas such as convergence to local optimum solutions, computation of improper segment-specific FIMIX-PLS results, and identification of suitable explanatory variables for a priori segmentation. These critical aspects have been discussed, for example, by Ringle [61] and Sarstedt [74]. By addressing these deficiencies, the effectiveness and precision of the approach could be extended, thus further extending the analytical ground of PLS path modeling.

References

1. Ritu Agarwal and Elena Karahanna. Time flies when you're having fun: Cognitive absorption and beliefs about information technology usage. *MIS Quarterly*, 24(4):665–694, 2000.
2. Eugene W. Anderson, Claes Fornell, and Donald. R. Lehmann. Customer satisfaction, market share and profitability: Findings from Sweden. *Journal of Marketing*, 58(3):53–66, 1994.
3. Richard P. Bagozzi. On the meaning of formative measurement and how it differs from reflective measurement: Comment on Howell, Breivik, and Wilcox (2007). *Psychological Methods*, 12(2):229–237, 2007.
4. Richard P. Bagozzi and Youjae Yi. Advanced topics in structural equation models. In Richard P. Bagozzi, editor, *Principles of Marketing Research*, pages 1–52. Blackwell, Oxford, 1994.
5. Thomas Bayes. Studies in the history of probability and statistics: IX. Thomas Bayes's essay towards solving a problem in the doctrine of chances; Bayes's essay in modernized notation. *Biometrika*, 45:296–315, 1763/1958.
6. David Biggs, Barry de Ville, and Ed Suen. A method of choosing multiway partitions for classification and decision trees. *Journal of Applied Statistics*, 18(1):49–62, 1991.
7. Kenneth A. Bollen. Interpretational confounding is due to misspecification, not to type of indicator: Comment on Howell, Breivik, and Wilcox (2007). *Psychological Methods*, 12(2):219–228, 2007.
8. Wynne W. Chin. The partial least squares approach to structural equation modeling. In George A. Marcoulides, editor, *Modern Methods for Business Research*, pages 295–358. Lawrence Erlbaum, Mahwah, NJ, 1998.
9. Sven F. Crone, Stefan Lessmann, and Robert Stahlbock. The impact of preprocessing on data mining: An evaluation of classifier sensitivity in direct marketing. *European Journal of Operational Research*, 173(3):781–800, 2006.
10. Fred D. Davis. Perceived usefulness, perceived ease of use, and user acceptance of information technology. *MIS Quarterly*, 13(3):319–340, 1989.
11. Adamantios Diamantopoulos. The C-OAR-SE procedure for scale development in marketing: A comment. *International Journal of Research in Marketing*, 22(1):1–10, 2005.
12. Adamantios Diamantopoulos, Petra Riefler, and Katharina P. Roth. Advancing formative measurement models. *Journal of Business Research*, 61(12):1203–1218, 2008.
13. Jens Dibbern and Wynne W. Chin. Multi-group comparison: Testing a PLS model on the sourcing of application software services across germany and the usa using a permutation based algorithm. In Friedhelm W. Bliemel, Andreas Eggert, Georg Fassott, and Jörg Henseler, editors, *Handbuch PLS-Pfadmodellierung. Methode, Anwendung, Praxisbeispiele*, pages 135–160. Schäffer-Poeschel, Stuttgart, 2005.
14. William Dumouchel, Chris Volinsky, Theodore Johnson, Corinna Cortes, and Daryl Pregibon. Squashing flat files flatter. In *Proceedings of the 5th ACM SIGKDD International Conference on Knowledge Discovery in Data Mining*, pages 6–15, San Diego, CA, 1999. ACM Press.
15. Jacob Eskildsen, Kai Kristensen, Hans J. Juhl, and Peder Østergaard. The drivers of customer satisfaction and loyalty: The case of Denmark 2000–2002. *Total Quality Management*, 15(5-6):859–868, 2004.

16. Vincenzo Esposito Vinzi, Christian M. Ringle, Silvia Squillacciotti, and Laura Trinchera. Capturing and treating unobserved heterogeneity by response based segmentation in PLS path modeling: A comparison of alternative methods by computational experiments. Working Paper No. 07019, ESSEC Business School Paris-Singapore, 2007.

17. Vincenzo Esposito Vinzi, Laura Trinchera, Silvia Squillacciotti, and Michel Tenenhaus. REBUS-PLS: A response-based procedure for detecting unit segments in PLS path modeling. *Applied Stochastic Models in Business and Industry*, 24(5):439–458, 2008.

18. Fabian Festge and Manfred Schwaiger. The drivers of customer satisfaction with industrial goods: An international study. In Charles R. Taylor and Doo-Hee Lee, editors, *Advances in International Marketing – Cross-Cultural Buyer Behavior*, volume 18, pages 179–207. Elsevier, Amsterdam, 2007.

19. Adam Finn and Ujwal Kayande. How fine is C-OAR-SE? a generalizability theory perspective on rossiter's procedure. *International Journal of Research in Marketing*, 22(1):11–22, 2005.

20. Claes Fornell and Fred L. Bookstein. Two structural equation models: LISREL and PLS applied to consumer exit-voice theory. *Journal of Marketing Research*, 19(4):440–452, 1982.

21. Claes Fornell, Michael D. Johnson, Eugene W. Anderson, Jaesung Cha, and Barbara Everitt Johnson. The American customer satisfaction index: Nature, purpose, and findings. *Journal of Marketing*, 60(4):7–18, 1996.

22. Claes Fornell and David F. Larcker. Evaluating structural equation models with unobservable variables and measurement error. *Journal of Marketing Research*, 18(1):39–50, 1981.

23. Claes Fornell, Peter Lorange, and Johan Roos. The cooperative venture formation process: A latent variable structural modeling approach. *Management Science*, 36(10):1246–1255, 1990.

24. Claes Fornell, William T. Robinson, and Birger Wernerfelt. Consumption experience and sales promotion expenditure. *Management Science*, 31(9):1084–1105, 1985.

25. David Gefen and Detmar W. Straub. Gender differences in the perception and use of e-mail: An extension to the technology acceptance model. *MIS Quarterly*, 21(4):389–400, 1997.

26. Oliver Götz, Kerstin Liehr-Göbbers, and Manfred Krafft. Evaluation of structural equation models using the partial least squares (PLS-) approach. In Vincenzo Esposito Vinzi, Wynne W. Chin, Jörg Henseler, and Huiwen Wang, editors, *Handbook of Partial Least Squares: Concepts, Methods and Applications in Marketing and Related Fields*, forthcoming. Springer, Berlin-Heidelberg, 2009.

27. Peter H. Gray and Darren B. Meister. Knowledge sourcing effectiveness. *Management Science*, 50(6):821–834, 2004.

28. Yany Grégoire and Robert J. Fisher. The effects of relationship quality on customer retaliation. *Marketing Letters*, 17(1):31–46, 2006.

29. Siegfried P. Gudergan, Christian M. Ringle, Sven Wende, and Alexander Will. Confirmatory tetrad analysis in PLS path modeling. *Journal of Business Research*, 61(12):1238–1249, 2008.

30. Peter Hackl and Anders H. Westlund. On structural equation modeling for customer satisfaction measurement. *Total Quality Management*, 11:820–825, 2000.

31. Carsten Hahn, Michael D. Johnson, Andreas Herrmann, and Frank Huber. Capturing customer heterogeneity using a finite mixture PLS approach. *Schmalenbach Business Review*, 54(3):243–269, 2002.

32. Jörg Henseler, Christian M. Ringle, and Rudolf R. Sinkovics. The use of partial least squares path modeling in international marketing. In Rudolf R. Sinkovics and Pervez N. Ghauri, editors, *Advances in International Marketing*, volume 20, pages 277–320. Emerald, Bingley, 2009.

33. Roy D. Howell, Einar Breivik, and James B. Wilcox. Is formative measurement really measurement? reply to Bollen (2007) and Bagozzi (2007). *Psychological Methods*, 12(2): 238–245, 2007.

34. Roy D. Howell, Einar Breivik, and James B. Wilcox. Reconsidering formative measurement. *Psychological Methods*, 12(2):205–218, 2007.

35. John Hulland. Use of partial least squares (PLS) in strategic management research: A review of four recent studies. *Strategic Management Journal*, 20(2):195–204, 1999.
36. Magid Igbaria, Nancy Zinatelli, Paul Cragg, and Angele L. M. Cavaye. Personal computing acceptance factors in small firms: A structural equation model. *MIS Quarterly*, 21(3): 279–305, 1997.
37. Kamel Jedidi, Harsharanjeet S. Jagpal, and Wayne S. DeSarbo. Finite-fixture structural equation models for response-based segmentation and unobserved heterogeneity. *Marketing Science*, 16(1):39–59, 1997.
38. Karl G. Jöreskog. Structural analysis of covariance and correlation matrices. *Psychometrika*, 43(4):443–477, 1978.
39. Richard E. Kihlstrom and Michael H. Riordan. Advertising as a signal. *Journal of Political Economy*, 92(3):427–450, 1984.
40. Philip Kotler and Kevin Lane Keller. *Marketing Management*. Prentice Hall, Upper Saddle River, NJ, 12 edition, 2006.
41. Kai Kristensen, Anne Martensen, and Lars Grønholdt. Customer satisfaction measurement at post denmark: Results of application of the European customer satisfaction index methodology. *Total Quality Management*, 11(7):1007–1015, 2000.
42. Wei-Yin Loh and Yu-Shan Shih. Split selection methods for classification trees. *Statistica Sinica*, 7(4):815–840, 1997.
43. Jan-Bernd Lohmöller. *Latent Variable Path Modeling with Partial Least Squares*. Physica, Heidelberg, 1989.
44. Scott B. MacKenzie, Philip M. Podsakoff, and Cheryl B. Jarvis. The problem of measurement model misspecification in behavioral and organizational research and some recommended solutions. *Journal of Applied Psychology*, 90(4):710–730, 2005.
45. Ranjan Maitra. Clustering massive data sets with applications in software metrics and tomography. *Technometrics*, 43(3):336–346, 2001.
46. Geoffrey J. McLachlan and Kaye E. Basford. *Mixture Models: Inference and Applications to Clustering*. Dekker, New York, NY, 1988.
47. Geoffrey J. McLachlan and Thriyambakam Krishnan. *The EM Algorithm and Extensions*. Wiley, Chichester, 2004.
48. Geoffrey J. McLachlan and David Peel. *Finite Mixture Models*. Wiley, New York, NY, 2000.
49. Christopher Meek, Bo Thiesson, and David Heckerman. Staged mixture modeling and boosting. In *In Proceedings of the 18th Annual Conference on Uncertainty in Artificial Intelligence*, pages 335–343, San Francisco, CA, 2002. Morgan Kaufmann.
50. Marina Meilă and David Heckerman. An experimental comparison of model-based clustering methods. *Machine Learning*, 40(1/2):9–29, 2001.
51. Paul Milgrom and John Roberts. Price and advertising signals of product quality. *Journal of Political Economy*, 94(4):796–821, 1986.
52. Vikas Mittal, Eugene W. Anderson, Akin Sayrak, and Pandu Tadikamalla. Dual emphasis and the long-term financial impact of customer satisfaction. *Marketing Science*, 24(4): 531–543, 2005.
53. Neil Morgan, Eugene W. Anderson, and Vikas Mittal. Understanding firms' customer satisfaction information usage. *Journal of Marketing*, 69(3):121–135, 2005.
54. Bengt O. Muthén. Latent variable modeling in heterogeneous populations. *Psychometrika*, 54(4):557–585, 1989.
55. Linda K. Muthén and Bengt O. Muthén. *Mplus User's Guide*. Muthén & Muthén, Los Angeles, CA, 4th edition, 1998.
56. Jum C. Nunnally and Ira Bernstein. *Psychometric Theory*. McGraw Hill, New York, NY, 3rd edition, 1994.
57. Francesco Palumbo, Rosaria Romano, and Vincenzo Esposito Vinzi. Fuzzy PLS path modeling: A new tool for handling sensory data. In Christine Preisach, Hans Burkhardt, Lars Schmidt-Thieme, and Reinhold Decker, editors, *Data Analysis, Machine Learning and Applications – Proceedings of the 31st Annual Conference of the Gesellschaft für Klassifika-*

tion e.V., Albert-Ludwigs-Universität Freiburg, March 7–9, 2007, pages 689–696, Berlin-Heidelberg, 2008. Springer.

58. Venkatram Ramaswamy, Wayne S. DeSarbo, David J. Reibstein, and William T. Robinson. An empirical pooling approach for estimating marketing mix elasticities with PIMS data. *Marketing Science*, 12(1):103–124, 1993.

59. Edward E. Rigdon. Structural equation modeling. In George A. Marcoulides, editor, *Modern methods for business research*, Quantitative Methodology Series, pages 251–294. Lawrence Erlbaum, Mahwah, 1998.

60. Edward E. Rigdon, Christian M. Ringle, and Marko Sarstedt. Structural modeling of heterogeneous data with partial least squares. *Review of Marketing Research*, forthcoming, 2010.

61. Christian M. Ringle. Segmentation for path models and unobserved heterogeneity: The finite mixture partial least squares approach. Research Papers on Marketing and Retailing No. 035, University of Hamburg, 2006.

62. Christian M. Ringle and Karl-Werner Hansmann. Enterprise-networks and strategic success: An empirical analysis. In Theresia Theurl and Eric C. Meyer, editors, *Strategies for Cooperation*, pages 133–152. Shaker, Aachen, 2005.

63. Christian M. Ringle and Rainer Schlittgen. A genetic segmentation approach for uncovering and separating groups of data in PLS path modeling. In Harald Martens, Tormod Næs, and Magni Martens, editors, *PLS'07 International Symposium on PLS and Related Methods – Causalities Explored by Indirect Observation*, pages 75–78, Ås, 2007. Matforsk.

64. Christian M. Ringle, Sven Wende, and Alexander Will. Customer segmentation with FIMIX-PLS. In Tomàs Aluja, Josep Casanovas, Vincenzo Esposito Vinzi, and Michel Tenenhaus, editors, *PLS'05 International Symposium on PLS and Related Methods – PLS and Marketing*, pages 507–514, Paris, 2005. Decisia.

65. Christian M. Ringle, Sven Wende, and Alexander Will. SmartPLS 2.0 (beta), 2005.

66. Christian M. Ringle, Sven Wende, and Alexander Will. The finite mixture partial least squares approach: Methodology and application. In Vincenzo Esposito Vinzi, Wynne W. Chin, Jörg Henseler, and Huiwen Wang, editors, *Handbook of Partial Least Squares: Concepts, Methods and Applications in Marketing and Related Fields*, forthcoming. Springer, Berlin-Heidelberg, 2009.

67. Christian M. Ringle, Marko Sarstedt, and Rainer Schlittgen. Finite mixture and genetic algorithm segmentation in partial least squares path modeling. In Andreas Fink, Berthold Lausen, Wilfried Seidel, and Alfred Ultsch, editors, *Advances in Data Analysis, Data Handling and Business Intelligence. Proceedings of the 32nd Annual Conference of the German Classification Society (GfKl)*, Springer, Heidelberg and Berlin, forthcoming.

68. John R. Rossiter. The C-OAR-SE procedure for scale development in marketing. *International Journal of Research in Marketing*, 19(4):305–335, 2002.

69. John R. Rossiter. Reminder: A horse is a horse. *International Journal of Research in Marketing*, 22(1):23–25, 2005.

70. Gastón Sánchez and Tomàs Aluja. A simulation study of PATHMOX (PLS path modeling segmentation tree) sensitivity. In Harald Martens, Tormod Næs, and Magni Martens, editors, *PLS'07 International Symposium on PLS and Related Methods – Causalities Explored by Indirect Observation*, pages 33–36, Ås, 2007. Matforsk.

71. Marko Sarstedt and Christian M. Ringle. Treating unobserved heterogeneity in PLS path modelling: A comparison of FIMIX-PLS with different data analysis strategies. *Journal of Applied Statistics*, forthcoming, 2010.

72. Marko Sarstedt, Christian M. Ringle, and Manfred Schwaiger. Do we fully understand the critical success factors of customer satisfaction with industrial goods? Extending Festge and Schwaiger's model to account for unobserved heterogeneity. *Journal of Business Market Management*, 3(3):185, 2009.

73. Marko Sarstedt. Market segmentation with mixture regression models. *Journal of Targeting, Measurement and Analysis for Marketing*, 16(3):228–246, 2008.

74. Marko Sarstedt. A review of recent approaches for capturing heterogeneity in partial least squares path modelling. *Journal of Modelling in Management*, 3(2):140–161, 2008.

75. Michel Tenenhaus, Vincenzo Esposito Vinzi, Yves-Marie Chatelin, and Carlo Lauro. PLS path modeling. *Computational Statistics & Data Analysis*, 48(1):159–205, 2005.
76. Viswanath Venkatesh and Ritu Agarwal. Turning visitors into customers: A usability-centric perspective on purchase behavior in electronic channels. *Management Science*, 52(3): 367–382, 2006.
77. Michel Wedel and Wagner Kamakura. *Market Segmentation: Conceptual and Methodological Foundations*. Kluwer, London, 2nd edition, 2000.
78. Ron Wehrens, Lutgarde M. C. Buydens, Chris Fraley, and Adrian E. Raftery. Model-based clustering for image segmentation and large datasets via sampling. *Journal of Classification*, 21(2):231–253, 2004.
79. John Williams, Dirk Temme, and Lutz Hildebrandt. A Monte Carlo study of structural equation models for finite mixtures. SFB 373 Discussion Paper No. 48, Humboldt University Berlin, 2002.
80. Herman Wold. Soft modeling: The basic design and some extensions. In Karl G. Jöreskog and Herman Wold, editors, *Systems Under Indirect Observations*, pages 1–54. North-Holland, Amsterdam, 1982.
81. Jianan Wu and Wayne S. DeSarbo. Market segmentation for customer satisfaction studies via a new latent structure multidimensional scaling model. *Applied Stochastic Models in Business and Industry*, 21(4/5):303–309, 2005.

Part II
Knowledge Discovery from Supervised Learning

Chapter 3
Building Acceptable Classification Models

David Martens and Bart Baesens

Abstract Classification is an important data mining task, where the value of a discrete (dependent) variable is predicted, based on the values of some independent variables. Classification models should provide correct predictions on new unseen data instances. This accuracy measure is often the only performance requirement used. However, comprehensibility of the model is a key requirement as well in any domain where the model needs to be validated before it can be implemented. Whenever comprehensibility is needed, justifiability will be required as well, meaning the model should be in line with existing domain knowledge. Although recent academic research has acknowledged the importance of comprehensibility in the last years, justifiability is often neglected. By providing comprehensible, justifiable classification models, they become acceptable in domains where previously such models are deemed too theoretical and incomprehensible. As such, new opportunities emerge for data mining. A classification model that is accurate, comprehensible, and intuitive is defined as acceptable for implementation.

David Martens
Department of Decision Sciences and Information Management, K.U.Leuven, Naamsestraat 69, 3000 Leuven, Belgium, e-mail: david.martens@econ.kuleuven.be; Department of Business Administration and Public Management, Hogeschool Gent, Voskenslaan 270, Ghent 9000, Belgium, e-mail: david.martens@hogent.be

Bart Baesens
School of Management, University of Southampton, Highfield Southampton, SO17 1BJ, UK, e-mail: bart@soton.ac.uk; Department of Decision Sciences and Information Management, K.U.Leuven, Naamsestraat 69, 3000 Leuven, Belgium, e-mail: bart.baesens@econ.kuleuven.be

R. Stahlbock et al. (eds.), *Data Mining,* Annals of Information Systems 8,
DOI 10.1007/978-1-4419-1280-0_3, © Springer Science+Business Media, LLC 2010

3.1 Introduction

Different data mining tasks are discussed in the literature [4, 63, 50], such as regression, classification, association rule mining, and clustering. The task of interest here is classification, which is the task of assigning a data point to a predefined class or group according to its predictive characteristics. The goal of a classification technique is to build a model which makes it possible to classify future data points based on a set of specific characteristics in an automated way. In the literature, there is a myriad of different techniques proposed for this classification task [4, 22], some of the most commonly used being C4.5 [42], CART [10], logistic regression [24], linear and quadratic discriminant analysis [9, 24], k-nearest neighbor [1, 18, 24], artificial neural networks (ANN) [9], and support vector machines (SVM) [14, 49, 59].

Classification techniques are often applied for credit scoring [6, 51], medical diagnosis, such as for the prediction of dementia [40], classifying a breast mass as benign or malignant, and selecting the best in vitro fertilized embryo [37]. Many other data mining applications have been put forward recently, such as the use of data mining for bioinformatics [25], marketing and election campaigns [23], and counterterrorism [43].

Several performance requirements exist for a classification model: providing correct predictions, comprehensibility, and justifiability. The first requirement is that the model generalizes well, in the sense that it provides the correct predictions for new, unseen data instances. This generalization behavior is typically measured by percentage correctly classified (PCC) test instances. Other measures include sensitivity and specificity, which are generated from a confusion matrix. Also commonly used are the receiver operating curve (ROC) and the area under this curve (AUC).

There exist many different classification output types, the most commonly used are as follows:

- **Linear models**, built by, e.g., linear and logistic regression. A typical logistic regression formulation for a data set $D = \{x_i, y_i\}_{i=1}^n$ with input data $x_i \in \mathbb{R}$ and corresponding binary class labels $y_i \in \{0, 1\}$ is

$$y_{\text{logit}}(x) = \frac{1}{1 + \exp(-(w_0 + \mathbf{w}^T x))} \qquad (3.1)$$

- **Nonlinear models**, built by, e.g., ANNs and SVMs. The model formulation for a SVM with RBF kernel is

$$y_{\text{SVM}}(x) = \text{sgn}[\sum_{i=1}^{N} \alpha_i y_i \exp\{-\|\mathbf{x} - \mathbf{x}_i\|_2^2 / \sigma^2\} + b] \qquad (3.2)$$

- **Rule-based models**, built by, e.g., RIPPER [13], CN2 [12], and AntMiner+ [33].
- **Tree-based models**, built by, e.g., C4.5 [42] and CART [10].

Other common model types are nearest neighbor classifiers and Bayesian networks. Benchmarking studies have shown that, in general, nonlinear models provide the most accurate predictions [6, 56], as they are able to capture nonlinearities in the data. However, this strength is also their main weakness, as the model is considered to be a black box: as Equation 3.2 shows, trying to comprehend the logics behind the decisions made is very difficult, if not impossible. This brings us to the second performance requirement: comprehensibility.

Comprehensibility can be a key requirement as well, demanding that the user can understand the logic behind the model's prediction. In some domains, such as credit scoring and medical diagnosis, the lack of comprehensibility is a major issue and causes a reluctance to use the classifier or even complete rejection of the model. In a credit scoring context, when credit has been denied, the Equal Credit Opportunity Act of the United States requires that the financial institution provides specific reasons why the customer's application was rejected, whereby vague reasons for denial are illegal [19]. In the medical diagnosis domain as well, clarity and explainability are major constraints.

Whenever comprehensibility is required, it will be needed so as to check whether the model is in line with existing domain knowledge. So if comprehensibility is needed, justifiability is needed as well. For example, it cannot be that loan applicants with a high income are rejected credit, while similar applicants with a low income are granted credit. Such a model is counterintuitive and hence unacceptable for implementation.

A data mining approach that takes into account the knowledge representing the experience of domain experts is therefore much preferred and of great focus in current data mining research. This chapter is structured as follows: Section 3.2 describes the comprehensibility requirement in a data mining context and discusses how to define, measure, and obtain comprehensibility. A further discussion into the academically challenging justifiability requirement is provided in Section 3.3. Here as well, approaches to obtain justifiable models are reviewed. Finally, a metric to measure justifiability is proposed.

3.2 Comprehensibility of Classification Models

Comprehensibility can be a key requirement for a classification model, demanding that the user can understand the motivations behind the model's prediction. It is an absolute necessity in any domain where the model needs to be validated before it can actually be implemented for practical use. Typical domains are the highly regulated credit scoring and medical diagnosis. However, in other domains as well, comprehensibility and justifiability will lead to a greater acceptance of the provided model. The crucial importance of comprehensibility in credit scoring is demonstrated by the following quote by German Chancellor Angela Merkel. As acting president of the G8, she stated in response to the credit crisis that started in mid-2007 that the top credit rating agencies (such as Moody's and S&P) should be more open about the way they arrive at their credit ratings:

*"In the future it should be clear what
the basis of their ratings of companies is,"* Merkel said.
*"There can't be some **black box** from which something comes out
and which **no one understands**."*
– Reuters (August 18, 2007)

The importance of comprehensibility for any data mining application is argued by Kodratoff, who states in his *comprehensibility postulate* that "Each time one of our favorite ML [Machine Learning] approaches has been applied in industry, each time the comprehensibility of the results, though ill defined, has been a decisive factor of choice over an approach by pure statistical means, or by neural networks." [28]

Defining comprehensibility is close to being a philosophical discussion. Still, to get some clarity on what this requirement exactly entails, we will try to define when a classification model is comprehensible. As will be argued next, ***comprehensibility*** **measures the "mental fit" [35] of the classification model, and its main drivers are as follows:**

- **The type of output**. Although the comprehensibility of a specific output type is largely domain dependent, generally speaking, rule-based classifiers can be considered as the most comprehensible and nonlinear classifiers as the least comprehensible.
- **The size of the output.** Smaller models are preferred.

Comprehensibility is defined by the Oxford English Dictionary as "Quality of being comprehensible," with comprehensible being defined as "To seize, grasp, lay hold of, catch." Of course, within data mining context, this is a rather vague definition.

The first main criterion for comprehensibility is the model output, which can be rule based, tree based, linear, nonlinear, instance based (e.g., *k*-nearest neighbor), and many others. Which of these rule types is the most comprehensible is largely domain specific, as comprehensibility is a subjective matter, or put differently, comprehensibility is in the eye of the beholder. Michalski is one of the first to address the comprehensibility issue in Knowledge Discovery in Data [34]. In his *comprehensibility postulate*, he states "The results of computer induction should be symbolic descriptions of given entities, semantically and structurally similar to those a human expert might produce observing the same entities." Mainon and Rokach address this subjectivity issue as follows [35]: "The comprehensibility criterion (also known as interpretability) refers to how well humans grasp the classifier induced. While the generalization error measures how the classifier fits the data, comprehensibility measures the "*Mental fit*" of that classifier." ... "the accuracy and complexity factors can be quantitatively estimated, while the comprehensibility is more subjective." This concept of mental fit is very interesting and points out that if the user is more familiar with linear models, the mental fit with such models will be greater than the fit with tree-based classifiers. However, generally speaking, one can argue that the models that are more linguistic will give a better mental fit. From that point of view, we can say that rule (and tree)-based classifiers are considered the most comprehensible and nonlinear classifiers the least comprehensible, once again, keeping in mind that for some domains other output types (such as a linear model or 1NN classifier) can be regarded as the most comprehensible.

For a given rule output, the comprehensibility decreases with the size [3]. Domingos motivates this with Occam's razor, interpreting this principle as [17] "preferring simpler models over more complex." Speaking in a rule-based context, the more the conditions (terms), the harder it is to understand [26]. For a given number of conditions, it is better to have many rules with a low average number of conditions per rule than few rules with many conditions [48]: "a theory consisting of few long clauses is harder to understand than one with shorter clauses, even if the theories are of the same absolute size." This size concept can of course be extended to all rule outputs [17]: the number of nodes in a decision tree, the number of weights in a neural network, the number of support vectors in a support vector machine, etc.

Finally, though we consider model output and model size to be the main components determining comprehensibility, the concept can be deepened even further, as it also depends on aspects such as the number of variables and constants in a rule, the number of instances it covers [48], and even the consistency with existing domain knowledge [38], an aspect that will be addressed in Section 3.3.

3.2.1 Measuring Comprehensibility

With the previous discussion in mind, we can measure the comprehensibility in following ways. First of all, there seems to be a ranking in the comprehensibility of the different output types, such that we can state that rule-based models are more comprehensible than linear ones, which are again more comprehensible than nonlinear ones. However, this seemingly obvious conclusion is not always true: a linear model with just one variable will surely be more comprehensible than a rule-based model with over 20 rules. Also, comparing the comprehensibility of a nearest neighbor classifier is also very difficult; whereas a 1NN classifier might be very logical and comprehensible in one domain, it might be pretty meaningless in another, e.g., domains with high-dimensional data, where even the most similar training instance still differs in many of the variables. Generally speaking, however, typically this ranking in output types will be true and observable. The best way however to verify this is to ask application domain experts and users for their own ranking, to determine their "mental fit" with the different output types.

Within a given output type, the size can be measured as follows:

- Nonlinear model: number of terms in the final mathematical formulation (e.g., the number of support vectors for SVMs, the number of weights for ANNs).
- Linear model: number of terms in the final mathematical formulation (hence the number of included variables).
- Rule (or tree)-based model: number of rules/leaves, number of terms per rule. For a certain number of terms, there is a preference for more rules with less terms.

3.2.2 Obtaining Comprehensible Classification Models

3.2.2.1 Building Rule-Based Models

Obtaining comprehensible classifiers can be done in a variety of ways. As we consider rules and trees to be the most comprehensible format for a classification model (given their linguistic nature, and hence ease in understanding for nonexperts), techniques that induce such models are of course the best suited. In this category fall rule induction and extraction techniques.

Rule induction techniques induce rules from structured data. Such techniques are C4.5, CART, CN2, and AntMiner+. Comprehensibility can also be added to black box models by extracting symbolic rules from the trained model, rather than immediately from the data. Rule extraction techniques attempt to open up the black box and generate symbolic, comprehensible descriptions with approximately the same predictive power as the model itself. If the rules mimic the model closely enough, and thus is explained sufficiently, one might opt to use the black box model. Extraction techniques have been proposed from ANNs [5], as well as from SVMs [29, 30].

3.2.2.2 Combining Output Types

An incremental approach that combines several output types can be followed, so as to find a trade-off between simple techniques with good readability, but restricted model flexibility and complexity, and advanced techniques with reduced readability but extended flexibility and generalization behavior.

Setiono et al. combine rules with logistic regression [44, 45]. First, rules are induced for all the nominal variables. Hereafter, instead of providing a simple class label, a linear model is estimated with the remaining ordinal classes. As such, a set of rules is obtained, with a linear regression model as the final prediction. This approach has been successfully applied to credit scoring [44].

Van Gestel et al. combine the comprehensibility of linear models with the good generalization behavior of SVMs [53–55]. On top of a simple logistic regression model, extra SVM terms are added that try to model the residual errors, resulting in an increased accuracy.

3.2.2.3 Visualization

The final approach we mention to incorporate comprehensibility is visualization, which includes plots, self-organizing maps, decision tables, and decision diagrams.

Data visualization is the display of information in graphical or tabular format [50]. The goal of visualization is to allow for better interpretation, and thus validation, of the information by a person, in order to obtain an acceptable classifier. The most straightforward manner of visualizing the data and classification model is with plots. For higher dimensional data, with four variables or more, such plots are no longer feasible and should be visualized with other techniques, such as self-organizing maps (SOMs).

SOMs are a single-layer feedforward neural network, where the outputs are arranged in a low-dimensional grid (typically two or three dimensional) and provide a low-dimensional representation of the training data. This unsupervised data mining technique is typically used for clustering and data visualization [50].

Another way to add comprehensibility to the models is with the use of decision tables. As the proposed justifiability metric in Section 3.3 is based on this notion, we will look at this representation form in more detail. Decision tables are a tabular representation used to describe and analyze decision situations [58] and consists of four quadrants, separated by double-lines, both horizontally and vertically. The vertical line divides the table into a condition part (left) and an action part (right), while the horizontal line separates subjects (above) from entries (below). The condition subjects are the problem criteria (the variables) that are relevant to the decision-making process. The action subjects describe the possible outcomes of the decision-making process, i.e., the classes of the classification problem: applicant = good or bad. Each condition entry describes a relevant subset of values (called a state) for a given condition subject (variable) or contains a dash symbol ("–") if its value is irrelevant within the context of that row. Subsequently, every action entry holds a value assigned to the corresponding action subject (class).

Every row in the entry part of the decision table thus comprises a classification rule, indicating what action(s) apply to a certain combination of condition states. For example, in Table 3.1, the final row tells us to classify the applicant as good if owns property = no and savings amount = high.

3.3 Justifiability of Classification Models

Although many powerful classification algorithms have been developed, they generally rely solely on modeling repeated patterns or correlations which occur in the data. However, it may well occur that observations that are very evident to classify by the domain expert do not appear frequently enough in the data in order to be appropriately modeled by a data mining algorithm. Hence, the intervention and interpretation of the domain expert still remain crucial. A data mining approach that takes into account the knowledge representing the experience of domain experts is therefore much preferred and of great focus in current data mining research.

Table 3.1 Classification model visualized by decision table

1. Owns property?	2. Years client	3. Savings amount	1. Applicant=good	2. Applicant=bad
Yes	≤ 3	Low	–	×
		High	×	–
	> 3	–	×	–
No	–	Low	–	×
		High	×	–

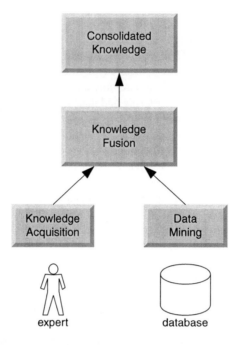

Fig. 3.1 The knowledge fusion process

In a data mining context, **a model is *justifiable* when it is in line with existing domain knowledge**. Therefore, for a model to be justifiable, it needs to be validated by a domain expert, which in turn means that the model should be comprehensible. The academically challenging problem of consolidating the automatically generated data mining knowledge with the knowledge reflecting experts' domain expertise constitutes the knowledge fusion problem (see Fig. 3.1).

3.3.1 Taxonomy of Constraints

Many types of constraints exist that a domain expert might want to incorporate. Although almost all research into this field is focused on the monotonicity constraint [2, 7, 21, 40, 47, 61], the taxonomy of possible domain constraints, shown in Table 3.2, indicates that this is a too limited view. It is, however, the most common one to be incorporated.

Each constraint can be either mandatory, which we name a *hard* constraint, or simply preferred, which we name a *soft* constraint. In what follows next, the potential domain constraints are listed and illustrated.

Univariate constraints are those that apply to one single variable.

Table 3.2 Taxonomy of possible constraints to incorporate

		Univariate			Multivariate
	Nominal	Monotone	Ordinal		
			Non-monotone		
			Piecewise monotone	Non-piecewise monotone	
Soft					
Hard					

- **Nominal** univariate constraints apply to a single, nominal variable. In a positive discrimination context, for instance, one might prefer women (for admittance, recruitment, etc.), thus the constraint being on the nominal variable *sex*.
- **Ordinal** univariate constraints apply to a single, ordinal variable.

  - **Monotone**, ordinal, univariate constraints are commonly referred to as monotonicity constraints, which are addressed in detail in the next section. A credit scoring example is that an increasing income, keeping all other variables equal, should yield a decreasing probability of loan default.
  - **Non-monotone**, ordinal, univariate constraints require a non-monotone relationship in the ordinal variable in question.

    · **Piecewise monotone** constraints allow a non-monotone constraint to be considered as several monotone constraints over the range of the variable. For example, the domain expert might demand a piecewise monotone constraint on the *age* variable, with decreasing probability of default for ages between 18 and 60, and an increasing probability of default for clients over 60 years old.
    · **Non-piecewise monotone** constraints are those constraints that are non-monotone and cannot be modeled as a piecewise monotone constraint. This constraint is the hardest to be fulfilled, as it cannot be based on the commonly researched monotonicity constraint.

Multivariate constraints are those that apply to multiple variables. All possible variations of two or more univariate constraints that need to be fulfilled simultaneously can be thought of. For example, a preference for young, highly educated customers is such a combination of a monotone, ordinal, univariate constraint and a nominal, univariate constraint. Although these clients will tend to have a rather low income and savings, considering the complete customer lifetime value, they can be very profitable to the bank. Incorporating such policies are of great importance as well, and to the best of our knowledge, not yet incorporated in data mining techniques. Additionally, the relative importance of a variable can also be included here. For credit scoring, one might state that income is more important than age, and this relative importance should be present in the model. Finally, most difficultly is arguably the introduction of interactions between predictive variables (see, e.g., [62]).

3.3.2 Monotonicity Constraint

The monotonicity is the most encountered domain constraint to be incorporated, and to the best of our knowledge, the only constraint researched so far. This constraint demands that an increase in a certain input(s) cannot lead to a decrease in the output. Such constraints exist in almost any domain, a few examples are as follows:

- For example, credit scoring context: increasing income should yield a decreasing probability of default; therefore
  Client A : Income = 1.000
  ⇨ Classified as GOOD
  Client B : Income = 2.000[1]
  ⇨ Classified as BAD   ➡ Not monotone
  ⇨ Classified as GOOD ➡ Monotone
- For example, medical diagnosis context: an increasing tumor size should yield an increasing probability of tumor recurrence; therefore
  Patient A : tumor size = 50
  ⇨ Tumor classified as RECURRING
  Patient B : tumor size = 100
  ⇨ Tumor classified as NOT RECURRING ➡ Not monotone
  ⇨ Tumor classified as RECURRING   ➡ Monotone
- For example, fuel consumption prediction: an increasing weight should yield an increasing fuel consumption; therefore
  Car A : weight = 1.000
  ⇨ Car classified as HIGH FUEL CONSUMPTION
  Car B : weight = 2.000
  ⇨ Car classified as LOW FUEL CONSUMPTION ➡ Not monotone
  ⇨ Car classified as HIGH FUEL CONSUMPTION ➡ Monotone

We can define such a monotonicity constraint more formally (similarly to [21]), given a data set $D = \{x^i, y^i\}_{i=1}^n$, with $x^i = (x_1^i, x_2^i, \ldots, x_m^i) \in X = X_1 \times X_2 \times \ldots X_m$, and a partial ordering $\leq$ defined over this input space X. Over the space Y of class values y^i, a linear ordering $\leq$ is defined. Then the classifier $f : x^i \mapsto f(x^i) \in Y$ is monotone if Equation 3.3 holds.

$$x^i \leq x^j => f(x^i) \leq f(x^j), \forall i, j \ \ (\text{or } f(x^i) \geq f(x^j), \forall i, j) \tag{3.3}$$

For instance, increasing income, keeping all other variables equal, should yield a decreasing probability of loan default. Therefore if client A has the same characteristics as client B, but a lower income, then it cannot be that client A is classified as a good customer and client B a bad one.

---

[1] Assuming the values for all other variables are equal.

3.3.3 Measuring Justifiability

Several adaptations to existing classification techniques have been proposed to cope with justifiability and will be discussed in Section 3.3.4, yet a measure to identify the extent to which the models conform to the required constraints is still lacking. In this section, a new measurement is proposed, where with the use of decision tables we provide a crisp performance measure for the critical justifiability measure.

In what follows, we propose a formal measurement for justifiability, where we assume that the data set consists of n variables V_i. We define a profile pr_i for variable V_i as the situation for which the variable settings differ only in variable V_i, with different classes assigned for the different settings. In the decision table view, this corresponds to the situation of having at least two rows with same values for all but the last column. A formal definition of our justifiability measure is given below, with n the number of variables, w_i a user-defined weight determining the relative importance of variable V_i (such that higher penalties are given to inconsistencies in variables with higher weights), and the $I(pr_{i,j})$ operator returning 1 if an inconsistency is present in the j_{th} profile of variable i, and 0 otherwise. The total number of profiles for variable V_i is denoted by $|pr_i|$. As we will demonstrate, such inconsistencies are easily detected with the use of decision tables.

$$\text{Justifiability} = 1 - \overbrace{\sum_{i=1}^{n} w_i \underbrace{\sum_{j=1}^{|pr_i|} \frac{1}{|pr_i|} \cdot I(pr_{i,j})}_{\text{penalty for } V_j}}^{\text{overall penalty}} \tag{3.4}$$

This justifiability metric is given by Equation 3.4, with $\sum_{i=1}^{n} w_i = 1$, such that $0 \leq$ justifiability ≤ 1. Each variable can bring about a penalty of maximum w_i, which ensures that the justifiability measure is limited between 0 and 1.

The weight parameter w_i can be set in the following manner:

1. Statistically based, with information theoretic measures, such as Information Value, Cramer's V and Gainratio [52]
2. Domain expert based

   - Use the statistically based measures and adapt them if necessary
   - Use the ranking of the variables, based on which the weights are determined
   - Immediately determine weights based on own expertise and intuition

Our measure can also be used for linear classifiers, without the use of decision tables: there is exactly one profile for each variable. An inconsistency takes place if the sign of variable V_i does not correspond to the expected sign. To determine the weights, a similar approach as for the rule-based classifiers can be followed. For the statistically proposed weights, one might use the (normalized) partial correlation coefficients, which is the correlation when all other variables are kept at fixed values [27], as statistically based suggestion. Other possibilities that determine the importance of a variable in the model include the (normalized) regression coefficients and the p-values of these coefficients.

Setting the weights is a difficult, yet necessary exercise in the justifiability calculations. Setting the weights completely statistically based and relying only on the domain expert's intuition are just two extremes of a spectrum. Using statistics only, one will mimic the relative importance of the variables in the data set exactly. Of course, due to the lack of perfect data quality (noise, limited data availability, etc.), one cannot completely rely on these measures alone. For example, if a variable has no correlation at all with the target variable, it does not mean the weight should therefore be 0. For the same reason, if one variable alone perfectly predicts the target variable, it should not be concluded that the weight of that variable is 1, and all others 0. On the other hand, relying on the domain expert's opinion alone is also not recommended, as the expert's intuition about the impact of different settings will be limited. From that perspective, approaches between these two extremes are sensible, where statistically based weights are adjusted according to the expert's opinion, or where an expert chooses among a set of possible weight configurations. To determine the impact of changing the weights, it is surely sensible to combine this with sensitivity analysis, to investigate what the impact is on the justifiability metric of (small) changes in the weight settings.

Before explaining the metric in detail, we should note that since we measure an inherently subjective concept, testing with human users and comparing the proposed metric with the user's values (referred to as "real human interest" [11]) can provide useful guidelines for the weight settings, and the metric as a whole. Such experiments are conducted in, e.g., [8, 36, 39, 41] but are beyond the scope of this chapter.

Our justifiability measure will be explained in more detail using two examples from, respectively, the credit scoring and medical diagnostic domains.

Credit Scoring Example

Table 3.3 reports the decision table, corresponding to the rule set inferred by AntMiner+ on the German credit scoring data set, as published in [32]. In this credit scoring example, we have a total of six profiles for the Credit History variable, which are numbered in Table 3.3 and are all inconsistent. For all other variables, no inconsistencies exist. Therefore, assuming the weights being taken as given by the normalized Gainratios in Table 3.4, the justifiability of this classifier is given by

$$\text{Justifiability} = 1 - 0.0741 \cdot \left[\frac{1}{6} + \frac{1}{6} + \frac{1}{6} + \frac{1}{6} + \frac{1}{6} + \frac{1}{6} \right] = 0.9259 = 92.59\%$$

Notice that this measure is very high, although all profiles of the Credit History variable are incorrect. This is the result of the very low weight given to this variable. An expert might decide to put some lower bound on the possible weight, or to simply adjust these weights.

Table 3.3 Discrepancies in the classification model, visualized by decision table

Purpose	Duration	Checking account	Savings account	Credit history	Bad	Good	
Furniture/business	≤15m	<0€	<250€	No credits taken/all credits paid back duly	×	–	1
				Critical account	–	×	
			≥250€	–	–	×	
		≥0€	–	–	–	×	
	>15m	<0€	<250€	–	×	–	2
			≥250 and <500€	No credits taken/all credits paid back duly	×	–	
				All credits at this bank paid back duly or Critical account	–	×	
			≥500€	–	–	×	
		≥0 and <100€	<500€	No credits taken/all credits paid back duly	×	–	3
				all credits at this bank paid back duly or Critical account	–	×	
			≥500€	–	–	×	
		≥100€	–	–	–	×	
Car/retraining or others	≤ 15m	–	<500€	No credits taken/all credits paid back duly	×	–	4
				Critical account	–	×	
			≥500€	–	–	×	
	>15m	<0€	<250€	–	×	–	5
			≥250 and <500€	No credits taken/all credits paid back duly	×	–	
				Critical account	–	×	
			≥500€	–	–	×	
		≥0€	<500€	No credits taken/all credits paid back duly	×	–	6
				Critical account	–	×	
			≥500€	–	–	×	

Table 3.4 Weights of variables, defined as normalized Gainratios

	Gainratio	w_i
Checking account	0.0526	0.1529
Duration	0.2366	0.6879
Credit history	0.0255	0.0741
Purpose	0.01258	0.03657
Savings account	0.01666	0.04843
Sum	0.34394	1

In the case of a linear classifier, a similar approach can be taken. Suppose we obtain the following (artificial) linear classification model for the credit scoring data set, with $z = +1$ denoting a good customer and $z = -1$ a bad customer:

$$z = sgn(0.7 \cdot \text{income} - 0.4 \cdot \text{savings} + 0.2 \cdot \text{checking} + 0.1 \cdot \text{age})$$

We expect that a higher income will yield a higher probability of being a good customer, and therefore the expected sign for the variable income is positive. Similarly we expect positive signs for savings and checking amount. For age we have no expectation, while the linear classification model shows an unwanted sign for savings. For simplicity reasons, we use the normalized regression coefficient as weight (e.g., $w_{\text{income}} = \frac{0.7}{0.7+0.4+0.2+0.1} = 0.5$) which results in the following justifiability:
Justifiability =

$$1 - \left[0.5 \cdot I(pr_{\text{income}}) + 0.29 \cdot I(pr_{\text{savings}}) + 0.14 \cdot I(pr_{\text{checking}}) + 0.07 \cdot I(pr_{\text{age}})\right]$$
$$= 1 - [0 + 0.29 + 0 + 0]$$
$$= 0.71 = 71\%$$

Medical Diagnosis Example

Our second example (Table 3.5) deals with the classification of patients as being
either demented or normal. For this, three variables are listed that are deemed relevant to the medical expert, being the number of years education the patient has had,
whether or not the patient can recall the name of the street he or she lives in, and
finally the age of the patient (slightly adjusted from [40]). First, we will measure the
justifiability of the example rule-based classifier.[2]

Table 3.5 Example dementia prediction rule set

> **if** (recalling name street = no **and** years education > 5)
> **then** patient = normal
>
> **else if** (recalling name street = yes and age > 80)
> **then** patient = normal
>
> **else if** (recalling name street = yes **and** years education > 5 **and** age ≤ 70)
> **then** patient = normal
>
> **else** patient = dementia diagnosed

Our rule-based classifier has a total of three rules to diagnose a patient's state
of mind. Our expectations for recalling the name of the street is rather obvious:
recalling the name speaks of the normal state of mind of the patient. Also for the
age variable straightforward expectations are made: older patients have an increased
likelihood of becoming demented. Finally, for the number of years education, a
medical expert might expect that more educated patients display higher degrees of
mental activities and are therefore less susceptible to dementia.

Table 3.6 shows the profiles for the three independent variables. For "recalling
name" there are two profiles: the first profile $pr_{recall,1}$ states that for two patients
with the same characteristics for age (> 70 and ≤ 80) and years education (> 5)
but different response to the recalling name variable, the one that recalls the name
of its street is demented, while the one that actually does not remember the name is
classified as normal. This is counterintuitive, and therefore $I(pr_{recall,1}) = 1$. For the
second profile $pr_{recall,2}$ the domain constraint is fulfilled, therefore $I(pr_{recall,2}) = 0$.
Similarly, the table illustrates that all profiles for the "years education" variable
satisfy our expectations, while none of the profiles for the "age" variable are in line
with the expectations, resulting in the following justifiability measure (assuming
equal weights):

---

[2] This example classifier is not generated from data, but humanly generated, as it only serves as an
example for the measure.

Table 3.6 Decision table format of dementia diagnosis classifier

(a) *Last variable "recalling name"*

Age	Years education	Recalling name	Impaired	Normal	
≤ 70	≤ 5	–	×	–	
	> 5	–	–	×	
> 70 and ≤ 80	≤ 5	–	×	–	
	> 5	yes	×	–	**1**
		no	–	×	
> 80	≤ 5	yes	–	×	**2**
		no	×	–	
	> 5	–	–	×	

(b) *Last variable "years education"*

Age	Recalling name	Years education	Impaired	Normal	
≤ 70	–	≤ 5	×	–	**1**
		> 5	–	×	
> 70 and ≤ 80	yes	–	×	–	
	no	≤ 5	×	–	**2**
		> 5	–	×	
> 80	yes	–	–	×	
	no	≤ 5	×	–	**3**
		> 5	–	×	

(c) *Last variable "age"*

Years education	Recalling name	Age	Impaired	Normal	
≤ 5	yes	≤ 80	×	–	**1**
		> 80	–	×	
	no	–	×	–	
> 5	yes	≤ 70	–	×	
		> 70 and ≤ 80	×	–	**2**
		> 80	–	×	
	no	–	–	×	

Justifiability =

$$
\begin{bmatrix}
\frac{1}{3} \cdot \left(\frac{1}{2}I(pr_{\text{recall},1}) + \frac{1}{2}I(pr_{\text{recall},2})\right) \\
+ \frac{1}{3} \cdot \left(\frac{1}{3}I(pr_{\text{education},1}) + \frac{1}{3}I(pr_{\text{education},2}) + \frac{1}{3}I(pr_{\text{education},3})\right) \\
+ \frac{1}{3} \cdot \left(\frac{1}{2}I(pr_{\text{age},1}) + \frac{1}{2}I(pr_{\text{age},2})\right)
\end{bmatrix}
$$

$$
= 1 - \left[\frac{1}{3} \cdot \left(\frac{1}{2} \cdot 1 + \frac{1}{2} \cdot 0\right) + \frac{1}{3} \cdot \left(\frac{1}{3} \cdot 0 + \frac{1}{3} \cdot 0 + \frac{1}{3} \cdot 0\right) + \frac{1}{3} \cdot \left(\frac{1}{2} \cdot 1 + \frac{1}{2} \cdot 1\right)\right]
$$

$$
= 0.5 = 50\%
$$

As for credit scoring, we can assess the justifiability performance of a linear classifier for medical diagnosis. Once again, we will consider the use of classification models for diagnosing dementia. Instead of considering $z = +1$ as a good customer and $z = -1$ a bad one, we will discriminate between patients that are demented ($z = -1$) and those that are not ($z = +1$). Such an example of linear classifier can be

$$
z = sgn(0.26 \cdot \text{recalling name} + 0.4 \cdot \text{age} + 0.3 \cdot \text{education})
$$

Since the variable recalling name can only be 0 (patient does not recall the name of the street) or 1 (patient does recall the name of the street), we expect a positive coefficient: recalling the name increases the probability of being normal. For age, we would also expect a negative coefficient: older patients have an increased chance of being demented. Finally, expecting more years of education has a positive influence on the state of mind of the patient; the justifiability of the linear classification model becomes

Justifiability =

$$1 - \left[\frac{1}{3} \cdot I(pr_{\text{recalling name}}) + \frac{1}{3} \cdot I(pr_{\text{age}}) + \frac{1}{3} \cdot I(pr_{\text{education}}) \right]$$
$$= 1 - \left[0 + \frac{1}{3} + 0 \right]$$
$$= 0.67 = 67\%$$

3.3.4 Obtaining Justifiable Classification Models

Although our taxonomy reveals that many types of domain constraints exist, to the best of our knowledge only the monotonicity constraint has been researched so far (with the exception of the AntMiner+ technique [32]). In this section, we will review existing approaches to incorporate monotonicity constraints. The application of such monotonicity constraints has been applied, among others, in the medical diagnosis [40], house price prediction [60], and credit scoring [16, 47] domains. The aim of all these approaches is to generate classifiers that are acceptable, meaning they are accurate, comprehensible, and justifiable.

We divided the different approaches in to three categories depending on where monotonicity is incorporated: in a preprocessing step, within the classification technique, and finally, in a postprocessing step. The different techniques are summarized in Table 3.7 and are shortly explained next.

Preprocessing

Daniels and Velikova propose an algorithm that transforms non-monotone data into monotone data by a relabeling process [16]. This procedure is based on changing the value of the target variable. The main idea is to remove all non-monotone data pairs, by iteratively changing the class of the data instance for which the increase in correctly labeled instances is maximal. They report an improved accuracy and comprehensibility by applying this relabeling procedure. A drawback of this approach is that, although the data will be completely monotone, there is still no guarantee that monotone classifiers are constructed.

Data Mining

The first attempt to incorporate monotonicity in classification trees combines both a standard impurity measure, such as entropy, and a non-monotonicity measure in the splitting criterion [7]. This measure is based on comparison of all possible leaf pairs and checking for monotonicity. The measure is defined as the ratio between actual number of non-monotonic pairs and the maximum number of possible non-monotonic pairs. The author reports a significant reduction in non-monotonicity in the classification trees, without a significant decrease in accuracy. However, a complete monotone classifier is not guaranteed.

Sill proposed a class of ANNs that can approximate any continuous monotone function to an arbitrary degree of accuracy [47]. This network has two hidden layers: the first hidden layer has linear activation functions. These hidden nodes are grouped and connected to a node in the second hidden layer, which calculates the maximum. Finally, the output unit computes the minimum over all groups. Monotonicity is guaranteed by imposing signs on the weights from the input to the first hidden layer. Although this approach is straightforward and shows good results, the comprehensibility of such an ANN classifier is limited.

Altendorf et al. build monotone Bayesian networks by imposing inequality constraints on the network parameters [2]. Monotonicity constraints on the parameter estimation problem are handled by imposing penalties in the likelihood function. The models constructed show to be as good or better in terms of predictive accuracy than when imposing no constraints.

In AntMiner+, monotonicity is obtained by imposing inequality signs. This approach is also able to include soft monotonicity constraints, by adapting a problem-dependent heuristic function [32].

Postprocessing

Instead of enforcing monotonicity constraints during model construction, a simple generate-and-test approach is applied by Feelders [20]. Many different trees are generated (each time on another randomization of the data) and the most monotonic one is used.

Another manner to achieve monotone classification models in a postprocessing step is by pruning classification trees [21]. This method prunes the parent of the non-monotone leaf that provides the largest reduction in the number of non-monotonic leaf pairs. Once more, similar accuracy is reported, with increased comprehensibility.

For linear models, checking monotonicity comes down to verifying the sign of the regression parameters. By removing those variables for which the sign is not as expected, and reestimating the parameters, a monotone model can be built (see, e.g., [53, 55]). Alternatively, adding variables might reverse the sign, or estimating the coefficients with nonnegative (positive) constraints can yield monotone linear models.

Finally, we want to point out the difference between *expressing* what is wanted, which should be done by the domain expert, and *demanding* these constraints, which is the focus of our research. Sometimes, data mining can reveal some interesting, though unexpected, patterns [46] which might not be detected when demanding

Table 3.7 Overview of existing approaches to incorporate monotonicity constraints

Preprocessing
Relabeling data to ensure monotonicity
Daniels and Velikova [16]
Data mining
Classification trees
Splitting criterion includes monotonicity metric
Ben-David [7]
Artificial neural networks
Linear activation functions and constrained weights
Sill [47]
Bayesian networks
Constrained Bayesian network parameters
Altendorf et al. [2]
Rule-based classification models
Constrained inequality signs
Martens et al. [32]
Postprocessing
Classification trees
Generate and test
Feelders [20]
Classification trees
Pruning non-monotone subtrees
Feelders and Pardoel [21]
Linear models
Constrained regression coefficients
Van Gestel [53]

justifiability constraints. Deciding when and for which variables to require such constraints is left entirely up to the domain expert.

3.4 Conclusion

Comprehensibility is an important requirement for classification models in many domains. Several approaches have been proposed in the literature to come to such models, from simple rule induction techniques to advanced incremental approaches. More recently, the importance of justifiability has been acknowledged by some researchers, in the form of the monotonicity constraint. Many other constraints exist that can be included in the learning algorithm, for which we introduced a taxonomy.

The newly introduced metric for justifiability allows not only to compare classifiers but also to demand a minimum justifiability threshold for acceptance before the classification model will be implemented in a decision support system, with threshold that can even go up to even 100%, requiring a model that is completely in line with business expectation.

By providing comprehensible, justifiable classification models, they become acceptable in domains where previously such models are deemed too theoretical and incomprehensible. We experienced this in previous case studies in domains such as audit mining (predicting the going concern opinion as issued by the auditor) [31], business/ICT alignment prediction [15], and software fault prediction [57]. This has also been confirmed in the research by, among others, Kodratoff [28] and Askira-Gelman [3]. As such, new opportunities emerge for data mining.

Acknowledgments We extend our gratitude to the guest editor and the anonymous reviewers, as their many constructive and detailed remarks certainly contributed much to the quality of this chapter. Further, we would like to thank the Flemish Research Council (FWO, Grant G.0615.05) for financial support.

References

1. D. W. Aha, D. F. Kibler, and M. K. Albert. Instance-based learning algorithms. *Machine Learning*, 6:37–66, 1991.
2. E. Altendorf, E. Restificar, and T.G. Dietterich. Learning from sparse data by exploiting monotonicity constraints. In *Proceedings of the 21st Conference on Uncertainty in Artificial Intelligence*, Edinburgh, Scotland, 2005.
3. I. Askira-Gelman. Knowledge discovery: Comprehensibility of the results. In *HICSS '98: Proceedings of the Thirty-First Annual Hawaii International Conference on System Sciences-Volume 5*, p. 247, Washington, DC, USA, 1998. IEEE Computer Society.
4. B. Baesens. *Developing intelligent systems for credit scoring using machine learning techniques*. PhD thesis, K.U. Leuven, 2003.
5. B. Baesens, R. Setiono, C. Mues, and J. Vanthienen. Using neural network rule extraction and decision tables for credit-risk evaluation. *Management Science*, 49(3):312–329, 2003.
6. B. Baesens, T. Van Gestel, S. Viaene, M. Stepanova, J. Suykens, and J. Vanthienen. Benchmarking state-of-the-art classification algorithms for credit scoring. *Journal of the Operational Research Society*, 54(6):627–635, 2003.
7. A. Ben-David. Monotonicity maintenance in information-theoretic machine learning algorithms. *Machine Learning*, 19(1):29–43, 1995.
8. D. Billman and D. Davila. Consistency is the hobgoblin of human minds: People care but concept learning models do not. In *Proceedings of the 17th Annual Conference of the Cognitive Science Society*, pp. 188–193, 1995.
9. C.M. Bishop. *Neural Networks for Pattern Recognition*. Oxford University Press, Oxford, UK, 1996.
10. L. Breiman, J. Friedman, R. Olshen, and C. Stone. *Classification and Regression Trees*. Chapman & Hall, New York, 1984.
11. D. R. Carvalho, A. A. Freitas, and N. F. F. Ebecken. Evaluating the correlation between objective rule interestingness measures and real human interest. In Alípio Jorge, Luís Torgo, Pavel Brazdil, Rui Camacho, and João Gama, editors, *PKDD*, volume 3721 of *Lecture Notes in Computer Science*, pp. 453–461. Springer, 2005.
12. P. Clark and T. Niblett. The CN2 induction algorithm. *Machine Learning*, 3(4):261–283, 1989.
13. W. W. Cohen. Fast effective rule induction. In Armand Prieditis and Stuart Russell, editors, *Proc. of the 12th International Conference on Machine Learning*, pp. 115–123, Tahoe City, CA, 1995. Morgan Kaufmann.

14. N. Cristianini and J. Shawe-Taylor. *An introduction to Support Vector Machines and Other Kernel-Based Learning Methods*. Cambridge University Press, New York, 2000.

15. B. Cumps, D. Martens, M. De Backer, S. Viaene, G. Dedene, R. Haesen, M. Snoeck, and B. Baesens. Inferring rules for business/ict alignment using ants. *Information and Management*, 46(2):116–124, 2009.

16. H. Daniels and M. Velikova. Derivation of monotone decision models from noisy data. *IEEE Transactions on Systems, Man and Cybernetics, Part C: Applications and Reviews*, 36(5):705–710, 2006.

17. P. Domingos. The role of occam's razor in knowledge discovery. *Data Mining and Knowledge Discovery*, 3(4):409–425, 1999.

18. R.O. Duda, P.E. Hart, and D.G. Stork. *Pattern Classification*. John Wiley and Sons, New York, second edition, 2001.

19. Federal Trade Commission for the Consumer. Facts for consumers: Equal credit opportunity. Technical report, FTC, March 1998.

20. A.J. Feelders. Prior knowledge in economic applications of data mining. In *Proceedings of the fourth European conference on principles and practice of knowledge discovery in data bases*, volume 1910 of *Lecture Notes in Computer Science*, pp. 395–400. Springer, 2000.

21. A.J. Feelders and M. Pardoel. Pruning for monotone classification trees. In *Advanced in intelligent data analysis V*, volume 2810, pp. 1–12. Springer, 2003.

22. D. Hand. Pattern detection and discovery. In D. Hand, N. Adams, and R. Bolton, editors, *Pattern Detection and Discovery*, volume 2447 of *Lecture Notes in Computer Science*, pp. 1–12. Springer, 2002.

23. D. Hand. Protection or privacy? Data mining and personal data. In *Advances in Knowledge Discovery and Data Mining, 10th Pacific-Asia Conference, PAKDD 2006, Singapore, April 9-12*, volume 3918 of *Lecture Notes in Computer Science*, pp. 1–10. Springer, 2006.

24. T. Hastie, R. Tibshirani, and J. Friedman. *The Elements of Statistical Learning, Data Mining, Inference, and Prediction*. Springer, New York, 2001.

25. J. Huysmans, B. Baesens, D. Martens, K. Denys, and J. Vanthienen. New trends in data mining. In *Tijdschrift voor economie en Management*, volume L, pp. 697–711, 2005.

26. J. Huysmans, C. Mues, B. Baesens, and J. Vanthienen. An empirical evaluation of the comprehensibility of decision table, tree and rule based predictive models. 2007.

27. D.G. Kleinbaum, L.L. Kupper, K. E. Muller, and A. Nizam. *Applied Regression Analysis and Multivariable Methods*. Duxbury Press, North Scituate, MA, 1997.

28. Y. Kodratoff. The comprehensibility manifesto. KDD Nuggets (94:9), 1994.

29. D. Martens, B. Baesens, and T. Van Gestel. Decompositional rule extraction from support vector machines by active learning. *IEEE Transactions on Knowledge and Data Engineering*, 21(2):178–191, 2009.

30. D. Martens, B. Baesens, T. Van Gestel, and J. Vanthienen. Comprehensible credit scoring models using rule extraction from support vector machines. *European Journal of Operational Research*, 183(3):1466–1476, 2007.

31. D. Martens, L. Bruynseels, B. Baesens, M. Willekens, and J. Vanthienen. Predicting going concern opinion with data mining. *Decision Support Systems*, 45(4):765–777, 2008.

32. D. Martens, M. De Backer, R. Haesen, B. Baesens, C. Mues, and J. Vanthienen. Ant-based approach to the knowledge fusion problem. In *Proceedings of the Fifth International Workshop on Ant Colony Optimization and Swarm Intelligence*, Lecture Notes in Computer Science, pp. 85–96. Springer, 2006.

33. D. Martens, M. De Backer, R. Haesen, M. Snoeck, J. Vanthienen, and B. Baesens. Classification with ant colony optimization. *IEEE Transaction on Evolutionary Computation*, 11(5):651–665, 2007.

34. R.S. Michalski. A theory and methodology of inductive learning. *Artificial Intelligence*, 20(2):111–161, 1983.

35. O.O. Maimon and L. Rokach. *Decomposition Methodology For Knowledge Discovery And Data Mining: Theory And Applications (Machine Perception and Artificial Intelligence)*. World Scientific Publishing Company, July 2005.

36. M. Ohsaki, S. Kitaguchi, K. Okamoto, H. Yokoi, and T. Yamaguchi. Evaluation of rule inter-estingness measures with a clinical dataset on hepatitis. In *PKDD '04: Proceedings of the 8th European Conference on Principles and Practice of Knowledge Discovery in Databases*, pp. 362–373, New York, NY, USA, 2004. Springer-Verlag New York, Inc.
37. L. Passmore, J. Goodside, L. Hamel, L. Gonzales, T. Silberstein, and J. Trimarchi. Assessing decision tree models for clinical in-vitro fertilization data. Technical Report TR03-296, Dept. of Computer Science and Statistics, University of Rhode Island, 2003.
38. M. Pazzani. Influence of prior knowledge on concept acquisition: Experimental and computational results. *Journal of Experimental Psychology: Learning, Memory, and Cognition*, 17(3):416–432, 1991.
39. M. Pazzani and S. Bay. The independent sign bias: Gaining insight from multiple linear regression. In *Proceedings of the Twenty First Annual Conference of the Cognitive Science Society*, pp. 525–530., 1999.
40. M. Pazzani, S. Mani, and W. Shankle. Acceptance by medical experts of rules generated by machine learning. *Methods of Information in Medicine*, 40(5):380–385, 2001.
41. M. Pazzani. Learning with globally predictive tests. In *Discovery Science*, pp. 220–231, 1998.
42. J. R. Quinlan. *C4.5 Programs for Machine Learning*. Morgan Kaufmann Publishers Inc., San Francisco, CA, 1993.
43. J.W. Seifert. Data mining and homeland security: An overview. *CRS Report for Congress*, 2006.
44. R. Setiono, B. Baesens, and C. Mues. Risk management and regulatory compliance: A data mining framework based on neural network rule extraction. In *Proceedings of the International Conference on Information Systems (ICIS 2006)*, 2006.
45. R. Setiono, B. Baesens, and C. Mues. Recursive neural network rule extraction for data with mixed attributes. *IEEE Transactions on Neural Networks*, Forthcoming.
46. A. Silberschatz and A. Tuzhilin. On subjective measures of interestingness in knowledge discovery. In *KDD*, pp. 275–281, 1995.
47. J. Sill. Monotonic networks. In *Advances in Neural Information Processing Systems*, volume 10. The MIT Press, Cambridge, MA, 1998.
48. E. Sommer. An approach to quantifying the quality of induced theories. In Claire Nedellec, editor, *Proceedings of the IJCAI Workshop on Machine Learning and Comprehensibility*, 1995.
49. J. A. K. Suykens, T. Van Gestel, J. De Brabanter, B. De Moor, and J. Vandewalle. *Least Squares Support Vector Machines*. World Scientific, Singapore, 2002.
50. P.-N. Tan, M. Steinbach, and V. Kumar. *Introduction to Data Mining*. Pearson Education, Boston, MA, 2006.
51. L. Thomas, D. Edelman, and J. Crook, editors. *Credit Scoring and its Applications*. SIAM, Philadelphia, PA, 2002.
52. T. Van Gestel and B. Baesens. *Credit Risk Management: Basic concepts: financial risk components, rating analysis, models, economic and regulatory capital*. Oxford University Press, 2009.
53. T. Van Gestel, B. Baesens, P. Van Dijcke, J. Garcia, J.A.K. Suykens, and J. Vanthienen. A process model to develop an internal rating system: sovereign credit ratings. *Decision Support Systems*, 42(2):1131–1151, 2006.
54. T. Van Gestel, B. Baesens, P. Van Dijcke, J.A.K. Suykens, J. Garcia, and T. Alderweireld. Linear and nonlinear credit scoring by combining logistic regression and support vector machines. *Journal of Credit Risk*, 1(4), 2005.
55. T. Van Gestel, D. Martens, B. Baesens, D. Feremans, J Huysmans, and J. Vanthienen. Forecasting and analyzing insurance companies' ratings. *International Journal of Forecasting*, 23(3):513–529, 2007.
56. T. Van Gestel, J.A.K. Suykens, B. Baesens, S. Viaene, J. Vanthienen, G. Dedene, B. De Moor, and J. Vandewalle. Benchmarking least squares support vector machine classifiers. *Machine Learning*, 54(1):5–32, 2004.

57. O. Vandecruys, D. Martens, B. Baesens, C. Mues, M. De Backer, and R. Haesen. Mining software repositories for comprehensible software fault prediction models. *Journal of Systems and Software*, 81(5):823–839, 2008.

58. J. Vanthienen, C. Mues, and A. Aerts. An illustration of verification and validation in the modelling phase of KBS development. *Data and Knowledge Engineering*, 27(3):337–352, 1998.

59. V. N. Vapnik. *The nature of statistical learning theory*. Springer-Verlag New York, Inc., New York, 1995.

60. M. Velikova and H. Daniels. Decision trees for monotone price models. *Computational Management Science*, 1(3–4):231–244, 2004.

61. M. Velikova, H. Daniels, and A. Feelders. Solving partially monotone problems with neural networks. In *Proceedings of the International Conference on Neural Networks*, Vienna, Austria, March 2006.

62. M. P. Wellman. Fundamental concepts of qualitative probabilistic networks. *Artificial Intelligence*, 44(3):257–303, 1990.

63. I. H. Witten and E. Frank. *Data mining: practical machine learning tools and techniques with Java implementations*. Morgan Kaufmann Publishers Inc., San Francisco, CA, 2000.

Chapter 4
Mining Interesting Rules Without Support Requirement: A General Universal Existential Upward Closure Property

Yannick Le Bras, Philippe Lenca, and Stéphane Lallich

Abstract Many studies have shown the limits of support/confidence framework used in Apriori-like algorithms to mine association rules. There are a lot of efficient implementations based on the antimonotony property of the support. But candidate set generation is still costly and many rules are uninteresting or redundant. In addition one can miss interesting rules like nuggets. We are thus facing a complexity issue and a quality issue.

One solution is to get rid of frequent itemset mining and to focus as soon as possible on interesting rules. For that purpose algorithmic properties were first studied, especially for the confidence. They allow to find all confident rules without a preliminary support pruning.

Recently, in the case of class association rules, the universal existential upward closure property of confidence has been exploited in an efficient manner. Indeed, it allows to use a pruning strategy for an Apriori-like but top-down associative classification rules algorithm.

We present a new formal framework which allows us to make the link between analytic and algorithmic properties of the measures. We then apply this framework to propose a general universal existential upward closure. We demonstrate a necessary condition and a sufficient condition of existence for this property. These results are then applied to 32 measures and we show that 13 of them do have the GUEUC property.

Yannick Le Bras and Philippe Lenca
Institut Telecom; Telecom Bretagne, UMR CNRS 3192 Lab-STICC, Technopôle Brest-Iroise – CS 83818, 29238 Brest Cedex 3 – France; Université Européenne de Bretagne, Rennes, France,
e-mail: yannick.lebras@telecom-bretagne.eu and
philippe.lenca@telecom-bretagne.eu

Stéphane Lallich
Laboratoire ERIC, Université de Lyon, Lyon 2, France,
e-mail: stephane.lallich@univ-lyon2.fr

R. Stahlbock et al. (eds.), *Data Mining*, Annals of Information Systems 8,
DOI 10.1007/978-1-4419-1280-0_4, © Springer Science+Business Media, LLC 2010

4.1 Introduction

Rules-based data mining methods are very important approaches in supervised learning. They provide a concise and potentially useful knowledge that is understandable by the user. In addition, they allow to explain why a particular prediction is made for a particular case. This ability to justify is very important in domains such as credit scoring or medicine.

The most popular rule-based method is the supervised learning of decision tree [12, 47, 65], which generates an understandable predictive model in a reasonable time. One of the limitations of decision trees is that they proceed according to a greedy strategy. As a consequence, they do not explore all the combinations of attributes. Indeed, the standard decision tree algorithms do not call into question the earlier choice of splitting attributes during the construction of the tree. Another feature of decision trees is that they are mainly intended for categorical attributes. Thus at each node, the quality of a candidate splitting attribute is estimated by taking into account all subpopulations, which could be induced from the candidate.

More recently, in order to build predictive rules, [37, 36, 61] have developed a very interesting alternative approach, named associative classification. This approach is based on the research of classification rules which are association rules of which the consequent is a class label (that is to say class association rules). In this chapter we mainly focus on this category of rules.

Let us first recall the fundamentals of association rule mining. An association rule is a rule $A \rightarrow B$, where A and B are two sets of items (also called itemsets) such that $A \neq \emptyset$, $B \neq \emptyset$, and $A \cap B = \emptyset$, meaning that given a database $\mathcal{D}$ of transactions (where each transaction is a set of items) whenever a transaction T contains A, then T probably contains B also [3]. In order to generate class association rules these methods use the confidence–support framework developed by [3] with the APRIORI algorithm to extract unsupervised association rules from boolean databases. Support is defined as the proportion of transactions containing A and B in the entire database (noted $supp(AB)$ or $P(AB)$), while confidence is the proportion of transactions containing A and B inside the set of transactions containing A (noted $conf(A \rightarrow B)$ or $P(B|A)$). APRIORI extracts the rules for which the support and the confidence exceed some prefixed thresholds according to a two-step process. In a first step, the minimum support constraint is applied to find all frequent itemsets in the database. In a second step, these frequent itemsets and the minimum confidence constraint are used to form rules.

Thus, the restriction of the rules search space is no longer ensured by a greedy exploration, but by the support and confidence conditions. The condition of support plays a prominent role insofar as it reduces the search space of frequent itemsets through the antimonotony property of the support. The associative classification methods thus remain captive of the support threshold, which prevents them to find nuggets, namely the rules of low support but very strong confidence, even though these rules are very suited to the prediction task.

Two other differences should be noted between associative classification and decision trees. The first one is that association rules were built primarily for boolean data [5], where only the presence of attributes is of interest for the user. The second, linked to the previous one, is that the usual measures of the interest of a rule, such as the confidence or the lift, take into account only the cases covered by the rule. Thereafter, various algorithms have been developed to build association rules between categorical and numerical attributes (see, for example, [21, 25]).

The development of algorithms which extract predictive association rules without involving the support condition is currently an important issue in data mining and various tracks were explored. In the case of categorical attributes, which means that the columns of the data matrix correspond to the flags of each modality of each attribute, *universal existential upward closure* (UEUC) [56, 54] was introduced. UEUC is a property of the confidence which allows to develop a pruning strategy and free oneself from the support condition.

However, different works showed that confidence brings only an incomplete information about the quality of extracted rules and that other interestingness measures may be preferred, for example, the centered confidence, the lift, the conviction, or the Bayes factor [50, 17, 31]. As a result, it seems to us that it is crucial to determine under what conditions an interestingness measure checks the UEUC property. Indeed, it is possible to directly extract the best rules according to such a measure, without the support condition.

The rest of the chapter is organized as follows. In Section 4.2 we briefly review some related works. The core of our study, the *universal existential upward closure* property of the confidence, is presented in Section 4.3. We also present in this section our *general universal existential upward closure* property. We present in Section 4.4 our framework that leads to study interestingness measures with respect to various quantities. This framework is then used in Section 4.5 to propose a sufficient condition and a necessary condition of existence of the *general universal existential upward closure* property. These conditions are then used to study the case of 32 measures. We conclude in Section 4.6.

4.2 State of the Art

Since the initial work [3] and the famous APRIORI algorithm [5] many efforts have been done in order to develop efficient algorithms for mining frequent itemsets and/or association rules (see, for example, the works following APRIORI, like [4, 51, 13]). These methods use the support–confidence framework developed by [3] to extract unsupervised association rules in boolean databases. In [64] the reader can find a comparison of five well-known algorithms (APRIORI implementation [11], CHARM [63], FP-growth [23], CLOSET, [44], and MAGNUMOPUS [57]). Current trends and overviews on the frequent itemset mining approaches are discussed in [19, 18, 22].

In order to build predictive rules (e.g., in a supervised setting), [37] (with the CBA algorithm), [36] (with the CMAR algorithm), [61] (with the CPAR algorithm) have developed an interesting alternative approach, named associative classification. These approaches also use the confidence–support framework developed by [5].

The use of the support condition is effective for computational reasons. However, without specific knowledge, users will have difficulties in setting the support threshold. The number of itemsets may differ by an order of magnitude depending on the support thresholds (see, for example, [14, 46]). If the support threshold is too large, there may be only a small number of rules, if the threshold is too small, there may be too many rules and the user cannot examine them. Notice also that low support implies huge computational costs. However, rules with very high confidence but with very low support may be of a great interest. In consequence many works also focused on solutions to avoid the use of the support [35, 15, 54, 8].

To get rid of the support constraint several authors proposed different solutions at the itemsets level by adding constraints on items and at the measurement of interestingness level by exploiting intrinsic properties of the interestingness measures.

In this chapter we are interested in exploiting algorithmic properties of measures of interest and we are thus at the second level. For constraint-based mining technique the reader can refer to the state-of-the-art paper of [10] and, for example, to [41, 55, 43, 42, 33, 14].

In what follows, we briefly review some of the works that focus on interestingness measures. Indeed, the measures of interest may play different roles [60]. In particular we are here interested in measures that can reduce the search space and also be useful for the evaluation of the quality of mined patterns. We believe that this approach is the most promising one.

In [7] the authors present DENSEMINER that mines consequent-constrained rule with an efficient pruning strategy based on constraints such as minimum support, minimum confidence, and minimum improvement. They define the improvement of a rule as the minimum difference between its confidence and the confidence of any proper sub-rule with the same consequent. The algorithm uses multiple pruning strategies based on the calculation of bounds for the three measures using only the support information provided by the candidates. Experiments demonstrate the efficiency of DENSEMINER, in contrast with frequent itemset mining approaches, especially on dense data sets.

In [59] the authors propose to make the confidence antimonotone following the example of support. In this way, they introduce the *h-confidence* a new measure over itemsets mathematically identical to *all-confidence* proposed by [40]. Both of these measures are antimonotone. They also introduce the concept of cross-support patterns (i.e., uninteresting patterns involving items with substantially different support levels). The authors thus propose *hyperclique miner*, an efficient algorithm for association rule mining that uses both the cross-support and antimonotone properties of the *h*-confidence measure. In [29], we prove a necessary condition for the existence of such a property for other measures. We then study 27 measures and show that only 5 of them can be made antimonotone by this way.

In [34] the author introduces the notion of optimal rule set for classification rules. A rule set is optimal if it contains all rules, except those with no greater interestingness than one of its more general rules. Optimal rule sets have similar properties as closed itemsets and nonredundant rule sets [62]. The main advantage of optimal rules is that they allow to define a pruning strategy that could apply to a large set of measures. The author gives a proof, case by case, for 13 measures. In [28], we extend the result of the original article and give a necessary and sufficient condition of applicability of this pruning strategy for an objective measure. We thus apply this framework to 32 measures and show that this pruning strategy applies to 26 measures out of them.

Focusing on the Jaccard measure, [15] propose an approximation algorithm based on hash tables for classification. The apparition of false negatives and false positives can be controlled with the algorithm parameters. The efficiency of the algorithm depends of course on the strength of the control. However, this control cannot be complete and there is always a risk to obtain false negatives and false positives.

In [66] the authors adapt a technique proposed by [38] to construct the COR-CLASS algorithm for associative classification. It is fully based on an antimonotonic property of convexity of the χ^2 and a direct branch-and-bound exploration of candidates.

In [9], the authors introduce the *loose antimonotony* property. This new type of constraint means that if it is satisfied by an itemset of cardinality k then it is satisfied by at least one of its subsets of cardinality $k-1$. It is applied to statistical constraints. The authors also deeply study different kinds of constraints (antimonotonicity, succinctness, monotony, convertibility) and propose a framework that can exploit different properties of constraints altogether in a level-wise Apriori-like manner.

Another approach is to use an intrinsic characteristic of confidence. In [54] the authors introduce the *universal existential upward closure* property based on a certain monotonicity of the confidence. This property applies to classification rules and allows to examine only confident rules of larger size for generating confident rules of smaller size. The authors deduce from this property a top-down confidence-based pruning strategy. Their algorithm is then efficient in terms of candidate generation and emphasizes the problem of nuggets.

Most of the previous approaches do not use a preliminary and costly frequent itemset mining. However, they mainly apply to very few measures and most of the time to the confidence. It would be interesting to have an algorithmic property, at the same time efficient and general, applicable to a large panel of interestingness measures. This is one of the major dimensions for extending frequent pattern mining framework [58]. As pointed out by several studies, in particular by [50, 20, 32, 53, 17, 31, 2, 1], measures of interest may have very different properties. The user should first use a set of measures adapted to his needs in order to select the good rules [30].

We here focus on the *universal existential upward closure* property of confidence proposed by [54] and will show that several measures share this property.

4.3 An Algorithmic Property of Confidence

4.3.1 On UEUC Framework

Among the algorithmic properties of interestingness measures, and especially of the confidence, we find the property of *universal existential upward closure* (UEUC) defined in [54]. Let us now describe the context of this property. We use notations similar to those used in the original article.

We consider a database T containing m nonclass categorical attributes $A_1, \ldots, A_m$ and one class attribute C. We denote by $\mathcal{A}_i$ (resp. C) the set of the values that an attribute A_i (resp. C) can take. A transaction is an element of $\mathcal{A}_1 \times \cdots \times \mathcal{A}_m \times C$. A k-rule is written as $A_{i_1} = a_{i_1}, \ldots, A_{i_k} = a_{i_k} \rightarrow C = c$, where $a_{i_p} \in \mathcal{A}_{i_p}$ and $c \in C$. We take the liberty to make the shortcut $X = x \rightarrow c$ instead of $A_{i_1} = a_{i_1}, \ldots, A_{i_k} = a_{i_k} \rightarrow C = c$ when no confusion is possible, where X is the set of attributes of the rule, and x the set of values on these attributes. If r is the rule $X = x \rightarrow c$ and $A \notin X$ then we say that $X = x, A = a \rightarrow c$ is a A-specialization of r if $a \in \mathcal{A}$, and that r is more general. The notions of support and confidence are the same as usual.

This framework is close to association rule mining, but is more general. In the case of a categorical database, we are able to focus on negative rules too, as described in [38], and more precisely in [66]. However, such categorical databases can be binarized by creating as many binary attributes as couples (A, a) possible, where $a \in \mathcal{A}$, and where A is a (non)class attribute. Then, an Apriori-like strategy can be applied.

On the other hand, this framework with categorical attributes and a target variable is close to the decision tree context. Indeed, in [56], the authors show how to build a classifier from confident association rules and compare it with classical decision tree methods. In this chapter, they introduce a property of confidence, called *existential upward closure*. More attention is paid to this property, recalled *universal existential upward closure*, in [54].

4.3.2 The UEUC Property

We now explain this property by reusing the example of the original article.

Let us consider the three rules

$$\begin{aligned}
\text{r:} \qquad & Age = young \rightarrow Buy = yes \\
\text{r1:}~ & Age = young, Gender = M \rightarrow Buy = yes \\
\text{r2:}~ & Age = young, Gender = F \rightarrow Buy = yes
\end{aligned}$$

We are interested in the behavior of young people (r) with respect to the purchase of an article, especially in case they are a male (r1) or a female (r2). With our formalism, we can write for the attributes: $A_1 = Age$, $A_2 = Gender$, $C = Buy$, and for the values: $\mathcal{A}_1 = \{young, old\}$, $\mathcal{A}_2 = \{M, F\}$, and $C = \{yes, no\}$. Since $Gender = M$ and

Gender = *F* generate two fully separate specializations, we can write the following equalities:

$$conf(\mathbf{r}) = \frac{supp(Age=young,Buy=yes)}{supp(Age=young)}$$

$$= \frac{supp(Age=young,Gender=M,Buy=yes)}{supp(Age=young,Gender=M)+supp(Age=young,Gender=F)}$$

$$+ \frac{supp(Age=young,Gender=F,Buy=yes)}{supp(Age=young,Gender=M)+supp(Age=young,Gender=F)}$$

$$= \frac{supp(Age=young,Gender=M)}{supp(Age=young,Gender=M)+supp(Age=young,Gender=F)} \times conf(\mathbf{r1})$$

$$+ \frac{supp(Age=young,Gender=F)}{supp(Age=young,Gender=M)+supp(Age=young,Gender=F)} \times conf(\mathbf{r2})$$

$$= \alpha_1 \times conf(\mathbf{r1}) + \alpha_2 \times conf(\mathbf{r2})$$

where α_1 and α_2 are positive numbers such that $\alpha_1 + \alpha_2 = 1$. Thus, the confidence of r is a barycenter of confidences of r1 and r2. From this equality, one can deduce that either $conf(\mathbf{r1})$ or $conf(\mathbf{r2})$ is greater or equal to $conf(\mathbf{r})$. Then, if neither r1 nor r2 is confident, then r can be removed.

This result can be generalized in the following way, where X is a set of attributes and A is an attribute not in X:

$$\exists(\alpha_1,\dots,\alpha_{|\mathcal{A}|}) \in \quad \mathbb{R}^+, \alpha_1 + \cdots + \alpha_{|\mathcal{A}|} = 1, \text{ such that}$$
$$conf(\mathtt{X} = x \rightarrow \mathtt{c}) = \sum_{a_i \in \mathcal{A}} \alpha_i \times conf(\mathtt{X} = x, \mathtt{A} = a_i \rightarrow \mathtt{c}) \quad (4.1)$$

The coefficients can be explicitly described by $\alpha_i = \frac{supp(\mathtt{X}=x,\mathtt{A}=a_i)}{supp(\mathtt{X}=x)}$.

We then can straightaway write the following property of confidence [54].

Definition 4.1 (UEUC). For every attribute A_i not occurring in a rule $\mathtt{X} = x \rightarrow \mathtt{c}$, (i) some A_i-specialization of $\mathtt{X} = x \rightarrow \mathtt{c}$ has at least the confidence of $\mathtt{X} = x \rightarrow \mathtt{c}$, (ii) if $\mathtt{X} = x \rightarrow \mathtt{c}$ is confident, so is some A_i-specialization of $\mathtt{X} = x \rightarrow \mathtt{c}$. This property is called the *universal existential upward closure* (UEUC).

4.3.3 An Efficient Pruning Algorithm

The UEUC property clearly offers a pruning strategy, which is shown to be efficient by its authors, in comparison with the well-known algorithm DENSEMINER [7]. We will now describe quickly the algorithm based on this pruning property.

This property suggests a top-down approach. It starts with all the transactions, which can be seen as rules with antecedent of size m, and calculates their confidence. We thus generate a first level of confident rules of size m. Suppose now that we have all confident rules of size $k + 1$, we will be able to generate all confident rules of size k. The following steps are directly inspired from the Apriori strategy reviewed under a top-down point of view:

- First, generate *projection candidates*. This step answers to the "some" part of Definition 4.1. A rule $X = x \rightarrow c$ is candidate if for some attribute A not in X, one A-specialization of that rule is confident. We also have to project every $(k+1)$-rule over the space of k-rules (there are $(k+1)$ possible projections).
- Second, generate *intersection candidates*. This step answers to the "for every" part of Definition 4.1. If a rule has at least one confident specialization, we have to check if it has confident specializations for every $(m-k)$ possible attributes.
- Then, after generating the candidates, we can start the step of *candidate validation*, by checking their confidence. Every candidate with a confidence higher than the given threshold is kept as a k-rule and will be used later for generating the $(k-1)$ level.

So this is a three-step algorithm that can be very costly in terms of used memory. If the database has very high dimensions, one level of rules might be too large to fit in the memory. Reference [54] propose a disk-based implementation of the algorithm, efficient in terms of hard drive access. This implementation uses hash tables to store rules and candidates, and hypergraphs to calculate frequencies.

The advantage of this pruning strategy is that one should not have to first calculate frequent itemsets and then confident rules. While we are focusing on interesting rules, this first step is useless: we here delete it. However, if we care about a frequency minimum, one can introduce it in the algorithm, since supports are evaluated while calculating confidence. There is no loss of time, but frequent rules have to be stored separately.

If we do not care about support constraints, the algorithm permits to mine nuggets of knowledge that cannot be mined when using a support threshold, due to combinatorial explosion. Nuggets of knowledge represent a large part of the confident rules: in a database like the well-known Mushrooms, our experiments show that 80% of the confident rules (confident threshold 0.8) have a support less than 1%. Such rules are typically appearing in medical databases when focusing on orphan diseases or rare diseases [39, 52].

4.3.4 Generalizing the UEUC Property

One of the limits of the UEUC property is that it is defined only for confidence. It is legitimate to ask oneself if other measures verify the UEUC property. To answer this question, we first define a general UEUC property.

Definition 4.2 (General UEUC Property). An interestingness measure μ verifies the general UEUC property if, and only if, for every attribute A_i not occurring in a rule $r: X = x \rightarrow c$, some A_i-specialization of r has at least the value taken by r on the measure μ.

A consequence is that, for such a measure, if r is interesting (with respect to a given threshold), so is some A_i-specialization of r.

Confidence clearly verifies this property. We will in the following use the term of GUEUC property instead of general UEUC property. We are searching for other interestingness measures that verify this property too. In addition, we would like to have the power to say if a given measure verifies or not the property: we are looking for sufficient and/or necessary conditions of existence of the GUEUC property.

We first illustrate the GUEUC property with two interestingness measures. The first one has the GUEUC property and the second one does not have it.

Consider the Sebag and Shoenauer interestingness measure [48]. We can write it as

$$seb(X = x \to c) = \frac{supp(X = x, c)}{supp(X = x, \neg c)}.$$

Let now A be an attribute not in X. The following equalities hold:

$$
\begin{aligned}
seb(X = x \to c) &= \frac{\displaystyle\sum_{a \in \mathcal{A}} supp(X = x, A = a, c)}{supp(X = x, \neg c)} \\
&= \sum_{a \in \mathcal{A}} \frac{supp(X = x, A = a, c)}{supp(X = x, \neg c)} \\
&= \sum_{a \in \mathcal{A}} \frac{supp(X = x, A = a, \neg c)}{supp(X = x, \neg c)} \times \frac{supp(X = x, A = a, c)}{supp(X = x, A = a, \neg c)} \\
&= \sum_{a \in \mathcal{A}} \alpha_a \times \frac{supp(X = x, A = a, c)}{supp(X = x, A = a, \neg c)} \\
&= \sum_{a \in \mathcal{A}} \alpha_a \times seb(X = x, A = a \to c)
\end{aligned}
$$

where the α_a are positive numbers such that $\sum_{a \in \mathcal{A}} \alpha_a = 1$. Then the measure of Sebag and Shoenauer has the same barycenter property as confidence (Equation 4.1), and we can deduce that it verifies the GUEUC property.

On the contrary, consider the measure of Piatetsky-Shapiro [45]:

$$ps(X = x \to c) = supp(X = x, c) - supp(X = x) \times supp(c).$$

For this measure, we have the following inequalities:

$$
\begin{aligned}
ps(X = x \to c) &= \sum_{a \in \mathcal{A}} (supp(X = x, A = a, c) - supp(X = x, A = a) \times supp(c)) \\
&= \sum_{a \in \mathcal{A}} ps(X = x, A = a \to c)
\end{aligned}
$$

Then the measure of Piatetsky-Shapiro of a rule is the sum of its specifications and does not verify our GUEUC property.

Thus a question arises: Do other measures verify the GUEUC property? We will now introduce a new framework to study the measures that will allow us to say which ones do and which ones do not.

4.4 A Framework for the Study of Measures

4.4.1 Adapted Functions of Measure

In [24] a framework for the study of interestingness measures is presented. An interestingness measure of a rule A → B is a function that will help the user to evaluate the quality of the rule A → B, such as the well-known support and confidence. Most of the authors consider measures of interest as functions of $\mathbb{R}^3 \longrightarrow \mathbb{R}$. In [31], one can also find an advanced study of target domains, while [16, 49] restricts this domain by normalizing measures. Given a set of measures the authors prove that they have the same behavior and that they can be simultaneously optimized. First, we will explain the concept of associated measure introduced in [24]. This chapter focuses only on the parametrization of interestingness measures in function of the number of examples, antecedents, and consequents. Since the UEUC property was first introduced for the confidence, we will focus on the behavior of measures with respect to confidence. One similar approach can be found in [26, 17, 27, 31]. One can also need to write measures as a function of the number of counterexamples of the rule and study their behavior with respect to this quantity, like in [31].

We propose here to introduce a formal framework which enables an analytic study of interestingness measures. We focus only on objective interestingness measures. Such measures can be expressed with the help of the contingency table in relative frequencies and consequently with three parameters. The study of their variations with respect to these variables will allow us to make a link between the measures and their algorithmic properties, but they imply the description of a domain of definition in order to study only real cases.

4.4.1.1 Association Rules

A boolean database $\mathcal{DB}$ is described by a triplet $(\mathcal{A}, R, \mathcal{T})$, where $\mathcal{A}$ is a set of items, $\mathcal{T}$ is a set of attributes, and R is a binary relation on $\mathcal{A} \times \mathcal{T}$. An association rule is defined by a database $\mathcal{DB}$, a nonempty set A $\subset \mathcal{A}$ (A is an *itemset*) called antecedent, and a nonempty set B $\subset \mathcal{A}$ called consequent such that A∩B = ∅. We denote a rule by A $\xrightarrow{\mathcal{DB}}$ B, or simply A → B when there is no possible confusion.

The support of an itemset A is the frequency of appearance of this itemset in the database. We denote it by $supp_{\mathcal{DB}}(A)$ or, if there is no ambiguity, $supp(A)$. The support of the rule A → B is the support of itemset AB.

4.4.1.2 Contingency Tables

Let A and B be two itemsets on the database $\mathcal{DB}$. The contingency table in relative joined frequencies of A and B gives information about the simultaneous presence of these itemsets (Fig. 4.1).

	B	B̄	
A	$supp(AB)$	$supp(A\bar{B})$	$supp(A)$
Ā	$supp(\bar{A}B)$	$supp(\bar{A}\bar{B})$	$supp(\bar{A})$
	$supp(B)$	$supp(\bar{B})$	1

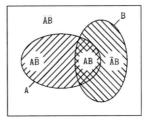

Fig. 4.1 Contingency table of A and B, as table and as graphic

The contingency table has three degrees of liberty: one needs at least three of its values to describe it, and three values are enough to find all other values. For example, the contingency table is fully described by the two marginal frequencies $supp(A)$ and $supp(B)$ and the joined frequency $supp(AB)$. An association rule on a given database is described by two itemsets. One can also speak about the contingency table of an association rule, which leads us to the notion of descriptor system.

Definition 4.3. We call *descriptor system* of the contingency table a triplet of functions (f, g, h) over the association rules which fully describes the contingency table of association rules.

Example 4.1. We define the following functions:

$$ant(A \rightarrow B) = supp(A); \quad cons(A \rightarrow B) = supp(B); \quad conf(A \rightarrow B) = \frac{supp(AB)}{supp(A)}$$

The triplet $(conf, ant, cons)$ is a descriptor system of the contingency table.

The same stands for the functions $ex(A \rightarrow B) = supp(AB)$ and $c\text{-}ex(A \rightarrow B) = supp(A\bar{B})$: the triplets $(ex, ant, cons)$ and $(c\text{-}ex, ant, cons)$ are descriptor systems.

An objective interestingness measure is a function from the space of association rules to the space of extended real numbers ($\mathbb{R} \cup \{-\infty, +\infty\}$). It is used to quantify the interest of a rule. There exists a large number of rules [50, 17, 60, 31], but most of them can be expressed with the contingency table and considered as three variable functions (a descriptor system of the contingency table). In this chapter, we focus only on this kind of measures.

4.4.1.3 Minimal Joint Domain

Like random variables in a probabilistic universe, one can define variables over the space of association rules. A descriptor system d of the contingency table is then a triplet of variables over this space, and an interestingness measure μ can be written with the help of a function ϕ_μ of this triplet. If we want to make an analytic study of this function, we need only to restrict the analysis to the joint variation domain of the triplet. Moreover, the study will have no sense out of this domain, where the points match no real situation.

Definition 4.4. We call the couple $(\phi_\mu, \mathcal{D}_d)$, made from this function and the joint variation domain (associated to a specific descriptor system), the *d-adapted function of measure* of the measure μ.

It is important to see that the form of the functional part of this function of measure depends on the descriptor system chosen. However, when this system is fixed, the adapted function of measure is uniquely defined. In the following, we voluntarily omit to mention the chosen descriptor system if there is no possible ambiguity.

If d is a descriptor system of the contingency table, the joint variation domain associated to this system is defined by the constraints laid down by the values of d between themselves.

Example 4.2. Let d_{conf} be the descriptor system based on confidence:

$$d_{conf} = (conf, ant, cons) \qquad (4.2)$$

For this system, we have the set of constraints:

$$
\begin{aligned}
0 &< ant < 1 \\
0 &< cons < 1 \\
0 &\leq conf \leq 1 \\
1 - \frac{1-cons}{ant} &\leq conf \leq \frac{cons}{ant}
\end{aligned}
$$

Indeed,

$$
\begin{aligned}
conf(A \to B) &= \frac{supp(AB)}{supp(A)} \\
&= \frac{supp(A) - supp(A\neg B)}{supp(A)} \\
&= 1 - \frac{supp(A\neg B)}{supp(A)} \\
&\geq 1 - \frac{supp(\neg B)}{supp(A)} \\
&\geq 1 - \frac{1 - supp(B)}{supp(A)}
\end{aligned}
$$

and

$$
\begin{aligned}
conf(A \to B) &= \frac{supp(AB)}{supp(A)} \\
&\leq \frac{supp(B)}{supp(A)}
\end{aligned}
$$

We thus define the following domain:

$$D = \left\{ \begin{pmatrix} c \\ y \\ z \end{pmatrix} \in \mathbb{Q}^3 \mid \begin{array}{ccc} 0 & < y < & 1 \\ 0 & < z < & 1 \\ \max\{0, 1 - \frac{1-z}{y}\} & \leq c \leq & \min\{1, \frac{z}{y}\} \end{array} \right\} \tag{4.3}$$

Because $\mathcal{D}_{d_{conf}}$ matches this definition, we know that $\mathcal{D}_{d_{conf}} \subset D$ (D is complete). To show the inverse inclusion (i.e., D is minimal), we have to prove that each element of D has a corresponding association rule (and then a database).

Proof. Consider an element (c, y, z) of D; we need to construct a database $\mathcal{DB}$ containing an association rule $A \rightarrow B$, such that $\mu(A \rightarrow B)$ equals $\phi_\mu(c, y, z)$. Since c, y, and z are rational numbers, we define n as an integer such that $(c \times y \times n, y \times n, z \times n) \in \mathbb{N}^3$. Our database should verify the following equalities:

$$conf(A \rightarrow B) = c, \; supp(A) = y, \; supp(B) = z \tag{4.4}$$

The constraints of the domain assure that $0 \leq (1 - c) \times y \leq y \leq y \times (1 - c) + z \leq 1$ holds. We can thus construct the database of Table 4.1, satisfying the equalities 4.4. Then, the second inclusion is verified. $\quad\square$

Table 4.1 Database for the domain $\mathcal{D}_{d_{conf}}$

Finally, we have identified the joint variation domain of the descriptor system d_{conf}: it is exactly D.

4.4.2 Expression of a Set of Measures of $\mathcal{D}_{d_{conf}}$

We now study in detail the case of 32 measures found in the literature [50, 17, 31] studied in the context of $\mathcal{D}_{d_{conf}}$ (Table 4.2). We can already make a remark about these results.

In fact, we can see here that most of the measures are increasing with respect to the confidence of the rule. The higher the confidence, the higher the measure.

Then we can see that most of the measures are decreasing with the number of consequents. This is a logical behavior: if the consequent has a high frequency, then it is nearly always true, and the rule is uninteresting. For example, it is not interesting in a market basket database to have "bread" in the consequent, since bread is present in almost all transactions.

Table 4.2 Expression and variations of measures over $\mathcal{D}_{d_{conf}}$.

measure	$(c,y,z) \in \mathcal{D}_{d_{conf}}$	c	y	z	measure	$(c,y,z) \in \mathcal{D}_{d_{conf}}$	c	y	z
SUPPORT	cy	↗	↗	→	KLOSGEN	$\sqrt{cy}(c-z)$	✕	✕	↘
CONFIDENCE	c	↗	→	→	ADDED VALUE	$\max(c-z,(\frac{cy}{z}-y))$	↗	✕	↘
COVERAGE	y	→	↗	→	CONVICTION	$\frac{1-z}{1-c}$	↗	→	↘
PREVALENCE	z	→	→	↗	ONE WAY SUPPORT	$c\log\frac{c}{z}$	✕	→	↘
RECALL	$\frac{cy}{z}$	↗	↗	↘	J_1-MEASURE	$cy\log\frac{c}{z}$	✕	✕	↘
SPECIFICITY	$1-\frac{z-cy}{1-y}$	↗	✕	↘	PIATETSKY-SHAPIRO	$y(c-z)$	↗	✕	↘
RELATIVE SPECIFICITY	$y-\frac{z-cy}{1-y}$	↗	✕	↘	COSINE	$c\sqrt{\frac{y}{z}}$	↗	↗	↘
PRECISION	$2cy+1-y-z$	↗	✕	↘	LOEVINGER	$1-\frac{1-c}{1-z}$	↗	→	↘
LIFT	$\frac{c}{z}$	↗	→	↘	INFORMATION GAIN	$\log\frac{c}{z}$	↗	→	↘
LEVERAGE	$c-yz$	↗	↘	↘	SEBAG SHOENAUER	$\frac{c}{1-c}$	↗	→	→
CENTERED CONFIDENCE	$c-z$	↗	→	↘	CONTRAMIN	$\frac{y(2c-1)}{z}$	↗	✕	✕
RELATIVE RISK	$c\times\frac{1-y}{z-cy}$	↗	✕	↘	BAYESIAN FACTOR	$\frac{c}{1-c}\times\frac{1-z}{z}$	↗	→	↘
JACCARD	$\frac{cy}{y+z-cy}$	↗	↗	↘	EX -COUNTEREX RATE	$1-\frac{1-c}{c}$	↗	→	→
GANASCIA	$2\times c-1$	↗	→	→	ZHANG	$\frac{(c-z)}{\max(c\times(1-z),z\times(1-c))}$	↗	→	↘
INTEREST	$|y(c-z)|$	✕	↗	✕	IMPLICATION INDEX	$\sqrt{n}\frac{\sqrt{y}(c-z)}{\sqrt{(1-z)}}$	↗	✕	↘
ODDS RATIO	$1+\frac{c-z}{(1-c)(z-cy)}$	↗	✕	↘	KAPPA	$2\frac{c-z}{1+\frac{c}{y}-2z}$	↗	✕	↘

The ✕ signs indicate that the variations depend on the situation in relation to the independency

The variations with respect to the number of antecedents are more diversified. The explanation can be the following. Considering the confidence and the number of consequents as fixed, one can say that when increasing the number of antecedents, the number of examples of the rule increases too, then the rule is more interesting. On the other hand, one can say that if the number of counterexamples increases too, the rule is less interesting.

Finally, one can find a compromise between these two possibilities by taking into account the situation in relation to the independency (Fig. 4.2(a)). We suppose that we have two rules r: A → B and r′: A′ → B such that

$$conf(A \rightarrow B) = conf(A' \rightarrow B), P(A') = (1+\delta)P(A), \delta > 0$$

We will compare the two quantities $\Delta = P(A'|B) - P(A|B)$ and $\bar{\Delta} = P(A'|\neg B) - P(A|\neg B)$. Δ represents the variation of examples (true positives) in the set of positive

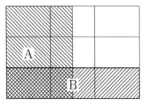
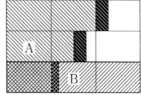
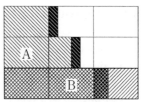

(a) Situation of independency   (b) Before the situation of in-   (c) After the situation of inde-
                                 dependency: $\bar{\Delta} \geq \Delta$ pendency: $\bar{\Delta} \leq \Delta$

Fig. 4.2 Different situations in relation to the independency for rules of the form A → B. The black parts correspond to the variation induced by a rule A′ → B, such that $P(A') > P(A)$ with same confidence and consequents. For each case, we compare the variations of false negatives ($\bar{\Delta}$) and examples (Δ)

cases, and $\bar{\Delta}$ represents the variation of false negatives in the set of negative cases.

$$P(A'B) - P(AB) = P(A') \times conf(A' \to B) - P(AB)$$
$$= (1+\delta)P(A) \times conf(A \to B) - P(AB)$$
$$= (1+\delta)P(AB) - P(AB)$$
$$= \delta \times P(AB)$$

In the same manner, we can prove that $P(A'\neg B) - P(A\neg B) = \delta \times P(A\neg B)$. Determining the sign of $\Delta - \bar{\Delta}$ is then the same as determining the sign of $P(A|B) - P(A|\neg B)$:

$$P(A|B) - P(A|\neg B) = \frac{P(\neg B)P(AB) - P(B)P(A\neg B)}{P(B)P(\neg B)}$$
$$= \frac{P(\neg B)P(AB) - P(B)P(A) + P(B)P(AB)}{P(B)P(\neg B)}$$
$$= \frac{P(AB) - P(B)P(A)}{P(B)P(\neg B)}$$

Then the sign of $\Delta - \bar{\Delta}$ depends on the situation in relation to the independency. Since r and r′ have the same confidence and the same consequent, they are both on the same side of independency. Suppose r is on the left side of the independency (Fig. 4.2(b)), that is, $P(AB) - P(B)P(A) < 0$, then $\Delta - \bar{\Delta} < 0$, which means that by taking the rule r′ instead of the rule r, we add more false negatives into the set of negative cases than true positives into the set of positive cases. On the contrary, when studying the situation after the independency (Fig. 4.2(c)), the opposite (and more acceptable) situation appears.

We now benefit of a rigorous formal framework for the study of interestingness measures. We will use it in the following to establish a necessary condition and a sufficient condition for existence of the GUEUC property.

4.5 Conditions for GUEUC

The framework we proposed in the previous section will help us to study the analytic properties of the interestingness measures and their behavior with respect to some given quantities. We will see in the following that the variations of adapted functions of measure can be a very useful information. We focus here on the descriptor system defined in Equation 4.2, and consequently, we note $\mathcal{D} = \mathcal{D}_{d_{conf}}$.

4.5.1 A Sufficient Condition

We first study a sufficient condition of existence of the GUEUC property for a given interestingness measure μ. We already discovered one, with the barycenter property of confidence.

Proposition 4.1 (Trivial Sufficient Condition for GUEUC). *Let μ be an interestingness measure for association rules and μ be an affine transformation of the confidence. Then μ does verify the barycenter property (with the same weights as for confidence) and thus μ has the GUEUC property.*

Six measures of Table 4.2 are in this case (confidence, prevalence, centered confidence, Ganascia, and the measure of Loevinger). In addition, we showed above that the measure of Sebag–Shoenauer verifies this condition too. The problem of this condition is that it is not automatically and systematically applicable, since it demands some possibly long calculations. We then formulate another sufficient condition.

Proposition 4.2 (Sufficient Condition for GUEUC). *Let μ be an interestingness measure for association rules and $(\phi_\mu, \mathcal{D})$ an adapted function of measure of μ. Let ϕ_μ verify the two following properties:*

(a) $\forall (y,z) \in \mathbb{Q}^2 \cap [0,1]^2$, the function $c \mapsto \phi_\mu(c,y,z)$, where c is such that $(c,y,z) \in \mathcal{D}$, is a monotone function (increasing or decreasing);

(b) $\forall (c,z) \in \mathbb{Q}^2 \cap [0,1]^2$, the function $y \mapsto \phi_\mu(c,y,z)$, where y is such that $(c,y,z) \in \mathcal{D}$, is a decreasing function (in the broad meaning of that term).

Then μ has the GUEUC property.

Proof. Let us first make a remark: If $(c,y,z) \in \mathcal{D}$ and if y' is a positive relative number such that $y' \leq y$ then we have $(c,y',z) \in \mathcal{D}$. This guarantees that we will remain in the domain.

Then, let $\mathtt{X} = x \rightarrow \mathtt{c}$ be a rule and $\mathtt{A}$ an itemset not in $\mathtt{X}$. Since the confidence has the GUEUC property, and moreover has the barycenter property, there exist $a_\downarrow$ and $a_\uparrow$ in $\mathcal{A}$ such that

$$conf(\mathtt{X} = x, \mathtt{A} = a_\downarrow \rightarrow \mathtt{c}) \leq conf(\mathtt{X} = x \rightarrow \mathtt{c}) \leq conf(\mathtt{X} = x, \mathtt{A} = a_\uparrow \rightarrow \mathtt{c})$$

Suppose now that, for $y = supp(X = x, A = a_\uparrow, c)$ (resp. $y = supp(X = x, A = a_\downarrow, c)$) the property (a) is an increasing (resp. decreasing) property. We have the following inequalities, where $a = a_\uparrow$ (resp. $a = a_\downarrow$). The letter over the inequality signs designates the property used.

$$
\begin{aligned}
\mu(X = x \rightarrow c) &= \phi_\mu(conf(X = x \rightarrow c), supp(X = x), supp(c)) \\
&\overset{(b)}{\leq} \phi_\mu(conf(X = x \rightarrow c), supp(X = x, A = a), supp(c)) \\
&\overset{(a)}{\leq} \phi_\mu(conf(X = x, A = a \rightarrow c), supp(X = x, A = a), supp(c)) \\
&\leq \mu(X = x, A = a \rightarrow c)
\end{aligned}
$$

Therefore the rule $X = x \rightarrow c$ has a more interesting A-specialization $X = x, A = a \rightarrow c$. μ verifies the GUEUC property. $\square$

This condition allows us to add two measures that are only function of the confidence (Sebag–Shoenauer and Ex-Counterex Rate) and five others measures (conviction, leverage, information gain, Bayesian factor, Zhang). The leverage case is interesting since it is the only decreasing function with $y = p_a$. One can so substitute one of these measures for the confidence or associate some of those to confidence [6].

This condition does not hold for all the measures presented above. We will now see that, in fact, most of the measures do not verify the GUEUC property.

4.5.2 A Necessary Condition

We will now give a necessary condition for the GUEUC property. First recall that if (c, y, z) is an element of $\mathcal{D}$ then so is $(c, \frac{y}{2}, z)$ (we made the remark above).

Proposition 4.3 (Necessary Condition for GUEUC). *If μ is an interestingness measure that verifies the GUEUC property, and $(\mathcal{D}, \phi_\mu)$ its adapted function of measure, then for every $(c, y, z) \in \mathcal{D}$, we have*

$$
\phi_\mu(c, y, z) \leq \phi_\mu(c, \frac{y}{2}, z) \tag{4.5}
$$

Suppose there exist two values of c and z such that the function $y \mapsto \phi_\mu(c, y, z)$ from $\{y | (c, y, z) \in \mathcal{D}\}$ into $\mathbb{R}$ is an increasing function. Then the inequality 4.5 never holds, and μ does not have the GUEUC property. This applies in particular to all the measures whose variations with respect to the second variable depends on the situation in relation to the independency. We now make a proof of this property.

Proof. We will construct our proof in Fig. 4.3. Consider in this figure that X, C, and A are categorical attributes whose values can be 0 or 1. For X and C, only the value 1 is represented. Suppose that the proportions are the following:

$$
supp(X = 1) = y, \; supp(C = 1) = z, \; supp(X = 1, C = 1) = cy
$$

It is here clear that, following the figure and the proportions, we have

$$supp(X = 1, A = 1) = \frac{y}{2}, \ supp(X = 1, A = 0) = \frac{y}{2}$$

$$conf(X = 1, A = 1 \rightarrow C = 1) = c, \ conf(X = 1, A = 0 \rightarrow C = 1) = c$$

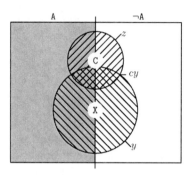

Fig. 4.3 Element of the proof of the necessary condition

Then since μ verifies the GUEUC property, at least one of the A-specializations of $X = 1 \rightarrow C = 1$ has a higher value with respect to μ. But both A-specializations are projected on the same point of $\mathcal{D}$, that is $(c, \frac{y}{2}, z)$, which implies $\mu(X = 1, A = 1 \rightarrow C = 1) = \mu(X = 1, A = 0 \rightarrow C = 1)$. Consequently, we obtain the inequality 4.5. $\square$

Therefore, we can exclude from the GUEUC property all the measures that increase with the number of antecedents. This is the case, for example, of support, coverage, and Jaccard. All the measures whose variations with respect to y depend on the situation in relation to the independency do not verify the property. They are not monotone around the independency. This is the case, for example, of specificity, relative risk, Piatetsky-Shapiro.

We now summarize the obtained results.

4.5.3 Classification of the Measures

In the last paragraph, we proved two conditions that let us classify a large panel of measures with respect to the GUEUC property. The GUEUC property gives an efficient algorithmic property which is a pruning strategy based on a top-down approach. Previously, this property had been studied only for confidence. However, as support, confidence is not satisfying, and there are many other measures available for a user. The choice of the measure will depend on his objectives [31], thus, it is interesting to know if these measures do verify an algorithmic property such as GUEUC, or not.

Table 4.3 gives an overview of the existence of GUEUC property for 32 measures found in the literature. An important point to see is that, with the 13 measures that

Table 4.3 Presence of the GUEUC property for the measures. The table recalls the variations of measures with respect to c and y. The last column indicates if the GUEUC property is present (YES) or not (NO)

Measure	$(c,y,z) \in \mathcal{D}$	c	y	GUEUC	Measure	$(c,y,z) \in \mathcal{D}$	c	y	GUEUC
SUPPORT	cy	↗	↗	NO	KLOSGEN	$\sqrt{cy}(c-z)$	×	×	NO
CONFIDENCE	c	↗	→	YES	ADDED VALUE	$\max(c-z,(\frac{cy}{z}-y))$	↗	×	NO
COVERAGE	y	→	↗	NO	CONVICTION	$\frac{1-z}{1-c}$	↗	→	YES
PREVALENCE	z	→	→	YES	ONE WAY SUPPORT	$c\log\frac{c}{z}$	×	→	?
RECALL	$\frac{cy}{z}$	↗	↗	NO	J_1-MEASURE	$cy\log\frac{c}{z}$	×	×	NO
SPECIFICITY	$1-\frac{z-cy}{1-y}$	↗	×	NO	PIATETSKY-SHAPIRO	$y(c-z)$	↗	×	NO
RELATIVE SPECIFICITY	$y-\frac{z-cy}{1-y}$	↗	×	NO	COSINE	$c\sqrt{\frac{y}{z}}$	↗	↗	NO
PRECISION	$2cy+1-y-z$	↗	×	NO	LOEVINGER	$1-\frac{1-c}{1-z}$	↗	→	YES
LIFT	$\frac{c}{z}$	↗	→	YES	INFORMATION GAIN	$\log\frac{c}{z}$	↗	→	YES
LEVERAGE	$c-yz$	↗	↘	YES	SEBAG SHOENAUER	$\frac{c}{1-c}$	↗	→	YES
CENTERED CONFIDENCE	$c-z$	↗	→	YES	CONTRAMIN	$\frac{y(2-c)}{z}$	↗	↗	NO
RELATIVE RISK	$\frac{z-cy}{1-y}$	↗	×	NO	BAYESIAN FACTOR	$\frac{c}{1-c}\times\frac{1-z}{z}$	↗	→	YES
JACCARD	$\frac{cy}{y+z-cy}$	↗	↗	NO	EX -COUNTEREX RATE	$1-\frac{1-c}{c}$	↗	→	YES
GANASCIA	$2\times c-1$	↗	→	YES	ZHANG	$\frac{(c-z)}{\max(c\times(1-z),z\times(1-c))}$	↗	→	YES
INTEREST	$|y(c-z)|$	×	↗	NO	IMPLICATION INDEX	$\sqrt{n}\frac{\sqrt{y}(c-z)}{\sqrt{(1-z)}}$	↗	×	NO
ODDS RATIO	$1+\frac{c-z}{(1-c)(z-cy)}$	↗	×	NO	KAPPA	$2\frac{c-z}{1+\frac{c}{y}-2z}$	↗	×	NO

verify the GUEUC property, one can characterize many situations, like the situation in relation to the independency with the lift or the information gain, the situation related to the counterexamples with Sebag–Shoenauer, the Bayesian factor, or the example–counterexample rate, and the size of the consequent with the prevalence, etc. Verifying the same property, these measures can be inserted altogether in an algorithm based on the GUEUC property and eventually combined with support. Without support, we can find all rules verifying multiple threshold constraints over multiples measures and particularly the nuggets of knowledge.

Figure 4.4 shows that the nuggets phenomenon can be highlighted with the classical database Mushroom. The algorithmic property of GUEUC, which allows a pruning strategy without a support threshold, let all these nuggets appear. The fact that many different situations can be characterized with measures verifying the GUEUC property make this property very interesting. Our framework gives also a systematic way to know if, for a given measure, the GUEUC property is respected. The only constraint is that one can write the measure as a function of confidence, support of antecedents, and support of consequents. Furthermore, for the moment, the measure

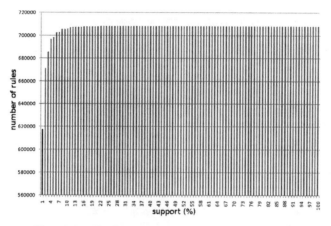

Fig. 4.4 Nuggets of knowledge. The figure shows the cumulated distribution of supports for all the confident rules ($conf \geq 0.8$) of the classical Mushroom database. Most of the rules have a support less than 1%: a too large support threshold would have missed those rules

of one way support cannot be categorized. It verifies neither the necessary condition nor the sufficient condition. Since we do not have a necessary condition and a sufficient condition, it is not surprising to have this kind of situation, whenever most of the measures are classified.

4.6 Conclusion

In this chapter, we focus on the universal existential upward closure property of confidence. We first review the original property and show that it is an interesting one since it permits to focus on interesting rule discovery without a first support pruning. In particular, it allows to focus on nuggets.

However, we recall that confidence does not offer many good characteristics, and that a lot of additional measures have been developed to bypass this problem. Unfortunately many of them are most of the time used in post-process phase (e.g., after the use of the support–confidence framework).

We here extend the universal existential upward closure property framework. We then establish a necessary condition and a sufficient condition of existence for our general UEUC property that cover a large number of measures. We apply these conditions to 32 interestingness measures and show that 13 of them do have the GUEUC property. As a consequence, they could be used in the mining phase in an efficient way.

All these measures that characterize many situations will select different kinds of patterns. Depending on his objectives, the user can prefer one or the other.

Our results give then an important and additional information about algorithmic properties of the measures.

References

1. Abe, H., Tsumoto, S.: Analyzing correlation coefficients of objective rule evaluation indices on classification rules. In: G. Wang, T. rui Li, J.W. Grzymala-Busse, D. Miao, A. Skowron, Y. Yao (eds.) 3rd International Conference on Rough Sets and Knowledge Technology, Chengdu, China, *Lecture Notes in Computer Science*, vol. 5009, pp. 467–474. Springer (2008)
2. Abe, H., Tsumoto, S., Ohsaki, M., Yamaguchi, T.: Finding functional groups of objective rule evaluation indices using pca. In: T. Yamaguchi (ed.) 7th International Conference on Practical Aspects of Knowledge Management, Yokohama, Japan, *Lecture Notes in Computer Science*, vol. 5345, pp. 197–206. Springer (2008)
3. Agrawal, R., Imieliski, T., Swami, A.: Mining association rules between sets of items in large databases. In: P. Buneman, S. Jajodia (eds.) ACM SIGMOD International Conference on Management of Data, Washington, D.C., United States, pp. 207–216. ACM Press, New York, NY, USA (1993)
4. Agrawal, R., Mannila, H., Srikant, R., Toivonen, H., Verkamo, A.I.: Fast discovery of association rules. In: U.M. Fayyad, G. Piatetsky-Shapiro, P. Smyth, R. Uthurusamy (eds.) Advances in Knowledge Discovery and Data Mining, pp. 307–328. AAAI/MIT Press, Menlo Park, CA, USA (1996)
5. Agrawal, R., Srikant, R.: Fast algorithms for mining association rules in large databases. In: J.B. Bocca, M. Jarke, C. Zaniolo (eds.) 20th International Conference on Very Large Data Bases, Santiago de Chile, Chile, pp. 478–499. Morgan Kaufmann (1994)
6. Barthélemy, J.P., Legrain, A., Lenca, P., Vaillant, B.: Aggregation of valued relations applied to association rule interestingness measures. In: V. Torra, Y. Narukawa, A. Valls, J. Domingo-Ferrer (eds.) 3rd International Conference on Modeling Decisions for Artificial Intelligence, Tarragona, Spain, *Lecture Notes in Computer Science*, vol. 3885, pp. 203–214. Springer (2006)
7. Bayardo Jr, R.J., Agrawal, R., Gunopulos, D.: Constraint-based rule mining in large, dense databases. In: 15th International Conference on Data Engineering, Sydney, Australia, pp. 188–197. IEEE Computer Society, Washington, DC, USA (1999)
8. Bhattacharyya, R., Bhattacharyya, B.: High confidence association mining without support pruning. In: A. Ghosh, R.K. De, S.K. Pal (eds.) 2nd International Conference on Pattern Recognition and Machine Intelligence, Kolkata, India, *Lecture Notes in Computer Science*, vol. 4815, pp. 332–340. Springer (2007)
9. Bonchi, F., Lucchese, C.: Pushing tougher constraints in frequent pattern mining. In: T.B. Ho, D.W.L. Cheung, H. Liu (eds.) 9th Pacific-Asia Conference on Knowledge Discovery and Data Mining, Hanoi, Vietnam, vol. 3518, pp. 114–124. Springer (2005)
10. Bonchi, F., Lucchese, C.: Extending the state-of-the-art of constraint-based pattern discovery. Data and Knowledge Engineering 60(2), 377–399 (2007)
11. Borgelt, C., Kruse, R.: Induction of association rules: APRIORI implementation. In: 15th Conference on Computational Statistics, Berlin, Germany, pp. 395–400. Physika Verlag, Heidelberg, Germany (2002)
12. Breiman, L., Friedman, J., Olshen, R., Stone, C.: Classification and Regression Trees. Wadsworth and Brooks, Monterey, CA (1984)
13. Brin, S., Motwani, R., Ullman, J.D., Tsur, S.: Dynamic itemset counting and implication rules for market basket data. In: J. Peckham (ed.) ACM SIGMOD International Conference on Management of Data, Tucson, Arizona, USA, pp. 255–264. ACM Press, New York, NY, USA (1997)
14. Cheung, Y.L., Fu, A.W.C.: Mining frequent itemsets without support threshold: With and without item constraints. IEEE Transaction on Knowledge and Data Engineering 16(9), 1052–1069 (2004)
15. Cohen, E., Datar, M., Fujiwara, S., Gionis, A., Indyk, P., Motwani, R., Ullman, J.D., Yang, C.: Finding interesting associations without support pruning. IEEE Transaction on Knowledge and Data Engineering 13(1), 64–78 (2001)

16. Diatta, J., Ralambondrainy, H., Totohasina, A.: Towards a unifying probabilistic implicative normalized quality measure for association rules. In: F. Guillet, H.J. Hamilton (eds.) Quality Measures in Data Mining, *Studies in Computational Intelligence*, vol. 43, pp. 237–250. Springer (2007)
17. Geng, L., Hamilton, H.J.: Interestingness measures for data mining: A survey. ACM Computing Surveys 38(3, Article 9) (2006)
18. Goethals, B.: Frequent set mining. In: O. Maimon, L. Rokach (eds.) The Data Mining and Knowledge Discovery Handbook, pp. 377–397. Springer, New York (2005)
19. Goethals, B., Zaki, M.J.: Advances in frequent itemset mining implementations: report on FIMI'03. SIGKDD Explorations 6(1), 109–117 (2004)
20. Gras, R., Couturier, R., Blanchard, J., Briand, H., Kuntz, P., Peter, P.: Quelques critères pour une mesure de qualité de règles d'association – un exemple : l'intensité d'implication. RNTI-E-1 (Mesures de qualité pour la fouille de données) pp. 3–31 (2004)
21. Guillaume, S.: Discovery of ordinal association rules. In: M.S. Cheng, P.S. Yu, B. Liu (eds.) 6th Pacific-Asia Conference on Knowledge Discovery and Data Mining, Taipei, Taiwan, *Lecture Notes in Computer Science*, vol. 2336, pp. 322–327. Springer-Verlag, London, UK (2002)
22. Han, J., Cheng, H., Xin, D., Yan, X.: Frequent pattern mining: current status and future directions. Data Mining and Knowledge Discovery 15(1), 55–86 (2007)
23. Han, J., Pei, J., Yin, Y.: Mining frequent patterns without candidate generation. In: W. Chen, J.F. Naughton, P.A. Bernstein (eds.) ACM SIGMOD International Conference on Management of Data, Dallas, Texas, pp. 1–12. ACM New York, NY, USA, Dallas, Texas, USA (2000)
24. Hébert, C., Crémilleux, B.: A unified view of objective interestingness measures. In: P. Perner (ed.) 5th International Conference on Machine Learning and Data Mining, Leipzig, Germany, *Lecture Notes in Computer Science*, vol. 4571, pp. 533–547. Springer (2007)
25. Karel, F.: Quantitative and ordinal association rules mining (qar mining). In: B. Gabrys, R.J. Howlett, L.C. Jain (eds.) 10th International Conference on Knowledge-Based Intelligent Information and Engineering Systems, Bournemouth,UK, *Lecture Notes in Computer Science*, vol. 4251, pp. 195–202. Springer (2006)
26. Lallich, S., Vaillant, B., Lenca, P.: Parametrised measures for the evaluation of association rule interestingness. In: J. Janssen, P. Lenca (eds.) 11th International Symposium on Applied Stochastic Models and Data Analysis, Brest, France, pp. 220–229 (2005)
27. Lallich, S., Vaillant, B., Lenca, P.: A probabilistic framework towards the parameterization of association rule interestingness measures. Methodology and Computing in Applied Probability 9, 447–463 (2007)
28. Le Bras, Y., Lenca, P., Lallich, S.: On optimal rules discovery: a framework and a necessary and sufficient condition of antimonotonicity. In: T. Theeramunkong, B. Kijsirikul, N. Cercone, H.T. Bao (eds.) 13th Pacific-Asia Conference on Knowledge Discovery and Data Mining, Bangkok, Thailand. Springer (2009)
29. Le Bras, Y., Lenca, P., Lallich, S., Moga, S.: Généralisation de la propriété de monotonie de la all-confidence pour l'extraction de motifs intéressants non fréquents. In: 5th Workshop on Qualité des Données et des Connaissances, in conjunction with the 9th Extraction et Gestion des Connaissances conference, Strasbourg, France (2009)
30. Lenca, P., Meyer, P., Picouet, P., Vaillant, B.: Aide multicritère à la décision pour évaluer les indices de qualité des connaissances – modélisation des préférences de l'utilisateur. In: M.S. Hacid, Y. Kodratoff, D. Boulanger (eds.) Revue des Sciences et Technologies de l'Information – série RIA ECA, vol. 17, pp. 271–282. Hermes Science Publications (2003)
31. Lenca, P., Meyer, P., Vaillant, B., Lallich, S.: On selecting interestingness measures for association rules: user oriented description and multiple criteria decision aid. European Journal of Operational Research 184(2), 610–626 (2008)
32. Lenca, P., Meyer, P., Vaillant, B., Picouet, P., Lallich, S.: Évaluation et analyse multicritère des mesures de qualité des règles d'association. RNTI-E-1 (Mesures de qualité pour la fouille de données) pp. 219–246 (2004)
33. Leung, C.K.S., Lakshmanan, L.V.S., Ng, R.T.: Exploiting succinct constraints using FP-trees. SIGKDD Explorations 4(1), 40–49 (2002)

34. Li, J.: On optimal rule discovery. IEEE Transaction on Knowledge and Data Engineering 18(4), 460–471 (2006)
35. Li, J., Zhang, X., Dong, G., Ramamohanarao, K., Sun, Q.: Efficient mining of high confidence association rules without support thresholds. In: J.M. Zytkow, J. Rauch (eds.) 3rd European Conference on Principles of Data Mining and Knowledge Discovery, Prague, Czech Republic, *Lecture Notes in Computer Science*, vol. 1704, pp. 406–411. Springer (1999)
36. Li, W., Han, J., Pei, J.: CMAR: Accurate and efficient classification based on multiple class-association rules. In: N. Cercone, T.Y. Lin, X. Wu (eds.) 1st IEEE International Conference on Data Mining, San Jose, California, USA, pp. 369–376. IEEE Computer Society, Washington, DC, USA (2001)
37. Liu, B., Hsu, W., Ma, Y.: Integrating classification and association rule mining. In: R. Agrawal, P.E. Stolorz, G. Piatetsky-Shapiro (eds.) 4th ACM SIGKDD International Conference on Knowledge Discovery and Data Mining, New York City, USA, pp. 80–86. AAAI Press (1998)
38. Morishita, S., Sese, J.: Transversing itemset lattices with statistical metric pruning. In: 19th ACM SIGMOD-SIGACT-SIGART Symposium on Principles of Database Systems, Dallas, Texas, United States, pp. 226–236. ACM, New York, NY, USA (2000)
39. Ohsaki, M., Kitaguchi, S., Okamoto, K., Yokoi, H., Yamaguchi, T.: Evaluation of rule interestingness measures with a clinical dataset on hepatitis. In: J.F. Boulicaut, F. Esposito, F. Giannotti, D. Pedreschi (eds.) 8th European Conference on Principles of Data Mining and Knowledge Discovery, Pisa, Italy, *Lecture Notes in Computer Science*, vol. 3202, pp. 362–373. Springer, New York, NY, USA (2004)
40. Omiecinski, E.: Alternative interest measures for mining associations in databases. IEEE Transaction on Knowledge and Data Engineering 15(1), 57–69 (2003)
41. Pei, J., Han, J.: Can we push more constraints into frequent pattern mining? In: 6th ACM SIGKDD International Conference on Knowledge Discovery and Data Mining, Boston, Massachusetts, United States, pp. 350–354. ACM, New York, NY, USA (2000)
42. Pei, J., Han, J.: Constrained frequent pattern mining: A pattern-growth view. SIGKDD Explorations 4(1), 31–39 (2002)
43. Pei, J., Han, J., Lakshmanan, L.V.: Mining frequent itemsets with convertible constraints. In: 17th International Conference on Data Engineering, Heidelberg, Germany, pp. 433–442. IEEE Computer Society, Washington, DC, USA (2001)
44. Pei, J., Han, J., Mao, R.: CLOSET: An efficient algorithm for mining frequent closed itemsets. In: ACM SIGMOD Workshop on Research Issues in Data Mining and Knowledge Discovery, pp. 21–30. ACM, Dallas, TX, USA (2000)
45. Piatetsky-Shapiro, G.: Discovery, analysis, and presentation of strong rules. In: Knowledge Discovery in Databases, pp. 229–248. AAAI/MIT Press (1991)
46. Plasse, M., Niang, N., Saporta, G., Villeminot, A., Leblond, L.: Combined use of association rules mining and clustering methods to find relevant links between binary rare attributes in a large data set. Computational Statistics & Data Analysis 52(1), 596–613 (2007)
47. Quinlan, J.R.: C4.5: Programs for Machine Learning. Morgan Kaufmann, San Mateo, CA (1993)
48. Sebag, M., Schoenauer, M.: Generation of rules with certainty and confidence factors from incomplete and incoherent learning bases. In: J. Boose, B. Gaines, M. Linster (eds.) European Knowledge Acquisition Workshop, pp. 28–1 – 28–20. Gesellschaft für Mathematik und Datenverarbeitung mbH, Sankt Augustin, Germany (1988)
49. Słowiński, R., Greco, S., Szczęch, I.: Analysis of monotonicity properties of new normalized rule interestingness measures. In: P. Brézillon, G. Coppin, P. Lenca (eds.) International Conference on Human Centered Processes, vol. 1, pp. 231–242. TELECOM Bretagne, Delft, The Netherlands (2008)
50. Tan, P.N., Kumar, V., Srivastava, J.: Selecting the right objective measure for association analysis. Information Systems 4(29), 293–313 (2004)
51. Toivonen, H.: Sampling large databases for association rules. In: T. Vijayaraman, A.P. Buchmann, C. Mohan, N. Sarda (eds.) 22nd International Conference on Very Large Data Bases, Bombay, India, pp. 134–145. Morgan Kaufman (1996)

52. Tsumoto, S.: Clinical knowledge discovery in hospital information systems: Two case studies. In: D.A. Zighed, H.J. Komorowski, J.M. Zytkow (eds.) 4th European Conference on Principles of Data Mining and Knowledge Discovery, Lyon, France, pp. 652–656. Springer (2000)
53. Vaillant, B., Lenca, P., Lallich, S.: A clustering of interestingness measures. In: E. Suzuki, S. Arikawa (eds.) 7th International Conference on Discovery Science, Padova, Italy, *Lecture Notes in Computer Science*, vol. 3245, pp. 290–297. Springer (2004)
54. Wang, K., He, Y., Cheung, D.W.: Mining confident rules without support requirement. In: 10th International Conference on Information and Knowledge Management, Atlanta, Georgia, USA, pp. 89–96. ACM, New York, NY, USA (2001)
55. Wang, K., He, Y., Han, J.: Mining frequent itemsets using support constraints. In: A.E. Abbadi, M.L. Brodie, S. Chakravarthy, U. Dayal, N. Kamel, G. Schlageter, K.Y. Whang (eds.) 26th International Conference on Very Large Data Bases, Cairo, Egypt, pp. 43–52. Morgan Kaufmann (2000)
56. Wang, K., Zhou, S., He, Y.: Growing decision trees on support-less association rules. In: 6th ACM SIGKDD International Conference on Knowledge Discovery and Data Mining, Boston, Massachusetts, United States, pp. 265–269. ACM, New York, NY, USA (2000)
57. Webb, G.I.: Efficient search for association rules. In: 6th ACM SIGKDD International Conference on Knowledge Discovery and Data Mining, Boston, Massachusetts, United States, pp. 99–107. ACM, New York, NY, USA (2000)
58. Wu, X., Kumar, V., Quinlan, J.R., Ghosh, J., Yang, Q., Motoda, H., McLachlan, G.J., Ng, A., Liu, B., Yu, P.S., Zhou, Z.H., Steinbach, M., Hand, D.J., Steinberg, D.: Top 10 algorithms in data mining. Knowledge and Information Systems 14(1), 1–37 (2008)
59. Xiong, H., Tan, P.N., Kumar, V.: Mining strong affinity association patterns in data sets with skewed support distribution. In: 3rd IEEE International Conference on Data Mining, Melbourne, Florida, USA, pp. 387–394. IEEE Computer Society, Washington, DC, USA (2003)
60. Yao, Y., Chen, Y., Yang, X.D.: A measurement-theoretic foundation of rule interestingness evaluation. In: T.Y. Lin, S. Ohsuga, C.J. Liau, X. Hu (eds.) Foundations and Novel Approaches in Data Mining, *Studies in Computational Intelligence*, vol. 9, pp. 41–59. Springer (2006)
61. Yin, X., Han, J.: CPAR: Classification based on predictive association rules. In: D. Barbará, C. Kamath (eds.) 3dr SIAM International Conference on Data Mining, San Francisco, CA, USA, pp. 331–335. SIAM (2003)
62. Zaki, M.J.: Mining non-redundant association rules. Data Mining and Knowledge Discovery 9(3), 223–248 (2004)
63. Zaki, M.J., Hsiao, C.J.: CHARM: An efficient algorithm for closed itemset mining. In: R.L. Grossman, J. Han, V. Kumar, H. Mannila, R. Motwani (eds.) 2nd SIAM International Conference on Data Mining, Arlington, VA, USA, pp. 457–473. SIAM (2002)
64. Zheng, Z., Kohavi, R., Mason, L.: Real world performance of association rule algorithms. In: 7th ACM SIGKDD International Conference on Knowledge Discovery and Data Mining, San Francisco, California, pp. 401–406. ACM, New York, NY, USA (2001)
65. Zighed, D.A., Rakotomalala, R.: Graphes d'induction : apprentissage et data mining. Hermès, Paris (2000). 475 p.
66. Zimmermann, A., De Raedt, L.: CorClass: Correlated association rule mining for classification. In: E. Suzuki, S. Arikawa (eds.) 7th International Conference on Discovery Science, Padova, Italy, *Lecture Notes in Computer Science*, vol. 3245, pp. 60–72. Springer (2004)

Chapter 5
Classification Techniques and Error Control in Logic Mining

Giovanni Felici, Bruno Simeone, and Vincenzo Spinelli

Abstract In this chapter we consider *box clustering*, a method for supervised classification that partitions the feature space with particularly simple convex sets (boxes). *Box clustering* produces systems of logic rules obtained from data in numerical form. Such rules explicitly represent the logic relations hidden in the data w.r.t. a target class. The algorithm adopted to solve the *box clustering* problem is based on a simple and fast agglomerative method which can be affected by the initial choice of the starting point and by the rules adopted by the method. In this chapter we propose and motivate a randomized approach that generates a large number of candidate models using different data samples and then chooses the best candidate model according to two criteria: *model size*, as expressed by the number of boxes of the model, and *model precision*, as expressed by the error on the test split. We adopt a Pareto-optimal strategy for the choice of the solution, under the hypothesis that such a choice would identify simple models with good predictive power. This procedure has been applied to a wide range of well-known data sets to evaluate to what extent our results confirm this hypothesis; its performances are then compared with those of competing methods.

Giovanni Felici
Istituto di Analisi dei Sistemi ed Informatica 'Antonio Ruberti', Consiglio Nazionale delle Ricerche, Viale Manzoni, 30, 00185 Rome, Italy, e-mail: felici@iasi.cnr.it

Bruno Simeone
Dipartimento di Statistica, Probabilità e Statistiche Applicate, Università 'La Sapienza', Piazzale Aldo Moro 5, 00185 Rome, Italy, e-mail: bruno.simeone@uniroma1.it

Vincenzo Spinelli
ISTAT – Istituto Nazionale di Statistica, Via Tuscolana, 1788, 00173 Rome, Italy, e-mail: vispinel@istat.it

R. Stahlbock et al. (eds.), *Data Mining,* Annals of Information Systems 8,
DOI 10.1007/978-1-4419-1280-0_5, © Springer Science+Business Media, LLC 2010

5.1 Introduction

One of the main problems in the identification of a good interpretative and predictive
model is the trade-off between the precision that a model exhibits on training data
and its ability to correctly predict data that are not in the training set. Such aspect
is particularly important for those models that are loosely constrained in their size
and can thus be adapted to the training data in a myopic way (e.g., neural networks,
support vector machines, decision trees). For this reason data mining techniques
should be equipped with methods and tools to deal with the problem of *overfitting*,
or *overtraining*, particularly serious when (a) the data are affected by noise, (b) it is
not possible to define clearly an objective function for the model, or (c) the problem
of finding an optimal solution for such a function is computationally intractable.

The problem of overfitting control can be considered within a formal framework
for those methods that are based on a treatable mathematical formulation of the
learning problem: for example, in *support vector machines* (SVM) and *neural networks* a regularization parameter that weights some measure of model complexity
is often used to balance the objective function and contrast any potential overfitting
behavior, see [1–3].

Such methods, being based on mathematical optimization algorithms, appear to
be well suited to those learning problems where the main objective is the correct
classification of training and verification data, but where the extraction of usable
knowledge from data plays a secondary role w.r.t. to predictive power.

Nevertheless, in certain settings it is important to find good predictive models that
are also able to be understood and verified by users, and, last but not least, integrated
with other knowledge from different data sources and experience. The most popular
method in this class is decision trees, see [4], which build a classification model
based on the iterative partition of the training data. The leaves of this tree represent
logic rules on the input variables and can be viewed as clauses of a *disjunctive normal form* (DNF) formula in propositional logic, see [5] for further details on DNFs.

Other interesting methods in this class are those that are based on the formulation
of combinatorial and integer optimization problems, see [6, 7], whose solution can
again be put in relation with DNF formulas that classify the training data, such as
the *LAD* methods (logical analysis of data), see [8, 9], the *Lsquare* method, see
[10], and the *OCAT* method, see [11, 12]. Such methods assume that the data are
described directly by logic variables and thus may require proper transformations
of numerical data into logic data, often referred to as *discretization* or *binarization*,
see [13, 14]. This means that the logic-based method is applied to the transformed
data. What eventually characterizes logic mining is the nature of the classification
model (expressed as a logical formulas in the input variables) rather than the nature
of the input data, which can, in principle, always be transformed into logic form.

Logic mining methods are typically based on models and algorithms where it
is difficult to formalize the trade-off between precision and generalization capabilities, i.e., to control overfitting. The introduction of a regularization term in the
objective function is not straightforward and moreover it may generate significant

computational problems. In decision trees, for example, overfitting is controlled by pruning techniques that are applied to the tree according to a user-specified confidence parameter; in *LAD*, see [15–18], several methods to control the format of the resulting formula are combined with the use of an objective function that minimizes the dimension of the solution formula; in *Lsquare*, a greedy approach that minimizes the cost of a conjunctive clause is one of the techniques by which overfitting control is achieved.

The method under analysis in this chapter – *box clustering (BC)* – can be considered a logic mining method in the sense that the system of boxes that forms the classification model can be mapped into a DNF formula, but operates in a slightly different way, as the identification of threshold values for numerical variables (the above mentioned *binarization*) is performed within the classification algorithm itself, and thus it extends the logic approach without resorting to any transformation of the original data, see [19].

BC presents two interesting properties for the control of the error and of the overtraining behavior: (a) its complexity can be easily put in relation with the number of boxes that compose the final solution; (b) alternative solutions of different complexity can be identified efficiently according to different random choice of the parameters by which the solution algorithm is initialized. In addition, *BC* can deal directly with classification problems with more than two classes, see Example 5.2, and it manages the missing values inside the model itself, (see data sets *ictus, annealing*, and *page* in Table 5.1).

It is therefore natural to try and combine these two features to derive a method to choose a solution among those obtained by a randomized application of the BC algorithm, with the specific aim of controlling prediction error and finding the right balance between precision and overfitting on the training data. In this sense the proposed approach evaluates the solution space w.r.t. to the specific problem under analysis and is able to capture the complexity of the latent model and the amount of noise present in the data. The best candidate model is selected according to a bi-criterion problem: *model size*, expressed by the number of boxes of the model, and *prediction accuracy*, expressed by the error on the test split. We propose a choice criterion for the model based on Pareto-optimality under the hypothesis that such a choice would identify simple models with high accuracy.

The chapter is organized as follows.

Section 5.2 defines the main terminology used in this chapter. In Section 5.3, we show how the *BC* approach can be used in supervised classification problems. In Section 5.4, we define a bi-criterion procedure to select the *best BC* solution for our classification problem, based on the error distribution on the test set. Here we formulate a hypothesis on the link between the classification performance on the test set and the validation one. From these considerations, we define, in Section 5.5, an iterative procedure to choose the best *BC* solution for our classification problem. Finally, in Section 5.6, we apply this procedure to obtain *BC* models for benchmark data sets used in machine learning and check the validity of our hypothesis.

5.2 Brief Introduction to Box Clustering

In logic mining it is assumed that a number of observations are given in the form of n-dimensional vectors; with each of these vectors a binary outcome is associated; according to its values the observations are usually termed either *positive* (or *true*) or *negative* (or *false*), see [9, 20, 21]. Here we assume that the components are not constrained in type (i.e., they can be binary, ordinal, or numeric[1]) and are represented in $\mathbb{R}^n$. We define a *box* as a multidimensional interval defined by a pair of points (L, U):

$$I(L, U) = \prod_{i=1}^{n} [l_i, u_i] = \{X \in \mathbb{R}^n \mid l_i \leq x_i \leq u_i \ i = 1, \ldots, n\}$$

where L and U are called the *lower* and *upper* bounds of I. A box is called positive (or negative) if it includes only positive (respectively, negative) observations. Positive and negative boxes will also be called *homogeneous* boxes. For any finite set S of points, we define its *box closure* $[S]$ as the intersection of all boxes containing S. Given two boxes $A = [S]$ and $B = [T]$, we define their join $A \vee B$ to be the box $[S \cup T]$. The *BC* model can be easily extended from 2-Class to N-Class supervised classification, and, from now on, we consider the general case: $N \geq 2$. A decision tree can be seen as a set of logical formulas, defined by its leaves, and each logical formula can be considered a special box, see Example 5.1.

Example 5.1. Let us consider a decision tree defined on three numerical variables: $\bar{x} = (x_1, x_2, x_3)$. If a leaf of the tree is defined by the logical formula

$$F(\bar{x}) \equiv (x_1 < 1.2) \wedge (x_2 > 0.0) \wedge (x_3 \geq 0) \wedge (x_3 \leq 1)$$

then we can consider the equivalent box

$$B =]-\infty, 1.2] \times]3, +\infty[\times [0, 1]$$

It is easy to check that $F(\bar{x})$ is true $\Leftrightarrow \bar{x} \in B$.

This example also shows that a decision tree can be always seen as a set of logical formulas in DNF format.

When considering a set of observations, the *BC* problem amounts to finding a finite set of boxes as a solution of a specific optimization problem, see [19]. Such problem may involve different types of constraints, generally based on the geometrical properties of the boxes:

- *coverage*: every observation is included in at least a box, and every box includes at least one observation.

---

[1] We can further extend this approach to N-value variables and manage the presence of missing values by including them inside the *BC* model itself.

- *homogeneity*: all the boxes are homogeneous, i.e., all the points inside a single box are from the same class. This implies that we must have at least N boxes, where N is the number of classes, if *coverage* and *homogeneity* do hold.
- *spanning*: every box is the *box-closure* of the set of observations inside it.
- *overlapping*: four types of overlapping conditions may be considered. If *homogeneity* does not hold, then we have (a) *none*, i.e., no overlapping configuration is allowed, and (b) *strong* otherwise. If *homogeneity* does hold, then we can define (c) *mild*, i.e., only homogeneous boxes can overlap, and (d) *weak* otherwise.
- *consistency*: every box must contain at least an observation not included in another box, and this implies that we must have at most n boxes, where $n \leq |S|$.
- *saturation*: if we consider a subset C of the above-mentioned constraints, we say that a set of boxes $\hat{B}$ are saturated if and only if they satisfy C and we cannot join any pair of boxes in $\hat{B}$ without violating any constraints in C.

It is worth noticing the link between spanning sets and set coverage of the parameter space. In the literature we can find *the maximum patterns* approach to get a set coverage of the parameter space by the minimum number of patterns, see [22]. In our approach, we find a saturated system of boxes that is spanning and has the coverage property for the training set, and this means that every training point is inside a box (*coverage*) and all the boxes have the minimum dimension to include the training points inside them (*spanning*). But the two constraints do not necessarily imply that the system of boxes covers all the points of the feature space.

If the constraints *coverage* and *homogeneity* hold then the *BC* approach is an exact method as it performs an exact separation of training data, see Proposition 5.1. This fact motivates the analysis of methods to select, among different *BC* models, those that appear less biased by overfitting.

The identification of a set of boxes that satisfies a set of constraints is not an easy problem to formulate and solve; a restricted version known as *the maximum box problem*, where one wants to find the homogeneous box including the largest number of positive (or negative) points, has been studied in [23]. For a fixed number of points, the (weighted) maximum box problem is in class $\mathcal{P}$, but it can be shown that the maximum box problem is in class $\mathcal{NPC}$ for any number of points, by using a polynomial reduction of the maximum clique problem on the graph $G = (V, E)$, known to be $\mathcal{NP}$-complete, to the maximum box problem. From the practical standpoint we use an agglomerative approach based on greedy or random choice, a well-known algorithmic framework in *BC* for the search of a saturated box system satisfying an input set of constraints, see [19, 24, 25].

The simplest version of the algorithm is mainly based on two steps:

- *starting box set*: all the points in the training set form initial singleton boxes.
- *main loop*: in each step of the loop, we search for two boxes that can be joined in the current box set. If these boxes do not exist then we exit the loop, otherwise, we make a new box by joining the two boxes and update the box set.

The maximum number of iterations for the search loop is no greater than the number of points in the training set, and the last box set is saturated.

5.3 *BC*-Based Classifier

A *BC*-based classifier is a function (shortly referred to as *classification* in the following), with three input parameters: (a) a set of boxes *BS*, (b) the weight/distance function $w()$, and (c) the point p to classify. This function first assigns the weight $w_i = w(B_i, p)$ to each $B_i \in BS$, then chooses the index of the box having the best, i.e., the minimum, weight; generally, there is a subset $\overline{B} \subseteq BS$ such that each box $B \in \overline{B}$ has the best value. There can be two exclusive conditions:

- the boxes in $\overline{B}$ belong to *only one class*: the function randomly chooses the index of one of these boxes.
- the boxes in $\overline{B}$ belong to *more than one class*: the most elementary classification functions return an undefined result; the most sophisticated ones try to find out a group of homogeneous boxes in $\overline{B}$ having *the best global weight*. If this search has no result then the function returns an undefined result.

The output of the function is the estimated class of the point p. The weight $w(B, p)$ is a measure of *the attraction intensity* of B with respect to p. If *BS* is a *set coverage* of the observation space, then we can naturally define $w()$ as the characteristic function shown (*yes–no*) in Equation 5.1.

$$w(B, p) = \begin{cases} 0 & p \in B \\ 1 & \text{otherwise} \end{cases} \tag{5.1}$$

If *BS* is not a *set coverage* of the observation space, e.g., the *spanning* constraint holds, then we can define $w()$ as the *Manhattan distance*, see Equation 5.2, naturally compatible with the geometry of boxes defined in Section 5.2.

$$w(B, p) = \begin{cases} 0 & p \in B \\ d_1(B, p) & \text{otherwise} \end{cases} \tag{5.2}$$

It is worth noticing that $d_1(B, p) = \min_{q \in B} d_1(q, p)$, and it can be proven that $d_1()$ depends only on the size and position of B and p in $\mathbb{R}^n$, as shown in Fig. 5.1: let us consider a generic box B and four points, P_1, P_2, P_3, and P_4, placed in generic positions outside B. The dashed lines are the broken paths whose lengths give us the values $d_1(B, P_i)$, $i = 1, 2, 3, 4$.

Consider the usual split of the available data into a *the training set* and *the test set*, and the complete data set $S = Tr \cup Ts$. We also assume that the target variables divide S into n classes.

With regard to *Tr*, we define $w()$ in a slightly different way:

$$d_1(B, Tr, p) = mean_{q \in Tr \cap B} d_1(q, p)$$

This new version of d_1 can be defined as a gravitational model, for $w()$ depending on the mutual position between the point p and all the points of Tr which are inside B. These three classes for the weight function have a similar behavior for every box: they take on positive values and get their minimum value inside each box.

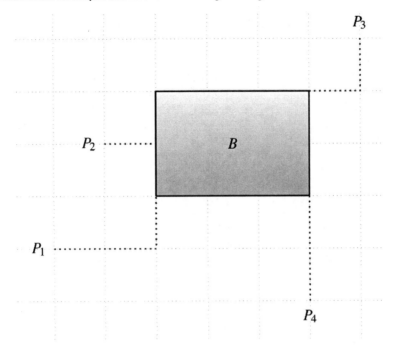

P_3

P_2 $\cdots\cdots\cdots\cdots\cdots$

B

P_1 $\cdots\cdots\cdots\cdots\cdots\cdots\cdots$

P_4

Fig. 5.1 The nearby weight function

We then apply the classification function to every point in Ts and define the confusion matrix $M_c = M_c(Ts) = \{m_{ij}\}$, where m_{ij} is defined as the number of points having the i-th class in Ts and classified in the j-th class by the BC classifier.[2] M_c is uniquely defined by the pair $(classification(), Ts)$, and thus it depends on $(BS, w(), Tr)$. It comes from the definition that $Ok(M_c) = \sum_{i=1,n} m_{ii}$ is the number of the correctly classified observations of Ts, and, for this reason, we can define the error function $Err(M_c) = \sum_{i \neq j} m_{ij}$.

Example 5.2. Let us consider a simple BS consisting of seven boxes in $[0,1]^2$; in BS there are two *green class* boxes, three *red class* boxes, and two *yellow class* boxes. There are three pairs of overlapping boxes: *red–red*, *yellow–red*, and *yellow–green* pairs. If we consider the *gray* color for the indefinite cases, we can classify a uniform point grid in $[0,1]^2$ by using the different weight functions, i.e., Equations 5.1 and 5.2. In Fig. 5.2, we can see the result we get for the function *yes–no*: every point outside the boxes is gray, and the same happens within the overlapping regions. In Fig. 5.3 we have the result for the function *nearby*: the gray regions are very narrow outside the boxes, and they can be more easily seen in Fig. 5.4. The edges are straight lines for the function *yes–no* in this two-dimensional problem (hyperplanes in $\mathbb{R}^n$),

---

[2] From the definition of M_c we can always consider $i, j = 1, \ldots, n+1$: when $i, j = n+1$ we can manage some extreme cases: $m_{n+1,j}$ is the number of the points having a class in Ts not defined in Tr, and $m_{i,n+1}$ is the number of the points that cannot be assigned to any class by the BC classifier.

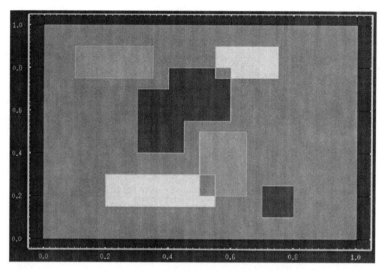

Fig. 5.2 *BC* classifier by characteristic function (see online version for color figures)

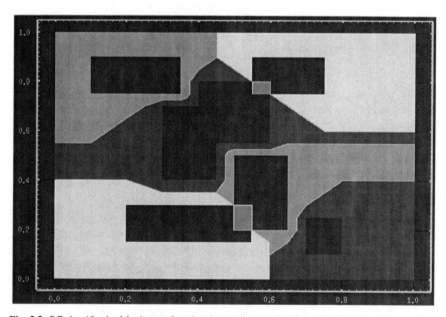

Fig. 5.3 *BC* classifier by Manhattan function (see online version for color figures)

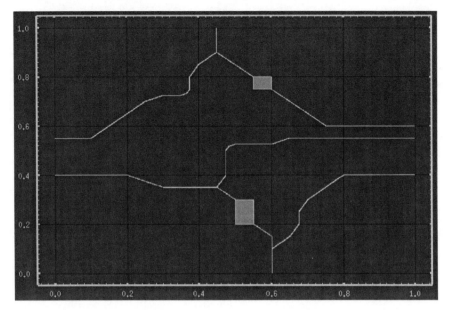

Fig. 5.4 *BC* class edges for Manhattan classifier

but this is not true for the function *nearby*. Moreover, it is worth considering the number of gray points we have in the two methods. In Figs. 5.2 and 5.4 we can see the different presence rate of the gray points in the sample grid. The function *yes–no* has many gray points, and this is coherent with the fact that *BS* is not a set coverage of $[0, 1]^2$. The *nearby* function returns a very small number of gray points and this is an important parameter in order to evaluate the classification power of this *BC* classifier.

From the definitions of the constraints {coverage, homogeneous} and $Err(M_c)$, we can check that the following propositions hold.

Proposition 5.1. *If the box set* BS *satisfies, at least, the constraints* {*coverage, homogeneous*} *on the point set* Tr, *then* $Err(M_c(\text{Tr})) = 0$ *holds for every* BC *classification technique, i.e.,* yes–no, nearby, *and* gravitational.

Proof. Let us consider $p \in Tr$. If *BS* has the coverage property then there exists $B \in BS$ such that $p \in B$; but *BS* is homogeneous, and this means that p and B are in the same class. If there is another $B' \in BS$ such that $p \in B'$ then B and B' must be in the same class. Since every *BC* classifier must correctly classify p, we can conclude that $Err(M_c(Tr)) = 0$. □

Proposition 5.2. *If* $S = \text{Tr} \cup \text{Ts}$ *and* BS *has the coverage property and it is homogeneous for* Tr, *then* $Err(M_c(S)) = Err(M_c(\text{Ts}))$.

Proof. For every *BC* classifier, we can check that

$$Err(M_c(S)) = Err(M_c(Tr)) + Err(M_c(Ts)).$$

From Proposition 5.1, we know that $Err(M_c(Tr)) = 0$, and this ends the proof. □

The above considerations lead to conclude that every *BC* classifier, based on *BS* with few elementary properties, is exposed to the risk of overtraining, a well-known situation where the learned model is adapted to noise or noninformative training data with excessive precision resulting in complex models and poor recognition performances on test data.

5.4 Best Choice of a Box System

Let us consider a *BC* approach for the supervised classification problem on *S*, and let $\hat{BS} = \{BS_i\}_{i=1,n}$ be a set of feasible solutions. Solutions in $\hat{BS}$ are created by the agglomerative algorithm with (a) different starting parameter values, (b) cross-validation, or (c) repeated percentage split on *S*.

We tackle the problem of choosing the *best* solution in $\hat{BS}$ according to two alternative objective functions:

$$\begin{cases} \min |BS| \\ s.t. \ \ BS \in \hat{BS} \end{cases} \tag{5.3}$$

$$\begin{cases} \min Err(M_c) \\ s.t. \ \ BS \in \hat{BS} \end{cases} \tag{5.4}$$

Problem 5.3 puts the focus on the simplest *BC* model we can consider for *S*; on the contrary, Problem 5.4 puts the focus on the reliability of the *BC* model in classification; the extent of trade-off between these two objectives is strongly depending on the data under analysis.

While the evaluation of the complexity of the model is directly measured by the objective function of Problem 5.3, the use of simple test accuracy for the evaluation of predictive power (see Problem 5.4) is not a straightforward choice. The literature proposes alternative and possibly more meaningful methods to evaluate the performance of a classifier, such as, among others, the area under the *ROC* (*receiver operating characteristic*) curve, or simply *AUC*, widely used to measure model performance in binary classification problems, see [26]. Huang and Ling, see [27], also show theoretically and empirically that *AUC* is a better measure for model evaluation than accuracy, as reflected in Problem 5.4. Nevertheless, the literature does not converge on a widely accepted performance method similar to *ROC* analysis for an *N*-class classifier ($N > 2$), and for this reason we make the choice to use accuracy as a performance evaluator. *BC* is by definition an *N*-class classifier, and many of the experimental tests presented later are based on data sets with $N > 2$ classes.

Propositions 5.1 and 5.2 suggest us not to solve just one problem to get a robust and reliable *BC* solution. For this reason we consider a bi-criterion problem:

$$f(BS, Ts) = \begin{pmatrix} |BS| \\ Err(M_c) \end{pmatrix} \tag{5.5}$$

$$\begin{cases} \min\ f(BS, Ts) \\ s.t.\ \ BS \in \hat{BS} \end{cases} \tag{5.6}$$

Optimization with multiple objective functions aims at a simultaneous improvement of the objectives. The goals in Problem 5.6 may easily conflict so that an optimal solution in the conventional sense does not exist. Instead we aim at, e.g., Pareto-optimality, i.e., we find the Pareto set in the plane π_{NE} as shown in Fig. 5.5, from which we can choose a promising BC solution.

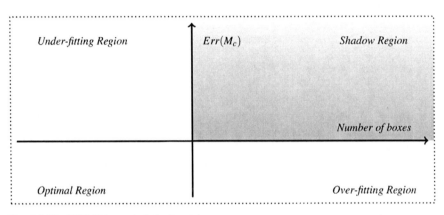

Fig. 5.5 The BCG-like matrix in logic mining

In the plane π_{NE}, see [28] we can place each BS_i according to its number of boxes and number of errors. We can classify the four quadrants of this plane as following:

- *Shadow region*: the worst BSs are in the upper right quadrant. They have many boxes and high error rate.
- *Overfitting region*: the BSs in the lower right quadrant have many boxes but low error rate; *model complexity* dominates *model generalization*.
- *Under-fitting region*: the BSs in the upper left quadrant have few boxes but high error rate; *model generalization* dominates *model complexity*.
- *Optimal region* or $Opt(\pi_{NE})$: the BSs in the lower left quadrant have few boxes and low error rate; for this reason they are the natural candidates to be considered as the best in our BS group. In general, we want to choose our best solution among the nondominated solutions in these regions.

We consider the point $(\bar{N}, \bar{E})$ as the origin of the axes, where $\bar{N}$ and $\bar{E}$ are the mean value of the number of boxes and the mean value of the error rate over all samples, respectively. Let us consider P^*, the lower left corner of π_{NE}, and the value $d^* = \min_{i=1,n} d(P_i, P^*)$. Furthermore, let us consider the set $\hat{P} = \{P_i | d^* = d(P_i, P^*)\}$:

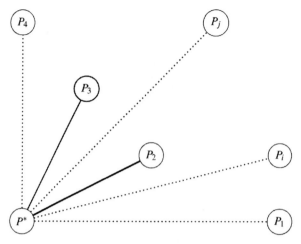

Fig. 5.6 Choice in Pareto set

- $|\widehat{P}| = 1$: this means that there is only one $P_i \in \widehat{P}$; this point is the best BS we can choose in $\hat{BS}$.
- $|\widehat{P}| > 1$: we have more than one BS at minimum distance, and we choose the unique BS having the minimum error rate (*model complexity* is preferred to *model generalization*).

  In Fig. 5.6 we have an example of this method: $\{P_1, P_2, P_3, P_4\}$ is the Pareto set of $\hat{BS} = \{P_1, P_2, P_3, P_4, \ldots, P_i, P_j, \ldots\}$, and P_2, P_3 are the BSs with the minimum distance to P^*. We prefer P_2 to P_3, because it has a lower error rate. Finally, P_4 is the *solution* for Problem 5.3, while P_1 is the *solution* for Problem 5.4.

In Fig. 5.7, we have the full scheme of this bi-criterion approach for $\hat{BS}$.
The solution BS_{i^*} is not generally the optimal solution to Problems 5.3 and 5.4. The point $\hat{P} = (N_{i^*}, E_{i^*})$ cannot be in the *shadow region* by definition. If we consider the equivalent definitions for the two-dimensional plane based on the validation set (i.e., $\pi_{NE'}$, $Opt(\pi_{NE'})$, and $\hat{P}'$), we can now informally formulate our hypothesis.

Hypothesis 1 *For n large $\hat{P} \in Opt(\pi_{NE}) \Rightarrow \hat{P}' \in Opt(\pi_{NE'})$.*

The hypothesis may also be proved under some probabilistic conditions on the distribution of the data, but we are more interested in its practical verification: probabilistic information is normally not available for real data and therefore its practical relevance is difficult to assess. On the other hand, it is of interest to check if the hypothesis is verified for reasonable values of n in real data sets with different sizes and nature.

```
function Pareto_choice(N, E, n)
{
1     P* = (min_i N_i, min_i E_i);
2     d* = +∞;
3     for(i = 1; i ≤ n; i++){
4         P_i = (N_i, E_i);
5         d_i = d(P*, P_i);
6     }
7     d* = min_i d_i;
8     E* = +∞;
9     i* = +∞;
10    for(i = 1; i ≤ n; i++){
11        if(d_i = d*){
12            if(E_i < E*){
13                E* = E_i;
14                i* = i;
15            }
16        }
17    }
18    return(i*);
}
```

Fig. 5.7 Pareto Choice schema for BC

5.5 Bi-criterion Procedure for *BC*-Based Classifier

The considerations of Section 5.4 can be used to design an algorithm for *BC* to control the validation set error. This algorithm is based on five basic functions: (a) *data-split*, (b) *local-search*, (c) *Pareto-choice*, (d) *normalization*, and (e) BC *classifier*.

The *data-split* function returns a percentage split of the input data set *S*. It uses a percentage of the total records as training data and the remaining percentage as test data. If 80% is shown as the split, then 80% of the data set is being used for training, while the remaining 20% is used for testing. The *local-search* function is the implementation of the well-known agglomerative algorithm for *BC*, see [19, 23, 25], as described at the end of Section 5.2, based on a random choice engine. The *Pareto-choice* function has been fully explained in Section 5.4. The input vectors, i.e., N and E, widely vary in size as a result of the units selected for representation: $N_i \leq |S|$ but $E_i \leq 1$. To avoid a higher influence of N with respect to E, when we consider the distance between two points, we need to normalize these vectors.

The BC *classifier* function is the general form of the function explained in Section 5.3; in this version, it returns the number of errors for the test set. In Figure 5.8 we have the full description of the algorithm.

function error-control(S, P)

```
  {
1   for(i = 1; i ≤ k; i++){
2     (Tr, Ts) = data-split(S, P);
3     BSᵢ = local-search(Tr, P);
4     Nᵢ = |BSᵢ|;
5     Eᵢ = BC-Classifier(BSᵢ, Tr, Ts, P);
6   }
7   E' = normalization(E, k);
8   N' = normalization(N, k);
9   i* = Pareto-choice(N', E', k);
10  return(BSᵢ*);
  }
```

Fig. 5.8 General schema for BC

The algorithm is based on a repeated percentage split strategy and a multi-criterion choice of one *BS* out of the k available. We summarize in the input parameter P the structure of specific parameters that we have used in the experiments. In particular, we have used four parameters for each run of the procedure: (a) $k = 100$, (b) 80% split, (c) *BC* random engine, and (d) overlapping = *none*. Several other low-level parameters needed in the different steps and not relevant to the scope of the experiments have been omitted here for brevity.

5.6 Examples

In this section we present some experimental results obtained with the proposed method. After a brief description of the data sets under consideration, we present the *BC* results and, in Section 5.6.3, a synthetic comparison with *decision trees* applied to the same problems.

5.6.1 The Data Sets

In order to empirically evaluate the efficiency of the procedure described in Section 5.5 and its effectiveness in data analysis, we have applied it to seven frequently used data sets, taken from the repository of the University of California, Irvine (UCI) (see [29]): Thyroid Domain (*thyroid*), Isolated Letter Speech Recognition (*isolet*), Blocks Classification (*page*), Multi-Spectral Scanner Image (*satimage*), Annealing Data (*annealing*), and Image Segmentation Data (*segment*). One further data set, *ictus*, is an unpublished medical archive related to the ictus disease.

Table 5.1 Summary of input data sets

Data set	Training	Validation	Percentage (%)	Features	Classes	Missing
ictus	698	0		37	2	y
	500	198	28.4			
annealing	798	0		18	6	y
	690	100	12.6			
page	5406	0		10	5	n
	4000	1308	24.6			
segment	210	2100	90.9	19	7	n
thyroid	3772	3428	47.6	21	3	n
satimage	4435	2000	31.1	36	6	n
isolet	6238	1559	20.0	617	26	n

These data sets have different values for size, number of features, and number of classes to classify. The main parameters of these six data sets are listed in Table 5.1: Columns 2–3 of the table report the number of records in training and validation sets that we have considered in each data set. For the data sets *ictus*, *annealing*, and *page* we do not have an explicit validation set, i.e., 0 in Column 3, and we have randomly split the training set in two parts as shown in the second row of the table for these data sets. Column 4 contains the percentage rate of the validation set. Similarly, Columns 5–6 of the table correspond to the number of features and classes in each data set. The last column indicates the presence or the absence of missing values.

5.6.2 Experimental Results with BC

For every data set, we now consider the results in the plane π_{NE}, as defined in Section 5.4, and display them in Figs. 5.7, 5.8, 5.9, and 5.10. The point P^* is represented as a black circle in each figure.

Table 5.2 gives us a summary of the minimal, maximal, and average values for the parameters N, E, and E'. Columns 2–4 are related to N, i.e., the number of boxes, while Columns 5–7 are related to E, i.e., the test errors. Columns 8–10 provide the same information on the validation errors. The distributions for N and E are evaluated inside the function *error-control* for $k = 100$, where we build 100 training and test sets (repeated percentage split). Even if there is one validation set for each data set and generally we do not know the distribution of E', we have modified the behavior of the function *error-control* to evaluate it, but only for a better comprehension of our approach. This means that we have also applied all the box sets to the validation set.

Table 5.2 must be compared to Table 5.3, where we show the performances of P^* when applied to the validation set and the difference with the test error for the same point. Columns 2 and 3 give us the coordinates of the points P^* in π_{NE}, Columns

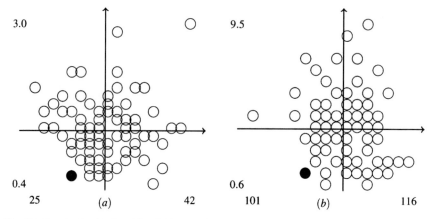

Fig. 5.9 Results for (**a**) *thyroid* and (**b**) *annealing*

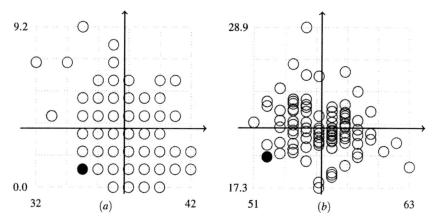

Fig. 5.10 Results for (**a**) *ictus* and (**b**) *isolet*

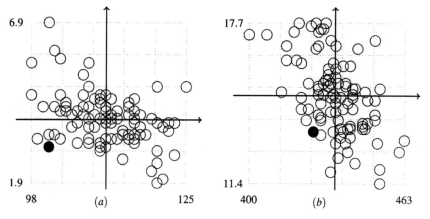

Fig. 5.11 Results for (**a**) *page* and (**b**) *satimage*

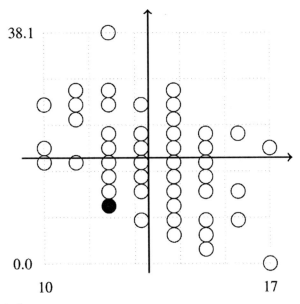

Fig. 5.12 Results for *segment*

2 and 4 those of the points P'^* in $\pi_{NE'}$. Column 5 shows an interesting measure of
the performance of our choice algorithm: it gives us the percentile of the validation
error rate of P'^* with respect to all the n points we could consider in our tests.[3] We
can check that for two data sets, i.e., *thyroid* and *annealing*, P'^* is really a very good
choice. For other two data sets, i.e., *page* and *satimage*, it has better performances
than 70.0% of all points, while, for the other three data sets, i.e., *ictus*, *isolet*, and
segment, it is better than 60.0% of all points. Finally, we must emphasize that even
if Hypothesis 1 is always satisfied in our tests (Column 6 in Table 5.3), there is one
case, i.e., *ictus*, where P'^* is on the border of $Opt(\pi_{NE'})$.

5.6.3 Comparison with Decision Trees

A decision tree can always be seen as a set of boxes, see Example 5.1. For this
reason, we have decided to use the performances of standard decision trees (a defi-
nitely well-established method in the literature) as the benchmark for the *BC* results
we present in Section 5.6.2. We use the Weka tool, see [30], to analyze the data
sets by the decision trees, and the results are in Table 5.4. These results have been
obtained by using the method *weka.classifiers.trees.J48*, based on *C4.5 approach*,
see [31]. We have used the default configuration in Weka, but have controlled the

---

[3] In this context, the known distribution of E' allows us to evaluate some statistical parameters we
use in these tables, i.e., mean, min, max, and percentile.

Table 5.2 Results in π_{NE} and $\pi_{NE'}$

| | Efficacy (#) | | | Efficiency (%) | | | | | |
| | Number of boxes | | | Err(Test set) | | | Err(Validation set) | | |
Data set	Min	Max	μ	Min	Max	μ	Min	Max	μ
thyroid	25	42	32.7	0.40	3.05	1.28	1.25	2.39	1.84
annealing	101	116	109.7	0.63	9.49	3.70	1.00	4.00	2.12
ictus	32	42	37.7	0.00	9.18	3.39	2.40	5.29	3.85
isolet	51	63	56.2	17.31	28.93	21.65	19.44	29.06	22.52
page	98	125	111.1	1.88	6.88	3.85	2.75	4.71	3.64
satimage	400	463	434.8	11.39	17.40	14.85	14.20	16.70	15.38
segment	10	17	13.2	0.00	38.10	17.48	9.95	15.71	12.56

Table 5.3 Best choices in π_{NE} and $\pi_{NE'}$

Data set	N^*	E^*	$Err(V)$	$Percentile(Err(V))$	$\hat{P}' \in Opt(\pi_{NE'})$
thyroid	29	0.53	1.34	P_3	y
annealing	106	1.27	1.00	P_1	y
ictus	35	1.02	3.85	P_{35}	y
isolet	52	19.55	22.13	P_{38}	y
page	101	3.00	3.33	P_{16}	y
satimage	426	13.42	15.05	P_{27}	y
segment	12	9.52	12.10	P_{35}	y

confidence parameter for building the classification models in the best possible way. In Table 5.4, the best results for 10-*cross-validation* (Columns 3–4), 20-*cross-validation* (Columns 5–6), and *percentage split (80%)* (Columns 7–8) are shown. The results for the data set *isolet* are not available because its size does not allow a complete set of trials by the Weka system.

Table 5.4 Decision tree results

| | | $k = 10$ | | $k = 20$ | | Percentage | |
Data Set	Confidence	Leaves	Error(%)	Leaves	Error(%)	Leaves	Error(%)
thyroid	0,20	16	0,4	16	0,3	16	0,6
annealing	0,20	49	6,3	49	7,4	49	7,3
ictus	0,30	12	12,0	12	11,0	12	11,2
isolet	–	–	–	–	–	–	–
page	0,10	29	2,7	29	2,7	29	2,9
satimage	0,05	121	12,9	121	12,9	121	13,2
segment	0,20	39	3,1	39	3,1	39	2,8

The comparison of Tables 5.3 and 5.4 helps us in assessing that the proposed *BC* approach provides good results with respect to decision trees – its more natural competitor. Its performance is better for the majority of the experiments, while the differences in solution sizes are not significant. In particular, we can highlight that

- The *BC* models are slightly more complex than the decision trees when we consider *annealing* and *ictus*, but they are more efficient, i.e., the difference of errors for the two approaches is greater than 5.0%.
- The *BC* models are more complex than the decision trees when we consider *thyroid* and *page*, but the efficiency is very similar, i.e., the difference of errors for the two approaches is less than 1.0%.
- The decision trees are better than *BC* models when applied to *satimage*.
- The *BC* models are simpler than the decision trees when we consider *segment*, but less efficient.

5.7 Conclusions

In this chapter we consider *box clustering*, a method designed to classify data described by variables in qualitative and numeric forms by a set of *boxes* that are equivalent to logical formulas on qualitative or discretized numerical variables. We adopt a standard agglomerative approach based on random choice to solve the *BC* problem and propose a method for the choice of the most interesting solution among those obtained by a sufficiently large number of different runs. Such method is based on the property of Pareto-optimality in the plane defined by the model complexity and the model training error (π_{NE}). Despite the evidence that the error obtained on test data is a good predictor of the error that the model will obtain on new data, we also claim that a particular nondominated solution in π_{NE} is still non-dominated in $\pi_{NE'}$. Such method of choice takes into account the complexity of the model and can play an important role in preventing the overtraining behavior of the model. We try to verify our hypothesis in several heterogeneous data sets from the literature and verify how it is experimentally confirmed for all of them. Moreover, the comparison of *BC* with a widely used *decision tree* technique w.r.t. to model size and test set accuracy provides positive evidence for the validation of the proposed method.

References

1. M.A. Davenport, R. G. Baraniuk, and C. D. Scott. *Learning minimum volume sets with support vector machines.* IEEE International Workshop on Machine Learning for Signal Processing (MLSP), Maynooth, Ireland, 2006.
2. M.A. Davenport, R. G. Baraniuk, and C. D. Scott. *Controlling false alarms with support vector machines.* IEEE International Conference on Acoustics, Speech, and Signal Processing (ICASSP), Toulouse, France, 2006.
3. Neyman-Pearson SVMs: *www.ece.rice.edu/md/np_svm.php/*.
4. L. Breiman, J. H. Friedman, R. A. Olshen, and C. J. Stone. *Classification and Regression Trees.* Wadsworth International, Belmont, CA, 1984.
5. S. Foldes, P.L. Hammer. Disjunctive and Conjunctive Normal Forms of Pseudo-Boolean Functions. *RUTCOR Research Report, RRR* 1-2000, Also available at *http://rutcor.rutgers.edu/pub/rrr/reports2000/01_2000.ps*.

6. G. Felici and K. Truemper. *A Minsat Approach for Learning in Logic Domains*. INFORMS Journal on Computing, 13 (3), 2001, 1–17.
7. G. Felici, F-S. Sun, and K. Truemper. *Learning Logic Formulas and Related Error Distributions*, in Data Mining and Knowledge Discovery Approaches Based on Rule Induction Techniques, G. Felici and E. Trintaphyllou eds., Springer Science, New York.
8. G. Alexe, P.L. Hammer, P.L. Kogan. *Comprehensive vs. Comprehensible Classifiers in Logical Analysis of Data*. RUTCOR Research Report, RRR 9-2002. Also available at *http://rutcor.rutgers.edu/pub/rrr/reports2002/40_2002.pdf*.
9. E. Boros, P.L. Hammer, T. Ibaraki, A. Kogan, E. Mayoraz, and I. Muchnik. *An implementation of logical analysis of data*. IEEE Transactions on Knowledge and Data Engineering, 12 (2): 292–306, November 2000.
10. K. Truemper. *Lsquare System for Learning Logic*. University of Texas at Dallas, Computer Science Program, April 1999.
11. E. Triantaphyllou. *The OCAT approach for data mining and knowledge discovery*. Working Paper, IMSE Department, Louisiana State University, Baton Rouge, LA 70803-6409, USA, 2001.
12. E. Triantaphyllou. *The One Clause At a Time (OCAT) Approach to Data Mining and Knowledge Discovery*, in Data Mining and Knowledge Discovery Approaches Based on Rule Induction Techniques, G. Felici and E. Trintaphyllou eds., Springer, Heidelberg, Germany, Chapter 2, pp. 45–87, 2005.
13. S. Bartnikowsi, M. Granberry, and J. Mugan. *Transformation of rational data and set data to logic data*, in Data Mining & Knowledge Discovery Based on Rule Induction Techniques. Massive Computing, Springer Science, 12 (5) : 253–278, November 2006.
14. P.L. Hammer, I.I. Lozina. *Boolean Separators and Approximate Boolean Classifiers*. RUTCOR Research Report, RRR 14-2006. Also available at *http://rutcor.rutgers.edu/pub/rrr/reports2006/14_2006.pdf*.
15. E. Boros, P.L. Hammer, T. Ibaraki, A. Kogan. *Logical Analysis of Numerical Data*. Mathematical Programming, 79: 163–190, 1997.
16. Y. Crama, P.L. Hammer, T. Ibaraki. *Cause-effect relationships and partially defined Boolean functions*. Annals of Operations Research, 16 : 299–325, 1988.
17. P.L. Hammer. *Partially defined Boolean functions and cause-effect relationships*. Lecture at the International Conference on Multi-Attribute Decision Making Via Or-Based Expert Systems, University of Passau, Germany, April 1986.
18. O. Ekin, P.L. Hammer, A. Kogan. *Convexity and Logical Analysis of Data*. RUTCOR Research Report, RRR 5–1998. Also available at *http://rutcor.rutgers.edu/pub/rrr/reports1998/05.ps*.
19. B. Simeone and V. Spinelli. *The optimization problem framework for box clustering approach in logic mining*. Book of Abstract of Euro XXII – 22nd European Conference on Operational Research, page 193. The Association of European Operational Research Societies, July 2007.
20. E. Boros, T. Ibaraki, L. Shi, M. Yagiura. *Generating all 'good' patterns in polynomial expected time*. Lecture at the 6th International Symposium on Artificial Intelligence and Mathematics, Ft. Lauderdale, Florida, January 2000.
21. P.L. Hammer, A. Kogan, B. Simeone, S. Szedmak. *Pareto-optimal patterns in logical analysis of data*. Discrete Applied Mathematics 144: 79–102, 2004. Also available at *http://rutcor.rutgers.edu/pub/rrr/reports2001/07.pdf*.
22. T.O. Bonates, P.L. Hammer, P.L. Kogan. *Maximum Patterns in Datasets*. RUTCOR Research Report, RRR 9-2006. Also available at *http://rutcor.rutgers.edu/pub/rrr/reports2006/9_2006.pdf*.
23. J. Eckstein, P.L. Hammer, Y. Liu, M. Nediak, and B. Simeone. *The maximum box problem and its application to data analysis*. Computational Optimization and Application, 23: 285–298, 2002.
24. P.L. Hammer, Y. Liu, S. Szedmák, and B. Simeone. *Saturated systems of homogeneous boxes and the logical analysis of numerical data*. Discrete Applied Mathematics, Volume 144, 1–2: 103–109, 2004.

25. B. Simeone, G. Felici, and V. Spinelli. *A graph coloring approach for box clustering techniques in logic mining*. Book of Abstract of Euro XXII – 22nd European Conference on Operational Research, page 193. The Association of European Operational Research Societies, July 2007.
26. S. Wu and P. Flach. *A scored AUC Metric for Classifier Evaluation and Selection*. Second Workshop on ROC Analysis in ML, Bonn, Germany, August 11, 2005.
27. J. Huang and C.X. Ling. *Using AUC and Accuracy in Evaluating Learning Algorithms*. IEEE Transactions on Knowledge and Data Engineering vol. 17, no. 3, pp. 299–310, 2005.
28. B. Henderson. *The experience curve reviewed: IV the growth share matrix or product portfolio*, 1973. Also available at *URL http://www.bcg.com/publications/files/, Experience_Curve_IV_Growth_Share_Matrix_1973.pdf*.
29. C.L. Blake and C.J. Merz. *UCI repository of machine learning databases*. University of California, Irvine, Department of Information and Computer Sciences, 1998. Also available at *http://www.ics.uci.edu/ mlearn/MLRepository.html*.
30. I.H. Witten and E. Frank. *Data Mining: Practical Machine Learning Tools and Techniques*. 2nd Edition, Morgan Kaufmann, San Francisco, 2005. *URL http://www.cs.waikato.ac.nz/ml/*.
31. R. Quinlan. *C4.5: Programs for Machine Learning*. Morgan Kaufmann Publishers, San Mateo, CA.

Part III
Classification Analysis

Chapter 6
An Extended Study of the Discriminant Random Forest

Tracy D. Lemmond, Barry Y. Chen, Andrew O. Hatch, and William G. Hanley

Abstract Classification technologies have become increasingly vital to information analysis systems that rely upon collected data to make predictions or informed decisions. Many approaches have been developed, but one of the most successful in recent times is the random forest. The discriminant random forest is a novel extension of the random forest classification methodology that leverages linear discriminant analysis to perform multivariate node splitting during tree construction. An extended study of the discriminant random forest is presented which shows that its individual classifiers are stronger and more diverse than their random forest counterparts, yielding statistically significant reductions in classification error of up to 79.5%. Moreover, empirical tests suggest that this approach is computationally less costly with respect to both memory and efficiency. Further enhancements of the methodology are investigated that exhibit significant performance improvements and greater stability at low false alarm rates.

6.1 Introduction

One of the greatest emerging assets of the modern technological community is *information*, as the computer age has enhanced our ability to collect, organize, and analyze large quantities of data. Many practical applications rely upon systems that are designed to assimilate this information, enabling complex analysis and inference. In particular, classification technologies have become increasingly vital to systems that learn patterns of behavior from collected data to support prediction and informed decision-making. Applications that benefit greatly from these methodologies span a broad range of fields, including medical diagnostics, network analysis (e.g., social, communication, transportation, and computer networks), image analy-

Tracy D. Lemmond · Barry Y. Chen · Andrew O. Hatch · and William G. Hanley
Lawrence Livermore National Laboratory, Systems and Decision Sciences, Livermore, CA, USA,
e-mail: lemmond1@llnl.gov, chen52@llnl.gov, hatch8@llnl.gov, hanley3@llnl.gov

R. Stahlbock et al. (eds.), *Data Mining,* Annals of Information Systems 8,   123
DOI 10.1007/978-1-4419-1280-0_6, © Springer Science+Business Media, LLC 2010

sis, natural language processing (e.g., document classification), speech recognition, and numerous others.

Many effective approaches to classification have been developed, but one of the most successful in recent times is the random forest. The random forest (RF) is a nonparametric ensemble classification methodology whose class predictions are based upon the aggregation of multiple decision tree classifiers. In this chapter, we present an in-depth study of the discriminant random forest (DRF) [14], a novel classifier that extends the conventional RF via a multivariate node-splitting technique based upon a linear discriminant function.

Application of the DRF to various two-class signal detection tasks has demonstrated that this approach achieves reductions in classification error of up to 79.5% relative to the RF. Empirical tests suggest that this performance improvement can be largely attributed to the enhanced strength and diversity of its base tree classifiers, which, as demonstrated in [3], lead to lower bounds on the generalization error of ensemble classifiers. Moreover, experiments suggest that the DRF is computationally less costly with respect to both memory and efficiency.

This chapter is organized as follows: Section 6.2 summarizes the motivation and theory behind the random forest methodology. We present the discriminant random forest approach in detail in Section 6.3 and contrast the performance of the RF and DRF methods for two signal detection applications in Section 6.4. This study incorporates an assessment of statistical significance of the observed differences in algorithm performance. Finally, our conclusions are summarized in Section 6.5.

6.2 Random Forests

The random decision forest concept was first proposed by Tin Kam Ho of Bell Labs in 1995 [12, 13]. This method was later extended and formalized by Leo Breiman, who coined the more general term random forest to describe the approach [3].

In [3], Breiman demonstrated that RFs are not only highly effective classifiers, but they readily address numerous issues that frequently complicate and impact the effectiveness of other classification methodologies leveraged across diverse application domains. In particular, the RF requires no simplifying assumptions regarding distributional models of the data and error processes. Moreover, it easily accommodates different types of data and is highly robust to overtraining with respect to forest size. As the number of trees in the RF increases, the generalization error, PE^*, has been shown in [3] to converge and is bounded as follows:

$$PE^* \leq \frac{\bar{\rho}(1-s^2)}{s^2} \tag{6.1}$$

$$s = 1 - 2PE^*_{tree} \tag{6.2}$$

where $\bar{\rho}$ denotes the mean correlation of tree predictions, s represents the strength of the trees, and PE^*_{tree} is the expected generalization error for an individual tree classifier. From Equation 6.1, it is immediately apparent that the bound on general-ization error decreases as the trees become stronger and less correlated. To reduce the mean correlation, $\bar{\rho}$, among trees, Breiman proposed a bagging approach [1, 2], in which each tree is trained on a bootstrapped sample of the original training data, typically referred to as its bagged training set [5, 9]. Though each bagged training set contains the same number of samples as the original training data, its samples are randomly selected with replacement and are representative of approximately two-thirds of the original data. The remaining samples are generally referred to as the *out-of-bag* (OOB) data and are frequently used to evaluate classification perfor-mance.

At each node in a classification tree, m features are randomly selected from the available feature set, and the single feature producing the "best" split (according to some predetermined criterion, e.g., Gini impurity) is used to partition the training data. As claimed in [3], small values of m, relative to the total number of features, are often sufficient for the forest to approach its optimal performance. In fact, large values of m, though they may increase the strength of the individual trees, induce higher correlation among them, potentially reducing the overall effectiveness of the forest. The quantity m is generally referred to as the *split dimension*.

Each tree is grown without pruning until the data at its leaf nodes are homoge-neous or until some other predefined stopping criterion is satisfied. Class predictions are then performed by propagating a test sample through each tree and assigning a class label, or *vote*, based upon the leaf node that receives the sample. Typically, the sample is assigned to the class receiving the majority vote. Note, however, that the resulting votes can be viewed as approximately i.i.d. random variables, and thus, the laws of large numbers imply that their empirical frequency will approach their true frequency as the number of trees increases. Moreover, the empirical distribution function from which they are drawn will converge to the true underlying distribu-tion function [15]. Ultimately, we can treat the resulting vote frequencies as class-specific probabilities and threshold upon this distribution to make a classification decision.

6.3 Discriminant Random Forests

Since its inception, the random forest has inspired the development of classifiers that exploit the flexibility and effectiveness of the ensemble paradigm to enhance performance. Recent work by Prinzie and Van den Poel [19], for example, utilizes logistic regression as the base learner for the ensemble and demonstrates a signif-icant increase in model accuracy relative to the single classifier variant. Like [19], the discriminant random forest leverages a linear model to strengthen its classifica-tion performance. In contrast, however, this model is combined with the tree-based classifiers of the RF to produce an ensemble classifier sharing the same theoretical foundation that affords the RF its remarkable effectiveness. The key distinction be-tween the RF and the DRF lies in the prescribed method for splitting tree nodes, for which the DRF leverages the parametric multivariate discrimination technique called linear discriminant analysis. The following sections describe the discriminant random forest methodology in greater detail.

6.3.1 Linear Discriminant Analysis

Linear discriminant analysis (LDA), pioneered by R.A. Fisher in 1936, is a discrimination technique that utilizes dimensionality reduction to classify items into distinct groups [8, 11, 17]. The LDA is an intuitively appealing methodology that makes class assignments by determining the linear transformation of the data in feature space that maximizes the ratio of their between-class variance to their within-class variance, achieving the greatest class separation, as illustrated in Fig. 6.1. The result is a linear decision boundary, identical to that determined by maximum-likelihood discrimination, which is optimal (in a Bayesian sense) when the underlying assumptions of multivariate normality and equal covariance matrices are satisfied [16]. It can be shown that, in the two-class case, the maximum class separation occurs when the vector of coefficients, $\mathbf{w}$, and intercept, b, used to define the linear transformation are as follows:

$$\mathbf{w} = \Sigma^{-1}(\mu_1 - \mu_0) \tag{6.3}$$

$$b = -0.5(\mu_1 + \mu_0)^T \Sigma^{-1}(\mu_1 - \mu_0) + \log \frac{\pi_1}{\pi_0} \tag{6.4}$$

where Σ is the common covariance matrix, μ_k is the mean vector for class k, and π_k is the prior probability of the kth class. Typically, when data are limited, we estimate Σ with the pooled covariance estimate, $\mathbf{S}_W$, given by

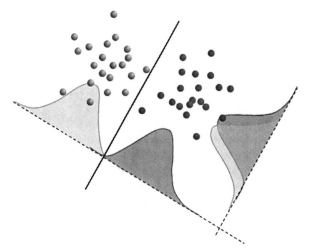

Fig. 6.1 LDA transformation and the optimal linear decision boundary (see online version for color figures)

$$S_W = \sum_{k=1}^{N} S_k \tag{6.5}$$

$$S_k = \sum_{i=1}^{N_k} (\mathbf{x}_{ki} - \bar{\mathbf{x}}_k)(\mathbf{x}_{ki} - \bar{\mathbf{x}}_k)^T \tag{6.6}$$

In the above equations, $\mathbf{x}_{ki}$ and $\bar{\mathbf{x}}_k$ denote the ith training sample of class k and the corresponding class sample mean, respectively.

6.3.2 The Discriminant Random Forest Methodology

Numerous variations of the random forest methodology have been proposed and documented in the literature, most of which address node-splitting techniques [4, 6]. Many of these are based upon an assessment of *node impurity* (i.e., heterogeneity) and include entropy-based methods, minimization of the Gini impurity index, or minimization of misclassification errors. Additional forest-based methods that focus upon alternative aspects of the algorithm include supplementing small feature spaces with linear combinations of available features [3], variations on early stopping criteria, selecting the split at random from the n best splits [6], and PCA transformation of random feature subsets [20].

Our discriminant random forest is a novel approach to the construction of a classification tree ensemble in which LDA is employed to split the feature data. Bagging and random feature selection are preserved in this approach, but unlike other forest algorithms, we apply LDA to the data at each node to determine an "optimal" linear decision boundary. By doing so, we allow decision hyperplanes of any orientation in multidimensional feature space to separate the data, in contrast to the conventional random forest algorithm, whose boundaries are limited to hyperplanes orthogonal to the axis corresponding to the feature yielding the best split. We have illustrated this effect in Fig. 6.2, which depicts decision lines in two-dimensional space for the RF (left) and the DRF (right).

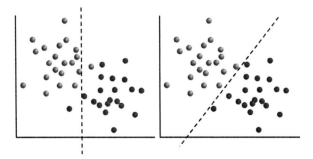

Fig. 6.2 Single feature splitting (*left*); LDA splitting (*right*) (see online version for color figures)

RF Decision Region
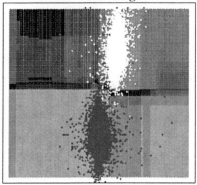

DRF Decision Region

Fig. 6.3 Posterior probability over a two-dimensional decision region for the RF and DRF; the training set is overlaid, where white/dark gray represent the positive/negative samples (see online version for color figures)

These two approaches to node splitting give rise to highly distinctive decision regions. Figure 6.3 shows an example of a two-dimensional decision region created by each forest, in which bright blue and bright gold areas represent regions of high posterior probability for the positive and negative classes, respectively. Darker areas indicate regions of greater uncertainty. The decision region produced by the DRF is notably more complex, and its boundaries are fluid and highly intricate, fitting more closely to the training data.

Pseudocode for training a DRF on a data set of size N is provided in Fig. 6.4. As previously discussed, the growth of the forest proceeds in a manner similar to the conventional random forest, with the notable exception of the decision boundary computation, which proceeds as described in Section 6.3.1. Note that the termination criterion for the leaf nodes in the given algorithm relies upon homogeneity of the node data. An alternative to this approach will be presented and discussed in Section 6.4.4.

6.4 DRF and RF: An Empirical Study

In the following suite of experiments, we compare the classification performance of the RF (utilizing the misclassification minimization node-splitting criterion) and DRF for two signal detection applications: (1) detecting hidden signals of varying strength in the presence of background noise and (2) detecting sources of radiation. Both tasks represent two-class problems, and like many other real-world applications, the costs for distinct types of error are inherently unequal. Thus, we evaluate the performance of the RF and DRF methodologies in terms of *false positives*, also known as false alarms or type I errors, and *false negatives*, also known as misses or type II errors. The false alarm rate (*FAR*), false negative rate (*FNR*), and true positive rate (*TPR* or detection rate) are defined as follows:

```
Train_DRF (Data, m, NumTrees):
   for (i = 0; i < NumTrees; i++)
       Dᵢ = bootstrap sample of size N from Data
       Train_DRF_Tree (Dᵢ, m)
   end for
end Train_DRF
```

```
Train_DRF_Tree (D, m):
   level = 0
   create root node at level with data D
   while not (all nodes at level are terminal)
      for (non-terminal node j at level)
         Fⱼ = sample m features w/o replacement
         Dⱼ' = project Dⱼ onto Fⱼ
         compute wⱼ and bⱼ using Dⱼ'; store in j
         Dₗ, Dᵣ = split Dⱼ such that
             Dₗ = xⱼ if wⱼᵀxⱼ'+bⱼ>0, ∀ xⱼ∈ Dⱼ, xⱼ'∈ Dⱼ'
             Dᵣ = xⱼ otherwise
         create left_child at level+1 with Dₗ
         if (Dₗ is homogeneous)
             assign class(Dₗ) to left_child
         end if
         create right_child at level+1 with Dᵣ
         if (Dᵣ is homogeneous)
             assign class(Dᵣ) to right_child
         end if
      end for
   end while
   increment level
end Train_DRF_Tree
```

Fig. 6.4 Pseudocode for the discriminant random forest algorithm (see online version for color figures)

$$FAR = \frac{\text{\# negative samples misclassified}}{\text{\# negative samples}}$$

$$FNR = \frac{\text{\# positive samples misclassified}}{\text{\# positive samples}}$$

$$TPR = 1 - FNR \qquad (6.7)$$

6.4.1 Hidden Signal Detection

The goal in the *hidden signal detection* application is to detect the presence of an embedded signal. In this application, it is assumed that each detection event requires a considerable amount of costly analysis, making false alarms highly undesirable. Hence, we have computed the *area under the receiver operating characteristic (ROC) curve*, or *AUC*, integrated over the *FAR* interval [0, 0.001] and scaled so that a value of 100% represents a perfect detection rate over this low *FAR* interval. The resulting quantity provides us with a single value that can be used to compare the prediction performance of the classifiers.

The data for these experiments are composed of two separate sets. The training data set, *T1*, consists of 7931 negative class samples (i.e., no embedded signal) along with two sets of positive class samples having 40 and 100% embedded signal strength (7598 and 7869 samples, respectively). The *J2* data set contains 9978 negative class samples and five positive classes having signal strengths of 20, 40, 60, 80, and 100% (7760, 9143, 9327, 9387, and 9425 samples, respectively). The training and testing data sets for each of the following experiments consist of the negative class combined with one of the available positive classes, as indicated in each case. All data samples consist of eight features useful for detecting the presence of embedded signals. We have applied both the RF and DRF forest methodologies at each split dimension ($m \in \{1, 2, ..., 8\}$) in an effort to assess the impact of this parameter on their performance.

6.4.1.1 Training on *T1*, Testing on *J2*

Figure 6.5 shows the plots of the *AUC* generated by training the forests on *T1* and testing on *J2* at signal strengths of 40 and 100%. Each RF or DRF was composed of 500 trees, a sufficient forest size to ensure convergence in *AUC*. In 14 of the 16

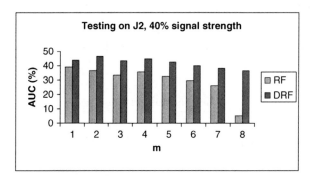

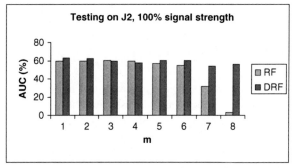

Fig. 6.5 *AUC* for RF and DRF: (*top*) trained on *T1*, tested on *J2* with 40% signal strength; (*bottom*) trained on *T1*, tested on *J2* with 100% signal strength. Each classifier is composed of 500 trees (see online version for color figures)

possible combinations of signal strength and split dimension, the DRF performance clearly exceeded that of the RF over the *FAR* region of interest. In the remaining two cases, the difference in the detection rate was negligible. Moreover, these results suggest that the DRF is more successful than the RF algorithm in detecting weaker signals and better utilizes more input features. As the split dimension increases, we would expect the trees to become more correlated for both methodologies, resulting in poorer prediction performance. Figure 6.5 suggests this trend, but the effect appears to be noticeably less severe for the DRF. The trade-off between tree strength and correlation with respect to m is discussed in greater detail in Section 6.4.2. ROC curves for both RF and DRF for the hidden signal detection application are plotted in Fig. 6.6, again indicating that the DRF exhibits superior performance across the low *FAR* region of interest.

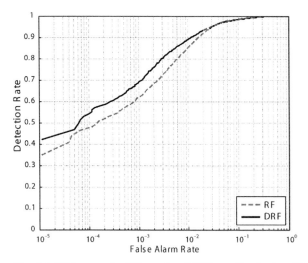

Fig. 6.6 ROC curves for RF $m = 1$ and DRF $m = 2$ trained on *T1*, tested on *J2* with 100% signal strength and 170 K additional negative samples (see online version for color figures)

6.4.1.2 Prediction Performance for *J2* with Cross-validation

To more thoroughly explore the impact of signal strength on prediction performance, we trained and tested the RF and DRF on all signal strengths of the *J2* data set. We used 5-fold cross-validation (CV) to evaluate the performance of both classifiers. In k-fold cross-validation, the data set is randomly partitioned into k equal-sized and equally proportioned subsets. For each run i, we set aside data subset i for testing, and we train the classifier on the remaining $k - 1$ subsets. We use the average of these k estimates to compute our performance estimate. Figure 6.7 shows the percentage increase in the *AUC* achieved by the DRF for split dimensionalities $m \in \{1, 2\}$ as compared to the best-performing RF (i.e., $m = 1$). As we observed when training on *T1*, the DRF yields substantially better detection performance on weaker signals.

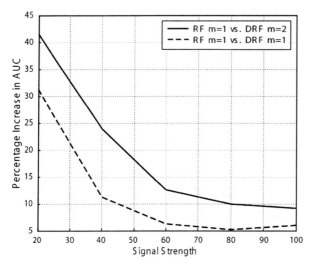

Fig. 6.7 Percentage increase in AUC on $J2$ for DRF $m = 1, 2$ over RF $m = 1$ at each signal strength. Each forest is composed of 500 trees (see online version for color figures)

6.4.2 Radiation Detection

The objective of the radiation detection effort is to detect the presence of a radiation source in vehicles traveling through a radiation portal monitoring system that measures the gamma ray spectrum of each vehicle quantized into 128 energy bins. The 128-dimensional normalized gamma ray spectra serve as the input features for the RF and DRF classifiers. The negative class is composed of radiation measurements from real vehicles containing no radiation source.

The data for the positive class were created by injecting a separate set of negative samples with spectra derived from two isotopic compositions of both uranium and plutonium in IAEA Category 1 quantities [18]. These sets of positive and negative samples were then partitioned into nonoverlapping, equally proportioned training and testing sets containing 17,000 and 75,000 samples, respectively.

The ROC curves in Fig. 6.8 show the *TPR* (i.e., detection rate) versus the *FAR* for RF and DRF. Over most of the low *FAR* range, the DRF maintains higher detection rates than the RF.

The large number of features available for this application presents an ideal opportunity to thoroughly explore the behavior of the RF and DRF methodologies for high split dimensions. In particular, we wish to investigate their relative computational efficiency, memory considerations, bounds on their generalization error (Equation 6.1), the interplay between tree strength and correlation, and the impact of each of these characteristics on overall algorithm performance. We have utilized the OOB data to compute these characteristics (see [3] for further details) at the minimal forest size required for each algorithm to achieve its peak performance. This can be readily observed in Fig. 6.9, which shows a plot of the minimum classification

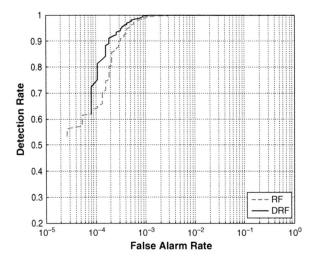

Fig. 6.8 ROC curves in the radiation detection task for RF and DRF (see online version for color figures)

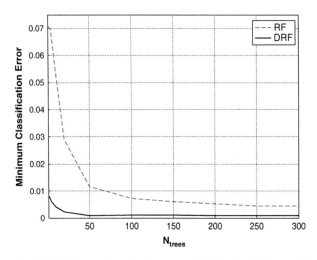

Fig. 6.9 Minimum classification error (*MCE*) for the RF and DRF as a function of forest size on the radiation detection data set (see online version for color figures)

error (*MCE*) achieved by both classifiers with respect to $m \in \{2^n | n = 0, 1, ..., 7\}$. The *MCE* is plotted as a function of the forest size and indicates that the peak DRF performance was achieved by a forest consisting of approximately 50 trees, far fewer than the 250 trees required by the RF.

In Table 6.1, performance statistics have been provided for the RF yielding the best *MCE* and for a DRF whose performance exceeded that of the RF with respect to error, computational requirements, and efficiency.[1] Both were trained on a dual-

---

[1] With respect to *MCE* alone, the best DRF achieved a 79.5% reduction relative to the RF.

Table 6.1 Performance summary for RF/DRF

	RF	DRF	Relative difference (%)
Dimension	4	8	–
Forest size	250	50	80
Avg. nodes/tree	1757.69	718.16	59.1
Classification error	4.37e-3	2.05e-3	53.0
Training time (s)	315	48	84.8
Memory usage (b)	439,423	323,172	26.5

core Intel 6600 2.4 GHz processor with 4 GB of RAM. To compute memory usage, we assumed that each RF node must store an integer feature ID and its corresponding floating-point threshold. Each DRF node must store m integers for its selected feature IDs along with $m + 1$ floating-point values for its weight vector, $\mathbf{w}$.

Table 6.1 indicates that the DRF was able to achieve a lower classification error rate than the RF while simultaneously reducing training time and memory usage by 84.8 and 26.5%, respectively. The smaller DRF trees clearly contribute to this improvement in efficiency, but their reduced size also suggests a dramatic increase in tree strength.

From an empirical standpoint, a node-splitting strategy that effects a better separation of the data, such as the multivariate LDA technique, naturally generates smaller classification trees as the split dimensionality increases, as shown in Fig. 6.10. Though such trees might exhibit superior prediction capabilities (i.e., greater strength), we would generally expect the variation among them to decrease (i.e., increased correlation), potentially leading to a reduction in overall performance.

The key to informative analysis of these two classification methodologies, as introduced in Section 6.2, lies in our ability to successfully characterize this interplay

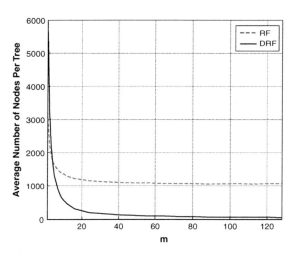

Fig. 6.10 Average decision tree size for the RF and DRF as a function of the dimensionality, m (see online version for color figures)

between the strength and correlation of individual trees. To provide further insight into these behaviors, Fig. 6.11 compares the OOB estimates of tree strength and correlation for the RF and DRF, along with their classification error and respective generalization error bounds plotted as a function of the split dimensionality, m. As expected, the strength of an individual DRF tree is, in general, significantly greater than that of its RF counterpart. Far more remarkable is the reduced correlation among the DRF trees for split dimensions up to $m \approx 90$. The relationship between strength and correlation is typically regarded as a trade-off [3], in which one is improved at the expense of the other, and the smaller DRF trees might naively be expected to exhibit greater correlation. However, [3] suggests that each base classifier is primarily influenced by the parameter vector representing the series of random feature selections at each node. In the multivariate setting, the number of potential feature subsets at each node increases combinatorially, dramatically enhancing the variability in the parameter vector that characterizes the classifier, which may explain the immediate drop in correlation as m increases. As m approaches the cardinality of the feature set, however, we observe a sudden and severe rise in correlation, behavior that is consistent with the reduced variation in the nodal features used for splitting.

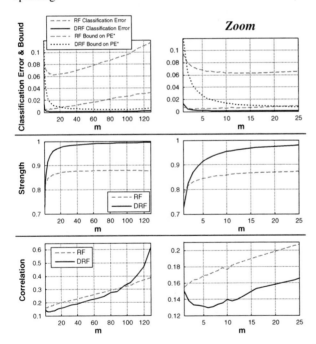

Fig. 6.11 OOB statistics as a function of the dimensionality, m, on the radiation detection data set (see online version for color figures)

Consistent with the generalization error bound (Equation 6.1), Fig. 6.11 shows that strength has a greater impact on the classifier performance than the correlation.

Even at the highest split dimensionality, the DRF classification error and error bound exhibit only minimal degradation. In contrast, the RF error steadily increases with the split dimensionality. Moreover, the DRF bound is far tighter than that of the RF, even surpassing the RF classification error at $m \approx 25$.

Interestingly, though the strength and correlation of both methodologies exhibit similar trends, their classification errors exhibit opposing behavior, suggesting that the relationship between strength and correlation is more complex than can be fully explained by our initial experiments.

6.4.3 Significance of Empirical Results

For the two signal detection applications discussed above, the performance of the DRF and RF methodologies was compared and contrasted via a series of experiments. The empirical evidence presented indicated that the DRF outperformed the RF with respect to each of the identified measures of performance. However, a natural question arises: *Are these observed differences statistically significant or simply an artifact of random fluctuations originating from the stochastic nature of the algorithms (e.g., bootstrap sampling of data and features)?* To more thoroughly investigate this issue, we revisited the hidden signal detection application introduced in Section 6.4.1. Specifically, using the original *T1* training data set and a new testing set called *J1*, we wish to build statistical confidence regions surrounding "average" DRF and RF ROC curves. The resultant confidence regions could then be used to determine whether the observed differences in performance are significant.

The *J1* data set is statistically equivalent to the original data set, *J2*. It contains 179,527 negative class samples and 9426 positive class samples with 100% embedded signal strength. As in the prior hidden signal detection experiments, all data samples consist of eight features useful for detecting the presence of embedded signals.

The *J1* data set, though equivalent to *J2*, was not utilized in any fashion during the development of the DRF algorithm; consequently, it is an ideal testing data set for independently assessing the performance of the methodology. This is a common practice in the speech recognition field and the broader machine learning community and is employed to prevent the subtle tuning of a methodology to a particular set of testing data.

For this study, we applied both the RF and DRF methodologies at all split dimensions $m \in \{1, 2, ..., 8\}$ and observed that the optimal RF occurred at $m = 2$, while the performance of the DRF peaked for $m = 1$. We will focus on these cases for the remainder of this discussion.

For each algorithm, 101 classifiers were trained and tested using variable random seeds. Based upon the resulting ROC curves, a "median" ROC and corresponding upper and lower confidence limits were computed for each methodology using a variant of the vertical averaging approach described by Fawcett [10]. Specifically, for each *FAR* value $\alpha \in [0.0, 1.0]$, the 101 corresponding detection rates were ranked, and their median detection rate $MDR(\alpha)$ was computed along with their 97.5 and 2.5 percentiles. Using these data, the median ROC, consisting of the collection of points $\{(\alpha, MDR(\alpha)) : \alpha \in [0.0, 1.0]\}$, and the 97.5 and 2.5 percentile bands were computed and are shown in Fig. 6.12.

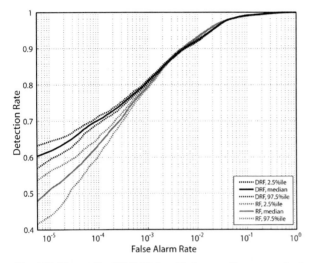

Fig. 6.12 The median ROC, 97.5 and 2.5 percentiles, curves for the DRF ($m = 1$) and RF ($m = 2$) classifiers using the hidden signal data sets (*T1* and *J1*) (see online version for color figures)

It is immediately apparent that the median ROC and associated percentile bands for the DRF and RF do not overlap for *FAR* values less than 10^{-3}, providing considerable evidence that the observed performance improvement exhibited by the DRF over this region is statistically significant.

We can also examine these results from the perspective of the *AUC*. Specifically, we computed the *AUC* values for the 101 classification results of each methodology using *FAR* cutoff levels of 10^{-3}, 10^{-4}, and 10^{-5}. Box plots of these results, with the medians and the 25th and 75th quantiles indicated, were computed and are shown in Fig. 6.13. Note that in each case, the interquartile ranges are nonoverlapping, further reinforcing our belief that the performance gain of the DRF over the RF is statistically significant in the lower *FAR* region of interest.

6.4.4 Small Samples and Early Stopping

Both the RF and DRF are tree-based ensemble classification methodologies whose construction relies fundamentally upon the processes of bagging and random feature selection. In fact, the DRF shares many similarities with the RF and, consequently, shares many of its most noteworthy advantages. However, a critical exception is the parametric node-splitting process utilized by the DRF. The conventional RF ap-

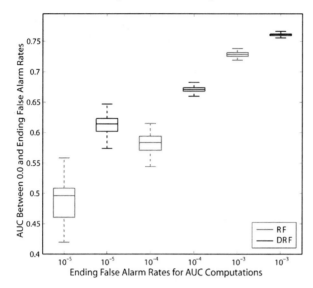

Fig. 6.13 Box plots of the *AUC* values for the DRF (*m* = 1) and RF (*m* = 2) classifiers using the hidden signal data sets (*T1* and *J1*) (see online version for color figures)

proach to node splitting is accomplished via a univariate thresholding process that is optimized relative to some predetermined criterion (e.g., Gini impurity). Generally, parameters are not estimated during this process. In contrast, node splitting under the DRF regime is performed by building an LDA model at each node in an effort to determine an "optimal" linear decision boundary. For the two-class problem, this requires the estimation of the class-specific mean vectors μ_k, $k = 0, 1$, and common covariance matrix Σ at each node in the forest.

This is an important distinction between the two methods that manifests itself in several ways. In particular, the training of the lower portions of all DRF trees (near the leaf nodes) must contend with progressively sparser data sets. Specifically, the LDA models are based upon point estimates (e.g., maximum-likelihood estimates) of the mean vectors and common covariance matrix. For a split dimension of $m \geq 1$, there are exactly $p = 3m + m(m-1)/2$ parameters to estimate at each node. Hence, as the value of m increases, the estimation problem becomes increasingly challenging at the more sparsely populated nodes in the forest. In severe cases, the common covariance matrix is not even estimable. This occurs exactly when a node is impure (i.e., both classes are represented) and the feature vectors within each class are identical. In these cases, the DRF splitting process defaults to a geometric method (i.e., the decision threshold is taken to be the perpendicular bisector of the chord connecting the sample means of the two classes). This tactic has proven relatively effective in practice, but the more subtle issue of small sample size parameter estimation remains.

Unfortunately, this is a common problem in statistics, and our study of the DRF methodology would be incomplete without investigating the role parameter estimation plays in its performance and reliability. A careful examination of the median ROC and the corresponding 97.5 and 2.5 percentile bands shown in Fig. 6.12 reveals behavior that may provide insight into this issue. Note that the distance between the percentile bands increases noticeably as the *FAR* drops from 10^{-4} to 5×10^{-6}, suggesting a considerable increase in the variability of the experimental ROC curves over this extreme interval. It is our conjecture that a contributing factor to this phenomenon is the sparseness of the data in the lower nodes of the DRF trees, which leads to greater instability in the parameter estimates.

This behavior is even more pronounced for the RF, appearing at first glance to contradict the above conjecture. Though its underlying cause is not entirely clear, this contradictory behavior may be at least partially explained by the fact that each DRF node-splitting decision utilizes two sources of information: the node data and the LDA model. The model may exert a dampening influence on the impact of the data, reducing variability at the leaf nodes and thus reducing variability in the ROC at low *FAR* values. In contrast, the RF is driven entirely by the data and hence may prove more vulnerable to data variation and sparseness.

In any case, enriching the data at the lower DRF nodes in an effort to improve parameter estimation and ultimately enhance performance (e.g., increase the median detection rate while reducing variability) is a challenge we would like to address. As a first attempt in this direction, we considered the optimal DRF ($m = 1$) for the hidden signal detection application problem first described in Section 6.4.1. In this case, the LDA model building exercise reduces to estimating three univariate Gaussian parameters. We have observed in our studies that the number of samples used to estimate these parameters near the leaf nodes is frequently very small – less than 10 in many cases. To more fully explore this issue, suppose we require that at least $n = 30$ samples be used to estimate the LDA parameters at any node in the forest. *What is the impact of this constraint on the detection performance?*

To address this question, we incorporated an early stopping criterion into our methodology whereby tree nodes continue to successively split until either purity is achieved or the number of node samples drops below a prescribed value, n. This version of the DRF methodology is called "early stopping" DRF and is denoted by DRF-ES.

For the following experiment, we once again randomly generated a collection of 101 forests, trained on *T1* and tested on *J1*, for both the DRF and DRF-ES methodologies. Figure 6.14 presents the median ROC curves and the corresponding 97.5 and 2.5 percentile bands for the standard DRF and the DRF-ES with $n = 30$. We observed that for *FAR* values less than approximately 5×10^{-2}, the DRF-ES classifier significantly outperforms the conventional DRF classifier. Figure 6.15 shows the corresponding box plots of the *AUC* values for *FAR* cutoff levels of 10^{-3}, 10^{-4}, and 10^{-5}. Over each of these intervals, the interquartile ranges are nonoverlapping, reinforcing the statement that the performance gain of the DRF-ES over the DRF is significant in this case ($m = 1$).

We then extended these studies and compared the performance of the DRF and DRF-ES classifiers for higher split dimensions of $m = 2$ and 3. Figure 6.16 shows the median ROC curves with percentile bands for the DRF and DRF-ES applied to

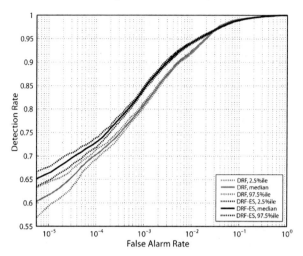

Fig. 6.14 The median ROC, 97.5 and 2.5 percentiles, curves for the DRF and DRF-ES classifiers for $m = 1$ using the hidden signal data sets (*T1* and *J1*) (see online version for color figures)

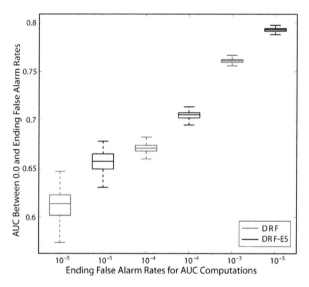

Fig. 6.15 Box plots of the *AUC* values for the DRF and DRF-ES classifiers at $m = 1$ using the hidden signal data sets (*T1* and *J1*) (see online version for color figures)

the hidden signal detection data sets. Note that in both cases the performance degrades as m increases, but the DRF-ES curves exhibit a "nesting" behavior (i.e., they have similar shape) that suggests greater performance stability relative to the DRF. Specifically, we see that the DRF-ES produces narrower percentile bands that are nonoverlapping, while those of the DRF are wider and repeatedly cross. In Fig. 6.17, the DRF and DRF-ES median ROC curves are plotted head to head, together with their 97.5 and 2.5 percentile bands, for each value of m. Note that the performance

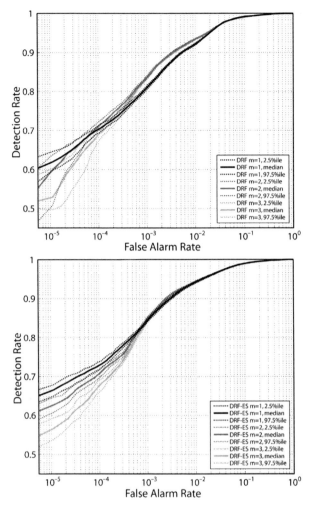

Fig. 6.16 Median ROC curves with 97.5 and 2.5 percentile bands for DRF (*top*) and DRF-ES (*bottom*) for the hidden signal detection problem at $m = 1, 2, 3$ (*T1* and *J1*) (see online version for color figures)

advantage enjoyed by the DRF-ES degrades as m increases from 1 to 2, evaporating entirely when m equals 3. In other words, as the number of parameters increases while holding the maximal sample size $n = 30$ constant, the DRF-ES performance gains are eroded. This behavior is consistent with our earlier conjecture that parameter estimation based upon sparse data, combined with increasing model dimensionality, adversely affects the performance and stability of the DRF methodology. However, it remains likely that the true mechanisms underlying these behaviors are quite complex and defy simple explanation. Issues such as model choice and misspecification may be contributing factors.

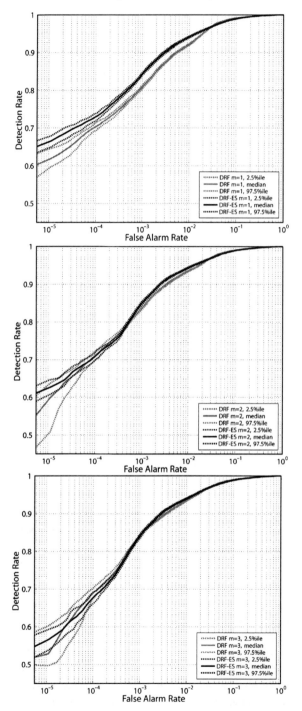

Fig. 6.17 Median ROC curves with 97.5 and 2.5 percentile bands for the DRF and DRF-ES for $m = 1, 2, 3$ (numbered top to bottom) (see online version for color figures)

6.4.5 Expected Cost

As we discussed in Section 6.1, information analysis systems are constructed for the purpose of compiling large stores of (potentially multisource) data to support informed analysis and decision-making. Though many algorithms in the classification field are designed to minimize the expected overall error in class predictions, it is common for real-world detection problems to be inherently associated with unequal costs for false alarms and missed detections (e.g., the hidden signal detection application). Thus, a natural performance metric that quantifies the expected cost of an incorrect decision in such cost-sensitive applications is given by

$$EC = p(+) \cdot (1 - DR) \cdot c(miss) + p(-) \cdot FAR \cdot c(falsealarm) \qquad (6.8)$$

where DR is the detection rate, $p(\cdot)$ is the prior probability for each class, and $c(\cdot)$ is the cost for each type of error. To enable visualization of the general behavior of this metric, Drummond and Holte developed "cost curves" that express expected cost as a function of the class priors and costs [7]. Specifically, cost curves plot the expected cost (normalized by its maximum value) versus the probability cost function (PCF), which is given by

$$PCF = \frac{p(+) \cdot c(miss)}{p(+) \cdot c(miss) + p(-) \cdot c(falsealarm)} \qquad (6.9)$$

Assuming equal priors, PCF is small when the cost of false alarms is large relative to that of missed detections. In the hidden signal detection application, a false alarm is considered to be at least 100 times more costly than a missed detection, making classifiers whose cost curves are lower at small values of PCF (e.g., $PCF < 0.01$) more desirable. In Fig. 6.18, we have plotted the median, 2.5 percentile, and 97.5 percentile cost curves for the RF, DRF, and DRF-ES. We immediately observe that the DRF and DRF-ES appear to be significantly more effective than the RF, with DRF-ES achieving the smallest expected cost across the low PCF range of interest.

6.5 Conclusions

The empirical results presented in Section 6.4 provide strong evidence that the discriminant random forest and its ES variant produce significantly higher detection rates than the random forest over the low FAR regions of interest. The superior strength and diversity of the trees produced by the DRF further support this observation. In addition, this methodology appears to be more successful in the detection of weak signals and may be especially useful for applications in which low signal-to-noise ratios are typically encountered.

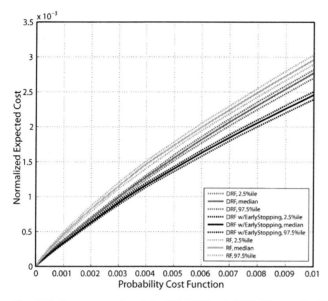

Fig. 6.18 Cost curves plotted for RF, DRF, and DRF-ES over the *PCF* range of interest. The discriminant-based classifiers outperform the RF over this range, with the DRF-ES achieving the lowest cost over all (see online version for color figures)

We have also found that our methodology achieves far lower prediction errors than the RF when high-dimensional feature vectors are used at each tree node. In general, we expect the performance of any forest to decline as the number of features selected at each node approaches the cardinality of the entire feature set. Under these conditions, the individual base classifiers that compose the forest are nearly identical, negating many of the benefits that arise from an ensemble-based approach. In such cases, the only variation remaining in the forest is due to the bagging of the input data. Though we observed the expected performance degradation for both forest methodologies at extremely high split dimensions, the effect was far less severe for the discriminant-based approach. This result suggests a versatility and robustness in the DRF methodology that may prove valuable for some application domains.

Our investigation into the effect of sparse data revealed that an early stopping strategy might help mitigate its impact on classification performance, ultimately increasing the detection rate over the lower *FAR* regions. This advantage, however, is diminished as split dimensionality increases. Though we did not thoroughly explore the sensitivity of the DRF performance to the early stopping parameter, *n*, we conjecture that increasing this parameter in response to increases in split dimensionality would be counterproductive. Such a strategy would eventually eliminate the fine-grained (i.e., small scale) class distinctions that are critical for effective classification. However, at split dimension $m = 1$, the DRF enjoys a significant performance improvement in low *FAR* regions via early stopping. In fact, for our applications, the DRF-ES at $m = 1$ outperformed the DRF over all dimensions.

Overall, our empirical studies provided considerable evidence that the DRF is significantly more effective than the RF over low *FAR* regions. Although computational efficiency may be adversely impacted by the more complex node splitting of the DRF at extremely high dimensions, its peak performance is typically achieved at much lower dimensions where it is more efficient than the RF with respect to memory and runtime.

The behavior of the discriminant random forest methodology is compelling and hints at complex internal mechanisms that invite further investigation. However, we have found statistically significant evidence supporting this technique as a highly robust and successful classification approach across diverse application domains.

Acknowledgments This work was performed under the auspices of the U.S. Department of Energy by Lawrence Livermore National Laboratory under Contract DE-AC52-07NA27344.

References

1. Breiman, L.: Bagging Predictors. Machine Learning. 26(2), 123–140 (1996)
2. Breiman, L.: Using Adaptive Bagging to Debias Regressions. Technical Report 547, Statistics Dept. UC Berkeley (1999)
3. Breiman, L.: Random Forests. Machine Learning. 45(1), 5–32 (2001)
4. Breiman, L., Friedman, J., Olshen, R.A., Stone, C.J.: Classification and Regression Trees. Boca Raton, FL: Chapman and Hall (1984)
5. Chernick, M.R.: Bootstrap Methods, A Practitioner's Guide. New York: John Wiley and Sons, Inc. (1999).
6. Dietterich, T.: An Experimental Comparison of Three Methods for Constructing Ensembles of Decision Trees: Bagging, Boosting, and Randomization. Machine Learning. 1–22 (1998)
7. Drummond, C., Holte, R.: Explicitly Representing Expected Cost: An Alternative to ROC Representation. Proc. of the Sixth ACM SIGKDD International Conference on Knowledge Discovery and Data Mining (2000)
8. Duda, R.O., Hart, P.E., Stork, D.H.: Pattern Classification, 2nd edition. New York: Wiley Interscience (2000)
9. Efron, B., Tibshirani, R.J.: An Introduction to the Bootstrap. New York: Chapman and Hall/CRC (1993)
10. Fawcett, T.: ROC Graphs: Notes and Practical Considerations for Researchers. Technical Report, Palo Alto, USA: HP Laboratories (2004)
11. Fisher, R.A.: The Use of Multiple Measurements in Taxonomic Problems. Annals of Eugenics. 7, 179–188 (1936)
12. Ho, T.K.: Random Decision Forest. Proc. of the 3rd International Conference on Document Analysis and Recognition. 278–282 (1995)
13. Ho, T.K.: The Random Subspace Method for Constructing Decision Forests. IEEE Trans. On Pattern Analysis and Machine Intelligence. 20(8), 832–844 (1998)
14. Lemmond, T.D., Hatch A.O., Chen, B.Y., Knapp, D.A., Hiller, L.J., Mugge, M.J., and Hanley, W.G.: Discriminant Random Forests. Proceedings of the 2008 International Conference on Data Mining (2008) *(to appear)*
15. Loeve, M.: Probability Theory II (Graduate Texts in Mathematics), 4th edition. New York: Springer-Verlag (1994)
16. Mardia, K., Kent, J., Bibby, J.: Multivariate Analysis. New York: Academic Press (1992)

17. McLachlan, G.J.: Discriminant Analysis and Statistical Pattern Recognition. New York, Wiley-Interscience (2004)
18. The Physical Protection of Nuclear Material and Nuclear Facilities. IAEA INF-CIRC/225/Rev.4 (Corrected).
19. Prinzie, A., Van den Poel, D.: Random Forests for multiclass classification: Random Multinomial Logit. Expert Systems with Applications. 34(3), 1721–1732 (2008)
20. Rodriguez, J.J., Kuncheva, L.I., et al.: Rotation forest: A New Classifier Ensemble Method. IEEE Transactions on Pattern Analysis and Machine Intelligence. 28(10), 1619–1630 (2006)

Chapter 7
Prediction with the SVM Using Test Point Margins

Süreyya Özöğür-Akyüz, Zakria Hussain, and John Shawe-Taylor

Abstract Support vector machines (SVMs) carry out binary classification by constructing a maximal margin hyperplane between the two classes of observed (training) examples and then classifying test points according to the half-spaces in which they reside (irrespective of the distances that may exist between the test examples and the hyperplane). Cross-validation involves finding the *one* SVM model together with its optimal parameters that minimizes the training error and has good generalization in the future. In contrast, in this chapter we collect *all* of the models found in the model selection phase and make predictions according to the model whose hyperplane achieves the maximum separation from a test point. This directly corresponds to the L_∞ norm for choosing SVM models at the testing stage. Furthermore, we also investigate other more general techniques corresponding to different L_p norms and show how these methods allow us to avoid the complex and time-consuming paradigm of cross-validation. Experimental results demonstrate this advantage, showing significant decreases in computational time as well as competitive generalization error.

7.1 Introduction

Data mining is the process of analyzing data to gather useful information or structure. It has been described as the *science of extracting useful information from large data sets or databases* [10] or *the nontrivial extraction of implicit, previously*

Süreyya Özöğür-Akyüz
Institute of Applied Mathematics, Middle East Technical University, 06531 Ankara, Turkey,
e-mail: sozogur@metu.edu.tr

Zakria Hussain · John Shawe-Taylor
Department of Computer Science, Centre for Computational Statistics and Machine Learning,
University College, London WC1E 6BT, UK, e-mail: z.hussain@cs.ucl.ac.uk,jst@cs.ucl.ac.uk

R. Stahlbock et al. (eds.), *Data Mining*, Annals of Information Systems 8,
DOI 10.1007/978-1-4419-1280-0_7, © Springer Science+Business Media, LLC 2010

unknown, and potentially useful information from data [7]. Computationally it is a highly demanding field because of the large amounts of experimental data in databases. Various applications of data mining are prevalent in areas such as medicine, finance, business. There are different types of data mining tools motivated from statistical analysis, probabilistic models, and learning theory.

In recent years, learning methods have become more desirable because of their reliability and effectiveness at solving real-world problems. In real-world situations, for instance, in the engineering or biological sciences, conducting experiments can be costly and time consuming. In such situations, accurate predictive methods can be used to help overcome these difficulties in more efficient and cost-effective ways. Large amounts of data are available through the internet in which data mining methods are needed to understand the structure and the pattern of the data. Different methodologies have been developed to tackle learning, including supervised and unsupervised learning.

Supervised learning is a learning methodology of searching for algorithms which are reasoned from given samples in order to generalize hypotheses and make predictions about future instances [13]. These algorithms learn functions based on training examples consisting of input–output pairs given to the learning system. These functions can subsequently be used to predict the output of test examples. Training sets are the main resource of supervised learning.

Popular data mining algorithms are ensemble methods [1]. Many researchers have investigated the technique which combines the predictions of multiple classifiers to produce a better classifier [3, 5, 18, 22]. The resulting classifier, which is an ensemble of functions, can be more accurate than a single hypothesis. Bagging [3] and boosting [8, 19] are among the most popular ensemble methods [16]. A second popular approach is the large margin algorithms [21, 6] known as the support vector machine (SVM). The algorithm looks to separate the two classes of training samples using a hyperplane that exhibits the maximum margin between the classes. Empirical and theoretical results of the SVM are very impressive and hence the algorithm is largely used in the data mining field.

For the majority of data mining tools, parameter selection is a critical question and attempts at determining the right model for data analysis and prediction. Different algorithms have been studied to choose the best parameters among the full set of functions, with cross-validation and leave one out being among the most popular.

In this research, we develop a fast algorithm for model selection that uses the benefit of all the models constructed during the parameter selection stage. We apply our model selection strategy to the SVM, as it is one of the most powerful methods in machine learning for solving binary classification problems. SVMs were invented by Vapnik [21] (and coworkers), with the idea to classify points by maximizing the distance between two classes [2].

More formally let $(\mathbf{x}, y)$ be an (input,output) pair where $\mathbf{x} \in \mathbb{R}^n$ and $y \in \{-1, 1\}$ and $\mathbf{x}$ comes from some input domain X and similarly y comes from some output domain Y. A training set is defined by m input–output pairs by $S = \{(\mathbf{x}_i, y_i)\}_{i=1}^{m}$. Given S and a set of functions $\mathcal{F}$ we would like to find a candidate function $f \in \mathcal{F}$ such that

$$f : \mathbf{x} \mapsto y$$

We refer to these candidate functions as *hypotheses* [6].

In this study, we will use the support vector machine (*SVM*) algorithm, a classification algorithm based on maximizing the margin γ between two classes of objects with some constraints. The classes are separated by an affine function, hyperplane $\langle \mathbf{w}, \mathbf{x} \rangle + b = 0$, where $\mathbf{w} \in \mathbb{R}^n$ is a normal vector (weight vector) helping to define the hyperplane, $b \in \mathbb{R}$ is the bias term [6], and $\langle \cdot, \cdot \rangle$ denotes the scalar product. Hence, given a set of examples S the SVM separates the two groups of points by a hyperplane.

In most real-world problems, data are not linearly separable. To use the facilities of the linear separable case, one can define a *nonlinear mapping* ϕ which transforms the input space into a higher dimensional *feature space* such that the points are separable in the feature space. But the mapping can be very high dimensional and sometimes infinite. Hence, it is hard to interpret decision (classification) functions which are expressed as $f(\mathbf{x}) = \langle \mathbf{w}, \phi(\mathbf{x}) \rangle + b$. Following the notation of [6], the *kernel function* is defined as an inner product of two points under the mapping ϕ, i.e., $\kappa(\mathbf{x}_i, \mathbf{x}_j) = \langle \phi(\mathbf{x}_i), \phi(\mathbf{x}_j) \rangle$ which can also be explained as the similarity between two points. The optimization problem for separating two classes is expressed as follows [6]:

Definition 7.1 (Primal Hard Margin Problem).

$$\min_{\mathbf{w},b} \langle \mathbf{w}, \mathbf{w} \rangle$$
$$\text{s.t.} \quad y_i \cdot (\langle \mathbf{w}, \phi(\mathbf{x}_i) \rangle + b) \geq 1 \quad (i = 1, 2, \ldots, m);$$

The dual allows us to work in kernel-defined feature space and reads as follows.

Definition 7.2 (Dual Hard Margin Problem).

$$\max_{\alpha} \sum_{i=1}^{m} \alpha_i - \frac{1}{2} \sum_{i=1}^{m} y_i y_j \alpha_i \alpha_j \kappa(\mathbf{x}_i, \mathbf{x}_j)$$
$$\text{s.t.} \quad \sum_{i=1}^{m} y_i \alpha_i = 0,$$
$$\alpha_i \geq 0 \quad (i = 1, 2, \ldots, m)$$

It is not satisfactory to apply strictly perfect maximal margin classifiers without any error term, since they will not be applicable to noisy real-world data. Therefore, variables are introduced that allow the maximal margin criterion to be violated; this classifier is called *a soft margin classifier*. Here, a vector ξ of some slack variables is inserted into the constraints and, equipped with a regularization constant C, into the objective function as well ($\|\cdot\|_2$ denotes Euclidean norm).

Definition 7.3 (Primal Soft Margin Problem).

$$\min_{\xi,\mathbf{w},b} \|\mathbf{w}\|_2^2 + C \sum_i \xi_i$$
$$\text{s.t.} \quad y_i \cdot (\langle \mathbf{w}, \phi(\mathbf{x}_i) \rangle + b) \geq 1 - \xi_i \quad (i = 1, 2, \ldots, m)$$

The dual problem in the soft margin case looks as follows.

Definition 7.4 (Dual Soft Margin Problem).

$$\max_\alpha \sum_{i=1}^m \alpha_i - \frac{1}{2}\sum_{i=1}^m y_i y_j \alpha_i \alpha_j \kappa(\mathbf{x}_i, \mathbf{x}_j)$$
$$\text{subject to } \sum_{i=1}^m y_i \alpha_i = 0,$$
$$0 \leq \alpha_i \leq C \quad (i = 1, 2, \ldots, m)$$

The solution of this optimization problem (Definition 7.4) yields a maximal margin hyperplane that we will refer to as the *support vector machine (SVM)*.

All machine learning algorithms require a model selection phase. This consists of choosing the best parameters for a particular data set and using them in order to make predictions. In the SVM (or ν-SVM) that uses a Gaussian kernel the number of parameters to tune is two – the C in the standard SVM (or the ν in the ν-SVM) and the kernel width parameter σ. Let us take the standard SVM and look at the model selection phase in detail. Firstly, given some data set S the most common model selection technique is to use k-fold cross-validation where $k > 0, k \in \mathbb{N}$. The idea is to split the data into k parts and use $k-1$ for training and the remaining for testing. The $k-1$ folds are trained with various values of C and σ and tested on each test set. The set of values that give the smallest validation error among all of the splits is used as the SVM model for the entire training set S. However, this can be very costly and time consuming. Alternative methods have been developed to find the best parameters such as a *gradient descent algorithm* where parameters are searched for by a gradient descent algorithm [4, 12]. Also when the number of parameters increases, grid search and CV become intractable and exhaustive. In such cases, while the error function is minimized over the hyperplane parameters, it can be maximized over kernel parameters simultaneously [4]. The intuition behind this algorithm is to minimize an error bound. Keerthi et al. [11] implemented a gradient-based solution for evaluating multiparameters for the SVM by using a radius/margin bound. However, it has some drawbacks; kernel functions may not be differentiable, causing a problem for gradient-based algorithms. To overcome this problem Friedrichs et al. developed evolutionary-based algorithms for searching multiple parameters of SVMs [9], capable of solving for nondifferentiable kernels. In [9], the proposed evolutionary algorithm is based on a *covariance matrix adaptation evolution strategy* that searches for an appropriate hyperparameter vector. Here, the fitness function corresponds to the generalization function performance.

In this chapter, we assume that the data sets consist of a small number of examples and applying cross-validation is costly as we will tend to use up a large proportion of points in the test set. We tackle this problem by using the full training set to construct all possible *SVM models* that can be defined using the list of parameter values. Hence, we benefit from all possible models for a given range of parameters and classify a test point by checking to see which SVM hyperplane (from the full list of models) the test point is furthest from. By this, different classifiers are matched to different test points in our test set. Rather than retraining the SVM using the best C and best σ value we simply store all of the SVM models and make predictions using any one of them. This speeds up the computational time of model selection when compared to the cross-validation model selection regime described above. The intuition of choosing parameters from the test phase is based on the studies of [17] and [20]. In [17], biological data are classified according to the output values defined with confidence levels by different classifiers being determined for each protein sequence. The second motivation for the work comes from the theoretical work of [20] that gives generalization error bounds on test points given that they achieve a large separation from the hyperplane. This suggests that we can make predictions once we receive test points, from the hyperplanes already constructed, and giving us a way of avoiding cross-validation. We would like to point out that this theoretical work motivates the L_∞ method we propose and NOT the L_1 and L_2 norm approaches also proposed, which are similar to ensemble methods. Finally we would like to make the point that with the test point margin methodology, imbalanced data sets can, we feel, be tackled more efficiently i.e., fraud detection. In this scenario the advantage of our approach is that we avoid using up too many of the smaller class of examples in the cross-validation splits and hence utilize all of the examples during training.

7.2 Methods

In this section, three different norms will be discussed for model selection at the testing phase. Given a set of functions $\{f_1(\mathbf{x}), \ldots f_\ell(\mathbf{x})\}$ output by the SVM with $\ell = |C| \times |\sigma|$ being the number of models that can be constructed from the set of parameter values $C = \{C_1, \ldots\}$ and $\sigma = \{\sigma_1, \ldots\}$ we can use some or a combination of them to make predictions. The first approach we propose uses the L_∞ norm for choosing which function to use. This is equivalent to evaluating the distance of a test point according to the function that achieves the largest (functional) margin. For example, assume we have three values for $C = \{C_1, C_2, C_3\}$ and two values for $\sigma = \{\sigma_1, \sigma_2\}$. Therefore, we have the following $\ell = 6$ SVM models together with their list of parameter values $\{C, \sigma\}$:

- $f_1 = \mathrm{SVM}_1\colon \{C_1, \sigma_1\}$
- $f_2 = \mathrm{SVM}_2\colon \{C_1, \sigma_2\}$
- $f_3 = \mathrm{SVM}_3\colon \{C_2, \sigma_1\}$
- $f_4 = \mathrm{SVM}_4\colon \{C_2, \sigma_2\}$
- $f_5 = \mathrm{SVM}_5\colon \{C_3, \sigma_1\}$
- $f_6 = \mathrm{SVM}_6\colon \{C_4, \sigma_4\}$

Now at evaluation, we would compute the functions for all test points. For instance, given a test example $\mathbf{x} \in X_{test}$, let us assume the following six functional values:

- $f_1(\mathbf{x}) = 1.67$
- $f_2(\mathbf{x}) = 0.89$
- $f_3(\mathbf{x}) = -0.32$
- $f_4(\mathbf{x}) = -0.05$
- $f_5(\mathbf{x}) = 1.1$
- $f_6(\mathbf{x}) = 1.8$

We assume here, without loss of generality, that the functions f compute the functional margins and not the geometrical margins (hence the reason that the example values we have presented are not bounded by 1 and −1). Finally, we would predict the class of $\mathbf{x}$ by looking for the maximum positive and the maximum negative values of all functions. This corresponds to f_6 and f_3. However, the distance of the test example $\mathbf{x}$ from the hyperplane is greater for the $f_6 = \text{SVM}_6$ function/model and therefore this example can be predicted as positive. Therefore, the L_∞ prediction function $F_\infty(\mathbf{x})$ given an example $\mathbf{x}$ can be expressed in the following way:

$$F_\infty(\mathbf{x}) = \text{sgn}\left(\max\{f_i(\mathbf{x})\}_{i=1}^\ell + \min\{f_i(\mathbf{x})\}_{i=1}^\ell\right) \qquad (7.1)$$

where $F = \{f_1, \ldots, f_\ell\}$ is the set of all the functions that can be constructed from the list of parameter values.

The L_∞ norm approach is also illustrated in Fig. 7.1 on a real-world data set. The figure gives the evaluations of 110 SVM models[1] (i.e., functional margin values), sorted in ascending order, for a particular test point. From the plot the maximum positive margin (the far rightmost bar) and minimum negative margin (far leftmost bar) are shown in black. Hence, the sign of the sum of these two function margin values will give us the prediction of the test point. This test point is classified positive.

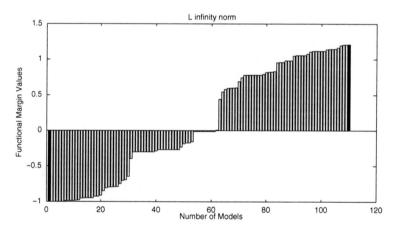

Fig. 7.1 Each stem corresponds to the functional margin value given for that particular SVM model f. The graph of L_∞ norm which predicts the example as $+1$ where actual class is $+1$

The second approach we introduce is for the L_1 norm where the decision depends on the sign of the Riemann sum of all outputs evaluated for a test point. This results in the following L_1 norm prediction function $F_1(\mathbf{x})$ given a test example $\mathbf{x}$:

---

[1] 11 C values $= \{2^{-5}, 2^{-3}, 2^{-1}, 2, 2^3, 2^5, 2^7, 2^9, 2^{11}, 2^{13}, 2^{15}\}$ and 10 σ values $= \{2^{-15}, 2^{-13}, 2^{-11}, 2^{-9}, 2^{-7}, 2^{-5}, 2^{-3}, 2^{-1}, 2, 2^3\}$

$$F_1(\mathbf{x}) = \text{sgn}\left(\sum_{i=1}^{\ell} f_i(\mathbf{x})\right) \tag{7.2}$$

This is illustrated in Fig. 7.2. It is clear that the prediction function looks at the integrals of the two areas (indicated in black) above and below the threshold of 0. Essentially, this equates to summing the above and below bars. In Fig. 7.2, it is clear that the summation will be positive since the area of the positive values (above 0) is bigger than the area of the negative values (below 0). This methodology corresponds to summing the weighted average of all the prediction functions with a uniform weighting of 1. This is closely related to taking a weighted majority vote.

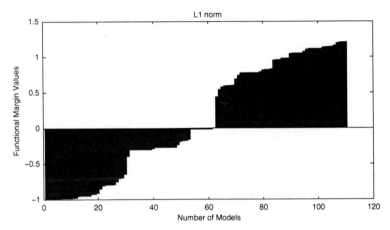

Fig. 7.2 Each stem corresponds to the functional margin value given for that particular SVM model f. The L_1 norm predicts $+1$ and the actual class of the example is $+1$

The final approach corresponds to the L_2 norm and is similar to the L_1 norm discussed above, but with a down-weighting if values are below 1 and an up-weighting if they are above 1. This means that we are giving a greater confidence to functions that predict functional values greater than 1 or -1 but less confidence to those that are closer to the threshold of 0. Another way of thinking about this approach is that it is equivalent to a weighted combination of functional margins with the absolute values of themselves. Therefore, given a test example $\mathbf{x}$, we have the following L_2 norm prediction function $F_2(\mathbf{x})$:

$$F_2(\mathbf{x}) = \text{sgn}\left(\sum_{i=1}^{\ell} f_i(\mathbf{x})|f_i(\mathbf{x})|\right) \tag{7.3}$$

The plot of Fig. 7.3 represents the L_2 norm solution for the same test point predicted by the L_∞ norm in Fig. 7.1 and the L_1 norm method shown in Fig. 7.2. As you can see the yellow region corresponds to the original values of the functions and the black bars are the down-weighted or up-weighted values of the 110 prediction

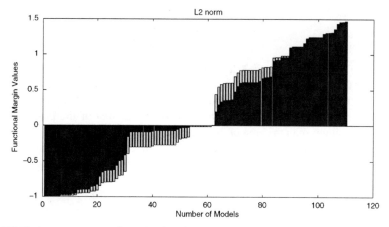

Fig. 7.3 Each stem corresponds to the functional margin value given for that particular SVM model f. The L_2 norm predicts $+1$ where the actual class of the example is $+1$

functions. The L_2 norm corresponds to summing the weights of the black bars only. It can be seen that the values that are smaller than 1 are down-weighted (decreased) and those greater than 1 are up-weighted (increased). Clearly, values that are close to 1 do not change significantly.

7.3 Data Set Description

In this study, we used the well-known standard UCI machine learning repository (can be accessed via http://archive.ics.uci.edu/ml/). From the repository, we used the Votes, Glass, Haberman, Bupa, Credit, Pima, BreastW, Ionosphere, Australian Credit, and the German Credit data sets. For the first seven data sets, we removed examples containing unknown values and contradictory labels (this is why the Votes data set is considerably smaller than the one found at the UCI website). The number of examples, attributes, and class distributions of all the data sets is given in Table 7.1.

7.4 Results

We call our methods the SVM-L_∞, SVM-L_1, and SVM-L_2 which correspond to using the L_∞, L_1, and L_2 methods we proposed in Section 7.2. We also test our methods against the SVM with cross-validation (CV), where we carry out 10-fold cross-validation to estimate the optimal C and σ values. Note that in the methods we propose we do not need to carry out this parameter tuning phase and hence achieve a 10-fold speedup against the SVM with CV.

Table 7.1 Data set description

Data set	# Instances	# Attributes	# Pos	# Neg
Votes	52	16	18	34
Glass	163	9	87	76
Haberman	294	3	219	75
Bupa	345	6	145	200
Credit	653	15	296	357
Pima	768	8	269	499
BreastW	683	9	239	444
Ionosphere	351	34	225	126
Australian	690	14	307	383
German	1000	20	300	700

Table 7.2 presents the results, including the standard deviation (STD) of the error over the 10-folds of cross-validation, the cumulative training, and testing time (time) in seconds for all folds of CV, the error as percentages (error %), as numbers (error #), the area under the curve (AUC), and the average over all data sets for the entire 10-fold cross-validation process.

The results of the SVM-L_p where $p = \infty, 2, 1$ show a significant decrease in computational time when compared to the SVM with CV. For example, we can see that the German data set takes approximately 4368 s to train and test and that our methods take between 544 and 597 s for training and testing purposes. This is approximately eight times faster than using cross-validation. We can also see from Table 7.2 that the L_∞ method seems to capture better prediction models compared to the other two L_p norm methods, but all three methods compare favorably with respect to test error against the SVM with CV. Since several data sets are imbalanced (see Table 7.1), we also report AUC results. It is well known that if the AUC tends to 1, the prediction accuracy will increase. In Table 7.2, we can see that the L_∞ norm has greater AUC values than the other L_p norm methods.

Finally, when comparing the three methods proposed it is clear that the most successful in terms of speed and accuracy is the L_∞ norm. This perhaps is less surprising when viewed from the theoretical motivation of this work, as [20] has proposed a bound that gives higher confidence of correct classification if the test point achieves a large separation from the hyperplane. This is exactly what the L_∞ norm method does. The other L_p norm methods do not have such theoretical justifications.

7.5 Discussion and Future Work

We proposed a novel method for carrying out predictions with the SVM classifiers once they had been constructed using the entire list of regularization parameters (chosen by the user). We showed that we could apply the L_p norms to help pick these classifier(s). Moreover, we introduced the SVM-L_∞, SVM-L_1, and SVM-L_2

Table 7.2 L_∞, L_1, and L_2 norm results against SVM with cross-validation

Data Set	SVM with CV					SVM-L_∞					SVM-L_1					SVM-L_2				
	STD	time	err %	err #	AUC	STD	time	err %	err #	AUC	STD	time	err %	err #	AUC	STD	time	err %	err #	AUC
Votes	0.2115	11.54	**8.33**	**5**	0.9417	0.0991	**0.94**	**8.33**	**5**	0.8333	0.095	1.84	27	14	0.4875	0.0979	0.9	19.17	10	0.6875
Glass	0.1222	89.93	34.85	57	0.7556	0.135	**16.29**	**31.99**	**52**	0.7095	0.1125	13.28	39.27	64	0.6401	0.1282	9.59	36.92	60	0.6363
Haberman	0.0437	169.51	**24.81**	**73**	0.7253	0.0363	**23.18**	25.17	74	0.7163	0.0127	23.44	25.49	75	0.7183	0.0133	13.62	25.83	76	0.7183
Bupa	0.0648	354.46	**28.95**	**100**	0.5437	0.0611	**82.87**	31.55	109	0.4535	0.0089	80.59	42.02	145	0.1632	0.0089	80.59	42.02	145	0.2117
Credit	0.1916	1812.86	**13.49**	**88**	0.9391	0.1439	370.21	18.09	118	0.9144	0.0984	245.43	18.52	121	0.9165	0.1036	199.31	18.22	119	0.9228
Pima	0.038	1300.82	**23.81**	**183**	0.7227	0.03	**265.9**	26.17	201	0.6649	0.0101	183.89	34.64	266	0.5336	0.0101	183.89	34.64	266	0.5791
BreastW	0.0192	414.18	**3.35**	**23**	0.9881	0.0361	111.26	4.81	33	0.9771	0.0421	72.35	4.8	33	0.9721	0.0363	105.88	3.78	26	0.9796
Ionosphere	0.0568	172.21	**6.21**	**22**	0.9507	0.0512	76.61	8.19	29	0.9211	0.0694	54.91	14.16	50	0.8673	0.0652	96.96	11.62	41	0.8980
Australian	0.0416	1868.16	**14.80**	**102**	0.7111	0.0333	**223.33**	14.97	103	0.6803	0.0195	225.98	17.29	119	0.5928	0.0186	236.45	16.71	114	0.6067
German	0.0454	4378.01	**23.80**	**238**	0.8678	0.0378	**544.93**	26.60	266	0.8488	0.0084	569.33	29.60	296	0.8187	0.0097	596.79	28.60	286	0.8245
Average	0.0834	1057.168	**18.12**	**89.1**	0.81	0.066	**171.552**	19.91	99	0.77	0.0477	147.104	25.19	118.3	0.67101	0.049	152.398	23.62	114.3	0.70645

strategies and discussed their attributes with real-world example. We showed that the L_∞ method would choose a single classifier for prediction, the one that maximally maximized the distance of a test point from its hyperplane. The L_1 and L_2 norms were similar to each other and gave predictions using a (weighted) sum of the prediction functions constructed by each SVM function. Finally, in Section 7.4 we gave experimental results that elucidated the methods described in this chapter.

The main benefit of the work proposed can be for imbalanced data sets. In such situations, such as fraud detection, we may have a very large number of examples but only a small number of fraud cases (positive examples). In this case using CV can be costly as we will tend to use up too many of the fraud cases within a large proportion of non-fraud cases and hence have a massive imbalance during training. However, in the models proposed, we can use all of the fraud cases and hence a larger proportion during training. We feel that this is an area that could greatly benefit from the work proposed in this chapter. Also, removing the CV dependency for finding parameters greatly improves training and testing times for the SVM algorithm.

A future research direction would be to use other methods for choosing the classifiers at testing. Perhaps a convex combination of the functions would yield better generalization capabilities. Such a combination of functions could be weighted by a factor in the following way:

$$F(\mathbf{x}) = \text{sgn} \left(\sum_{i=1}^{\ell} \beta_i f_i(\mathbf{x}) \right) \qquad (7.4)$$

$$\text{s.t} \quad \sum_i^{\ell} \beta_i = 1$$

where $\beta_i \in \mathbb{R}$.

Finally, we believe that tighter margin-based bounds would help to improve the selection of the SVM functions at testing. The bound proposed by [20] suggests the L_∞ method we proposed in this chapter. However, from the results section it is clear that this does not always create smaller generalization error than the SVM with CV. Therefore, a future research direction is to use a tighter bounding principle for the margin-based bound of [20], such as a PAC-Bayes analysis (due to [15], and extended to margins by [14]). Therefore, we could use the bounds to indicate which classifiers to use at testing. We believe that a tighter estimate of the bounds would yield improved generalization.

References

1. R. Berk. An introduction to ensemble methods for data analysis. In *eScholarship Repository, University of California. http://repositories.cdlib.org/uclastat/papers/2004072501*, 2004.
2. B.E. Boser, I.M. Guyon, and V. N. Vapnik. A training algorithm for optimal margin classifiers. In *In Fifth Annual Workshop on Computational Learning Theory , ACM.*, pages 144–152, Pittsburgh, 1992. ACM.

3. L. Breiman. Stacked regressions. *Machine Learning*, 24(1):49–64, 1996.
4. O. Chapelle and V. Vapnik. Choosing multiple parameters for support vector machines. *Machine Learning*, 46:131–159, 2002.
5. R. Clemen. Combining forecasts: A review and annotated bibliography. *Journal of Forecasting*, 5:559–583, 1989.
6. N. Cristianini and J. Shawe-Taylor. *An introduction to Support Vector Machines*. Cambridge University Press, Cambridge, UK, 2000.
7. W. Frawley, G. Piatetsky-Shapiro, and C. Matheus. Knowledge discovery in databases: An overview. In *AI Magazine*, pages 213–228, 1992.
8. Y. Freund and R. Schapire. Experiments with a new boosting algorithm. In *In Proceedings of the Thirteenth International Conference on Machine Learning*, pages 148–156, Bari, Italy, 1996.
9. F. Friedrichs and C. Igel. Evolutionary tuning of multiple svm parameters. *Neurocomputing*, 64:107–117, 2005.
10. D. Hand, H. Mannila, and P. Smyth. *An introduction to Support Vector Machines*. MIT Press, Cambridge, MA, 2001.
11. S.S. Keerthi. Efficient tuning of svm hyperparameters using radius/margin bound and iterative algorithms. *IEEE Transactions on Neural Networks*, 13:1225-1229, 2002.
12. S.S. Keerthi, V. Sindhwani, and O. Chapelle. An efficient method for gradient-based adaptation of hyperparameters in svm models. In *In Schölkopf, B.; Platt, J.C.; Hoffman, T. (ed.): Advances in Neural Informations Processing Systems 19*. MIT Press, 2007.
13. S.B. Kotsiantis. Supervised machine learning: A review of classification techniques. *Informatica*, 249, 2007.
14. J. Langford and J. Shawe-Taylor. PAC bayes and margins. In *Advances in Neural Information Processing Systems 15*, Cambridge, MA, 2003. MIT Press.
15. D.A. McAllester. Some pac-bayesian theorems. *Machine Learning*, 37(3):355–363, 1999.
16. D. Opitz. Popular ensemble methods: An empircal study. *Journal of Artificial Intelligence Research*, 11, 1999.
17. S. Özöğür, J. Shawe-Taylor, G.-W. Weber, and Z.B. Ögel. Pattern analysis for the prediction of fungal pro-peptide cleavage sites. *article in press in special issue of Discrete Applied Mathematics on Networks in Computational Biology, doi:10.1016/j.dam.2008.06.043*, 2007.
18. M. Perrone. *Improving Regression Estimation: Averaging Methods for Variance Reduction with Extension to General Convex Measure Optimization*. Ph.D. thesis, Brown University, Providence, RI, 1993.
19. R. Schapire, Y. Freund, P. Bartlett, and W. Lee. A new explanation for the effectiveness of voting methods. In *In Proceedings of the Fourteenth International Conference on Machine Learning*, Nashville, TN, 1997.
20. J. Shawe-Taylor. Classification accuracy based on observed margin. *Algorithmica*, 22: 157–172, 1998.
21. V. N. Vapnik. *Statistical Learning Theory*. John Wiley and Sons, New York, 1998.
22. D. Wolpert. Stacked generalization. *Neural Networks*, 5:241–259, 1992.

Chapter 8
Effects of Oversampling Versus Cost-Sensitive Learning for Bayesian and SVM Classifiers

Alexander Liu, Cheryl Martin, Brian La Cour, and Joydeep Ghosh

Abstract In this chapter, we examine the relationship between cost-sensitive learning and resampling. We first introduce these concepts, including a new resampling method called "generative oversampling," which creates new data points by learning parameters for an assumed probability distribution. We then examine theoretically and empirically the effects of different forms of resampling and their relationship to cost-sensitive learning on different classifiers and different data characteristics. For example, we show that generative oversampling used with linear SVMs provides the best results for a variety of text data sets. In contrast, no significant performance difference is observed for low-dimensional data sets when using Gaussians to model distributions in a naive Bayes classifier. Our theoretical and empirical results in these and other cases support the conclusion that the relative performance of cost-sensitive learning and resampling is dependent on both the classifier and the data characteristics.

8.1 Introduction

Two assumptions of many machine learning algorithms used for classification are that (1) the prior probabilities of all classes in the training and test sets are approximately equal and (2) mistakes on misclassifying points from any class should be penalized equally. In many domains, one or both of these assumptions are violated. Example problems that exhibit both imbalanced class priors and a higher

Alexander Liu · Cheryl Martin · Brian La Cour
Applied Research Labs, University of Texas at Austin, Austin, TX, USA,
e-mail: ayliu@mail.utexas.edu, cmartin@arlut.utexas.edu,
blacour@arlut.utexas.edu

Alexander Liu · Joydeep Ghosh
Department of Electrical and Computer Engineering, University of Texas at Austin, Austin, TX, USA,
e-mail: ayliu@mail.utexas.edu, ghosh@ece.utexas.edu

R. Stahlbock et al. (eds.), *Data Mining*, Annals of Information Systems 8,
DOI 10.1007/978-1-4419-1280-0_8, © Springer Science+Business Media, LLC 2010

misclassification cost for the class with fewer members include detecting cancerous cells, fraud detection [3, 21], keyword extraction [23], oil-spill detection [12], direct marketing [14], information retrieval [13], and many others.

Two different approaches to address these problems are resampling and cost-sensitive learning. Resampling works by either adding members to a class (oversampling) or removing members from a class (undersampling). Resampling is a classifier-agnostic approach and can therefore be used as a preprocessing step requiring no changes to the classification algorithms. In contrast, cost-sensitive learning approaches may modify classifier algorithms to minimize the total or expected cost of misclassification incurred on some test set. Some studies have shown that resampling can be used to perform cost-sensitive learning. However, in this chapter, we contrast resampling methods with cost-sensitive learning approaches that do not use resampling.

We examine the relationship between cost-sensitive learning and two oversampling methods: random oversampling and generative oversampling. We first introduce these concepts, including a new resampling method called "generative oversampling." We compare the performance both theoretically and empirically using a variety of classifiers and data with different characteristics. In particular, we compare low versus high-dimensional data, and we compare Bayesian classifiers and support vector machines, both of which are very popular and widely used machine learning algorithms. Since there is already an abundance of empirical studies comparing resampling techniques and cost-sensitive learning, the emphasis of this chapter is to examine oversampling and its relationship with cost-sensitive learning from a theoretical perspective and to analyze the reasons for differences in empirical performance.

For low-dimensional data, assuming a Gaussian event model for the naive Bayes classifier, we show that random oversampling and generative oversampling theoretically increase the variance of the estimated sample mean compared to learning from the original sample (as done in cost-sensitive naive Bayes). Empirically, using generative oversampling and random oversampling seems to have minimal effect on Gaussian naive Bayes beyond adjusting the learned priors. This result implies that there is no significant advantage for resampling in this context. In contrast, for high-dimensional data, assuming a multinomial event model for the naive Bayes classifier, random oversampling and generative oversampling change not only the estimated priors but also the parameter estimates of the multinomial distribution modeling the resampled class. This conclusion is supported both theoretically and empirically. The theoretical analysis shows that oversampling and cost-sensitive learning are expected to perform differently in this context. Empirically, we demonstrate that oversampling outperforms cost-sensitive learning in terms of producing a better classifier.

Finally, we present parallel empirical results for text classification with linear SVMs. We show empirically that generative oversampling used with linear SVMs provide the best results, beating any other combination of classifier and resampling/cost-sensitive method that we tested on our benchmark text data sets. We then discuss our hypothesis for why generative oversampling in particular works well with SVMs and present experiments to support this hypothesis.

8.2 Resampling

Resampling is a simple, classifier-agnostic method of rebalancing prior probabilities. Resampling has been widely studied, particularly with respect to two-class problems, where the class with the smaller class prior is called the minority class and the class with the larger prior is called the majority class. By convention, the positive class is set as the minority class and the negative class is set as the majority class.

Resampling creates a new training set from the original training set. Many resampling methods have been proposed and studied in the past. Resampling methods can be divided into two categories: oversampling and undersampling. Oversampling methods increase the number of minority class data points, and undersampling methods decrease the number of majority class data points. Some widely used approaches are random oversampling, SMOTE [4], random undersampling, and cost-proportionate rejection sampling [26].

This chapter focuses on two oversampling methods: random oversampling and generative oversampling, introduced below. Some empirical comparisons against SMOTE and random undersampling are also provided for context, but the empirical results in this chapter are primarily used to illustrate analytical results (see [18, 24, 9] for some empirical benchmarks of resampling techniques).

8.2.1 Random Oversampling

Random oversampling increases the number of minority class data points in the training set by randomly replicating existing minority class members. Random oversampling has performed well in empirical studies (e.g., [1]) even when compared to other, more complicated oversampling methods. However, random oversampling has been criticized since it only replicates existing training data and may lead to overfitting [18]. Both SMOTE [4] and generative oversampling address this criticism by creating artificial points instead of replicating existing points.

8.2.2 Generative Oversampling

The set of points in the minority class is characterized by some unknown, true distribution. Ideally, to resample a data set, one could augment the existing data with points drawn from this true distribution until the training set has the desired ratio of positive and negative class points.

Unfortunately, since the true distribution is typically unknown, one cannot usually draw additional points from this original distribution. However, one can attempt to model the distribution that produced the minority class and create new points based on this model. This is the motivation for generative oversampling [15], an oversampling technique that draws additional data points from an assumed distribution. Parameters for this distribution are learned from existing minority class data points. Generative oversampling could be effective in problem domains where there are probability distributions that model the actual data distributions well. Generative oversampling works as follows:

1. a probability distribution is chosen to model the minority class
2. based on the training data, parameters for the probability distribution are learned
3. artificial data points are added to the resampled data set by generating points from the learned probability distribution until the desired number of minority class points in the training set has been reached.

Generative oversampling is simple and straightforward. The idea of creating artificial data points through a probability distribution with learned parameters has been used in other applications (e.g., [19] for creating diverse classifier ensembles, [2] for model compression). Surprisingly, however, it has not previously been used to address imbalanced class priors.

8.3 Cost-Sensitive Learning

In a "typical" classification problem, there is an equal misclassification cost for each class. Cost-sensitive learning approaches, however, account for conditions where there are unequal misclassification costs. The goal of cost-sensitive learning is to minimize the total or expected cost of misclassification incurred on some test set.

Researchers have looked at a number of ways of modeling costs in cost-sensitive learning. Perhaps the most common approach is to define a cost for classifying a point from class $Y = y_j$ as a point from class $Y = y_i$ [6]. In this formulation, a cost matrix $\mathbf{C}$ can be defined where $\mathbf{C}(i, j)$ is the misclassification cost for classifying a point with true class $Y = y_j$ as class $Y = y_i$. Typically, $\mathbf{C}(i, i)$ is set to zero for all i such that correct classifications are not penalized. In this case, the decision rule is modified (as discussed in [6]) to predict the class that minimizes $\sum_j P(Y = y_j | \mathbf{X} = \mathbf{x}) C(i, j)$, where $\mathbf{x}$ is the data point currently being classified. When costs are considered in the two-class imbalanced data set problem, a two-by-two cost matrix can be defined, meaning that all points in the positive class share some misclassification cost and all points in the negative class share some misclassification cost.

Different classifiers can be modified in different ways in order to take costs into account. For example, a Bayesian classifier can be easily modified to predict the class that minimizes $\sum_j P(Y = y_j | \mathbf{X} = \mathbf{x}) C(i, j)$ as in Section 8.5.2. This modification involves shifting the decision boundary by some threshold. In comparison, SVMs can be modified as described in [20], where instead of a single parameter controlling the number of empirical errors versus the size of the margin, a separate parameter for false positives and a second parameter for false negatives are used.

8.4 Related Work

In [6], a direct connection between cost-sensitive learning and resampling is made. The author shows that, theoretically, one can resample points at a specific rate in order to accomplish cost-sensitive learning. In Section 4.1 of [6], the author describes the effect of resampling on Bayesian classifiers. In particular, the author claims that resampling only changes the estimates of the prior probabilities. Thus, an equivalent model can be trained either through resampling or a cost-sensitive Bayesian model. In this chapter, we further examine the assumptions required for this equivalence to hold. The remainder of this section describes additional related work.

Widely used resampling approaches include SMOTE [4], random oversampling, random undersampling, and cost-proportionate rejection sampling [26]. Random undersampling decreases the number of majority class data points by randomly eliminating majority class data points currently in the training set. Like random oversampling, random undersampling has empirically performed well despite its simplicity [27]. A disadvantage of undersampling is that it removes potentially useful information from the training set. For example, since it indiscriminately removes points, it does not consider the difference between points close to the potential decision boundary and points very far from the decision boundary.

A more sophisticated undersampling method with nice theoretical properties is cost-proportionate rejection sampling [26]. Cost-proportionate rejection sampling is based on a theorem that describes how to turn any classifier that reduces the number of misclassification errors into a cost-sensitive classifier. Given that each data point has a misclassification cost, each data point in the training set has a probability of being included in the resampled training set proportional to that point's misclassification cost. We limit the scope of analysis in this chapter to oversampling methods, but a discussion of the relationship between cost-proportionate rejection sampling and cost-sensitive learning is provided in [26].

SMOTE (Synthetic Minority Oversampling TEchnique), an oversampling method, attempts to add information to the training set. Instead of replicating existing data points, "synthetic" minority class members are added to the training set by creating new data points. A new data point is created from an existing data point as follows: find the k nearest neighbors to the existing data point ($k = 5$ in [4] and in this chapter); randomly select one of the k nearest neighbors; the new, synthetic point is a randomly chosen point on the line segment joining the original data point and its randomly chosen neighbor. Empirically, SMOTE has been shown to perform well against random oversampling [4, 1]. Compared to generative oversampling (a parametric oversampling method that adds synthetic points), SMOTE can be considered a non-parametric method.

Empirically, there seems to be no clear winner as to which resampling technique to use or whether cost-sensitive learning outperforms resampling [16, 18, 24, 9]. Many studies have been published, but there is no consensus on which approach is generally superior. Instead, there is ample empirical evidence that the best resampling method to use is dependent on the classifier [9]. Since there is already an abundance of empirical studies comparing resampling techniques and cost-sensitive learning, the emphasis of this paper is to examine oversampling and its relationship with cost-sensitive learning from a theoretical perspective and to analyze the reasons for differences in empirical performance.

8.5 A Theoretical Analysis of Oversampling Versus Cost-Sensitive Learning

In this section, we study the effects of random and generative oversampling on Bayesian classification and the relationship to a cost-sensitive learning approach. We begin with a brief review of Bayesian classification and discuss necessary background. We then examine two cases and analyze differences in the estimates of the parameters that must be calculated in each case to estimate the probability distributions being used to model the naive Bayes likelihood. For the first case, with a Gaussian data model, we show that there is little difference between random oversampling, generative oversampling, and cost-sensitive learning. When multinomial naive Bayes is used, however, there is a significant difference. In this case, we show that the parameter estimates one obtains after either random or generative oversampling differ significantly from the parameter estimates used for cost-sensitive learning.

8.5.1 Bayesian Classification

Suppose one is solving a two-class problem using a Bayesian classifier. Let us denote the estimated conditional probability that some (possibly multi-dimensional) data point $\mathbf{x}$ is from the positive class y_+ given $\mathbf{x}$ as $\hat{P}(Y = y_+|\mathbf{X} = \mathbf{x})$ and the estimated conditional probability that $\mathbf{x}$ is from the negative class y_- given $\mathbf{x}$ as $\hat{P}(Y = y_-|\mathbf{X} = \mathbf{x})$. According to Bayes rule:

$$\hat{P}(Y = y_+|\mathbf{X} = \mathbf{x}) = \frac{\hat{P}(\mathbf{X} = \mathbf{x}|Y = y_+, \hat{\theta}_+)\hat{P}(Y = y_+)}{\hat{P}(\mathbf{X} = \mathbf{x})} \quad (8.1)$$

and

$$\hat{P}(Y = y_-|\mathbf{X} = \mathbf{x}) = \frac{\hat{P}(\mathbf{X} = \mathbf{x}|Y = y_-, \hat{\theta}_-)\hat{P}(Y = y_-)}{\hat{P}(\mathbf{X} = \mathbf{x})} \quad (8.2)$$

where $\hat{P}(Y = y_+)$ and $\hat{P}(Y = y_-)$ are the class priors and $\hat{P}(\mathbf{X} = \mathbf{x}|Y = y_+, \hat{\theta}_+)$, $\hat{P}(\mathbf{X} = \mathbf{x}|Y = y_-, \hat{\theta}_-)$, $\hat{P}(Y = y_+)$, and $\hat{P}(Y = y_-)$ are estimated from the training set. $\hat{\theta}_+$ and $\hat{\theta}_-$ are the estimates of the parameters of the probability distributions being used to model the likelihoods $\hat{P}(\mathbf{X} = \mathbf{x}|Y = y_+, \hat{\theta}_+)$ and $\hat{P}(\mathbf{X} = \mathbf{x}|Y = y_-, \hat{\theta}_-)$. The decision rule is to assign $\mathbf{x}$ to y_+ if the posterior probability $\hat{P}(Y = y_+|\mathbf{X} = \mathbf{x})$ is greater than or equal to $\hat{P}(Y = y_-|\mathbf{X} = \mathbf{x})$. This is equivalent to classifying $\mathbf{x}$ as y_+ if

$$\frac{\hat{P}(Y = y_+|\mathbf{X} = \mathbf{x})}{\hat{P}(Y = y_-|\mathbf{X} = \mathbf{x})} \geq 1. \quad (8.3)$$

8.5.2 Resampling Versus Cost-Sensitive Learning in Bayesian Classifiers

More generally, one can adjust the decision boundary by comparing the ratio of the two posterior probabilities to some constant. That is, one can adjust the decision boundary by assigning $\mathbf{x}$ to y_+ if

$$\frac{\hat{P}(Y = y_+|\mathbf{X} = \mathbf{x})}{\hat{P}(Y = y_-|\mathbf{X} = \mathbf{x})} \geq \alpha, \tag{8.4}$$

where α is used to denote some constant. For example, one may use a particular value of α if one has known misclassification costs [6]. If one were to use a cost-sensitive version of Bayesian classification, α is based on the cost c_+ of misclassifying a positive class point and the cost c_- of misclassifying a negative class point. In this case, $\alpha = c_-/c_+$ based on the cost-sensitive decision rule given in Section 8.3.

One can also adjust the learned decision boundary by resampling (i.e., adding or removing points from the training set) which changes the estimated priors of the classes. Let $\hat{P}^{(rs)}(Y = y_+)$ and $\hat{P}^{(rs)}(Y = y_-)$ denote the estimated class priors after resampling (regardless of what resampling method has been used). Let $\hat{P}^{(rs)}(\mathbf{X} = \mathbf{x}|Y = y_+, \hat{\theta}_+^{(rs)})$ and $\hat{P}^{(rs)}(\mathbf{X} = \mathbf{x}|Y = y_-, \hat{\theta}_-^{(rs)})$ be estimated from the resampled training set. If $\hat{P}(\mathbf{X} = \mathbf{x}|Y = y_+, \hat{\theta}_+) = \hat{P}^{(rs)}(\mathbf{X} = \mathbf{x}|Y = y_+, \hat{\theta}_+^{(rs)})$ and $\hat{P}(\mathbf{X} = \mathbf{x}|Y = y_-, \hat{\theta}_-) = \hat{P}^{(rs)}(\mathbf{X} = \mathbf{x}|Y = y_-, \hat{\theta}_-^{(rs)})$, then the effect of using $\alpha \neq 1$ and the effect of adjusting the priors by resampling can be made exactly equivalent. That is, if resampling only changes the learned priors, then resampling at a specific rate corresponding to $\alpha = c_-/c_+$ is equivalent to cost-sensitive learning. In particular, one can show that if $\hat{P}(\mathbf{X} = \mathbf{x}|Y = y_+, \hat{\theta}_+) = \hat{P}^{(rs)}(\mathbf{X} = \mathbf{x}|Y = y_+, \hat{\theta}_+^{(rs)})$ and $\hat{P}(\mathbf{X} = \mathbf{x}|Y = y_-, \hat{\theta}_-) = \hat{P}^{(rs)}(\mathbf{X} = \mathbf{x}|Y = y_-, \hat{\theta}_-^{(rs)})$, then resampling to adjust the priors to correspond to $\alpha = c_-/c_+$ can be accomplished if

$$\alpha = \frac{c_-}{c_+} = \frac{\hat{P}^{(rs)}(Y = y_-)\hat{P}(Y = y_+)}{\hat{P}^{(rs)}(Y = y_+)\hat{P}(Y = y_-)}. \tag{8.5}$$

However, in practice, this equivalency may not be exact since resampling may do more than simply adjust the class priors. That is, the assumption that $\hat{P}(\mathbf{X} = \mathbf{x}|Y = y_+, \hat{\theta}_+) = \hat{P}^{(rs)}(\mathbf{X} = \mathbf{x}|Y = y_+, \hat{\theta}_+^{(rs)})$ and $\hat{P}(\mathbf{X} = \mathbf{x}|Y = y_-, \hat{\theta}_-) = \hat{P}^{(rs)}(\mathbf{X} = \mathbf{x}|Y = y_-, \hat{\theta}_-^{(rs)})$ may be invalid because the estimated parameters with and without resampling may change.

In the remainder of this section, we will theoretically examine the effect of two resampling techniques (random oversampling and generative oversampling) on probability estimation and Bayesian learning. In particular, we will examine the difference between learning from the resampled set and learning from the original training set when Gaussian and multinomial distributions are chosen to model the resampled class.

We will assume without loss of generality that the positive class is being over-sampled. In this case, since points are neither added nor removed from the negative class, $\hat{P}(\mathbf{X} = \mathbf{x} | Y = y_-, \hat{\theta}_-) = \hat{P}^{(rs)}(\mathbf{X} = \mathbf{x} | Y = y_-, \hat{\theta}_-^{(rs)})$. Thus, we will examine how $\hat{\theta}_+$ may differ from $\hat{\theta}_+^{(rs)}$ due to oversampling. Since we will only be discussing the positive class, we will omit the $+$ subscripts when it is obvious that we are referring to parameters estimated on the positive class.

In addition, we will also use the notation $\hat{\theta}^{(r)}$ and $\hat{\theta}^{(g)}$ to refer to the parameters estimated after random oversampling and generative oversampling, respectively, when such a distinction needs to be made, while $\hat{\theta}^{(rs)}$ will continue to refer to parameter estimates after either resampling technique has been used, and $\hat{\theta}$ will continue to refer to parameters estimated from the original training set when no resampling has occurred.

8.5.3 Effect of Oversampling on Gaussian Naive Bayes

In this section, we examine the effect of oversampling when $\hat{P}(\mathbf{X} = \mathbf{x} | Y = y_+, \hat{\theta}_+)$ and $\hat{P}^{(rs)}(\mathbf{X} = \mathbf{x} | Y = y_+, \hat{\theta}_+^{(rs)})$ are modeled by Gaussian distributions. In Gaussian naive Bayes, each feature is modeled by an independent Gaussian distribution. Thus, $\hat{P}(\mathbf{X} = \mathbf{x} | Y = y_+, \hat{\theta}_+) = \prod_{i=1}^{d} N(x_i | \mu_{+,i}, \sigma_{+,i})$ where d is the number of dimensions, $\mu_{+,i}$ and $\sigma_{+,i}$ are the mean and standard deviation estimated for the ith dimension of the positive class, and $N(x_i | \mu_{+,i}, \sigma_{+,i})$ is the probability that a normal distribution with parameters $\mu_{+,i}$ and $\sigma_{+,i}$ generated x_i. Since we are only discussing over-sampling the positive class, we will drop the $+$ subscripts and simply refer to the parameters as $\hat{\theta}$, μ_i, and σ_i.

For the sake of simplicity, we will limit our discussion to one-dimensional Gaussian distributions. However, since the parameters of each dimension are estimated independently in a naive Bayes classifier, our analysis can be extended to multiple dimensions if the features are indeed independent. Analysis when features are correlated will be left to future work. In the one-dimensional case, $\hat{\theta}$ corresponds to a single sample mean and sample standard deviation estimated from the positive class points in the original training set, while $\hat{\theta}^{(rs)}$ corresponds to the sample mean and sample standard deviation estimated after resampling.

We will first examine the theoretical effect of oversampling on estimating $\hat{\theta}$, $\hat{\theta}^{(r)}$, and $\hat{\theta}^{(g)}$. For the sake of brevity, we limit our discussion of these parameter estimates to the expected value and variance of the sample mean. In particular, we show that the expected value of the sample mean is always the same regardless of whether no resampling, random oversampling, or generative oversampling is applied.

Let the set of n points in the positive class be denoted as $X_1, \ldots, X_n$. These points are an i.i.d. set of random variables from a normal distribution with true mean μ and true variance σ^2. Both μ and σ^2 are unknown and must be estimated as parameters $\hat{\theta}$.

The sample mean estimated from the original training set will be denoted as $\bar{X}$ while the sample mean for the training set created after resampling will be denoted as either $\bar{X}_*^{(r)}$ for random oversampling or $\bar{X}_*^{(g)}$ for generative oversampling. Note that, in the Appendix, we make a differentiation between the sample mean estimated only on the newly resampled points (denoted as $\bar{X}^{(r)}$ for random over-sampling) versus the sample mean estimated for a training set comprised of both the resampled points and the original points (denoted as $\bar{X}_*^{(r)}$). Here, however, we will simply present results for the "pooled" training set consisting of both the resampled points and the original points.

Consider the sample mean of the original sample, $\bar{X}$. The expected value and variance of the sample mean are as follows:

$$E[\bar{X}] = \mu \tag{8.6}$$

$$\text{Var}[\bar{X}] = \frac{\sigma^2}{n}. \tag{8.7}$$

In the next sections, we will find the expected value and variance of the sample mean of the points generated by random oversampling and generative oversampling. Derivations of these equations can be found in the Appendix.

8.5.3.1 Random Oversampling

Consider a random sample $X_1^{(r)}, \ldots, X_m^{(r)}$ produced through random oversampling. Thus, each $X_j^{(r)} = X_{K_j}$, where $K_1, \ldots, K_m$ are i.i.d. uniformly distributed random variables over the discrete set $\{1, \ldots, n\}$.

We now seek the mean and variance of $\bar{X}_*^{(r)}$.

For the mean, we have

$$E[\bar{X}_*^{(r)}] = \mu \tag{8.8}$$

and

$$\text{Var}[\bar{X}_*^{(r)}] = \left[1 + \frac{m(n-1)}{(n+m)^2} \right] \frac{\sigma^2}{n}. \tag{8.9}$$

Thus, the expected value of the sample mean of the pooled training set is equal to the expected value of the sample mean without resampling. However, the variance of the estimated sample mean of the training set after resampling is greater.

8.5.3.2 Generative Oversampling

Now consider points $X_1^{(g)}, \ldots, X_m^{(g)}$ created via generative oversampling. These points are of the form

$$\mathbf{X}_j^{(g)} = \bar{\mathbf{X}} + s\mathbf{Z}_j, \tag{8.10}$$

where $X_j^{(g)}$ is the jth point created by generative oversampling, s is the original estimated sample standard deviation, and $Z_1, \ldots, Z_m$ are i.i.d. $N(0,1)$ and independent of $X_1, \ldots, X_n$ as well.

The expected value of the mean $\bar{X}_*^{(g)}$ is

$$E[\bar{X}_*^{(g)}] = \mu, \tag{8.11}$$

while the variance of the sample mean is

$$\text{Var}[\bar{X}_*^{(g)}] = \left[1 + \frac{mn}{(n+m)^2}\right] \frac{\sigma^2}{n}. \tag{8.12}$$

Thus, like random oversampling, the sample mean estimated from the resampled points created via generative oversampling has, on average, the same value as the sample mean estimated from the original points, but with greater variance.

8.5.3.3 Comparison to Cost-Sensitive Learning

A cost-sensitive naive Bayes classifier uses the parameter estimates $\hat{\theta}$ from the original set of points. Thus, in expectation, the estimated mean for a Gaussian naive Bayes classifier will be the same, regardless of whether random oversampling, generative oversampling, or cost-sensitive learning is used. When one resamples, one incurs additional overhead in terms of time required to create additional samples, memory needed to store the additional samples, and additional time required to train on the resampled points. Cost-sensitive naive Bayes is therefore preferable over resampling when a Gaussian distribution is assumed. We will support this claim empirically in Section 8.6.2.

8.5.4 Effects of Oversampling for Multinomial Naive Bayes

In multinomial naive Bayes (see [17] for an introduction to multinomial naive Bayes), there is a set of d possible features, and the probability that each feature will occur needs to be estimated. For example, in the case of text classification, each feature is a word in the vocabulary, and one needs to estimate the probability that a particular word will occur. Thus, the parameter vector θ for a multinomial distribution is a d-dimensional vector, where θ_k is the probability that the kth word in the vocabulary will occur.

Let F_i denote the number of times the ith word occurs in the positive class in the training set, and let $F_i^{(r)}$ represent the number of times that a word occurs in only the randomly oversampled points (we will use $F_i^{(g)}$ when discussing generative oversampling). In addition, let n represent the number of words that occur in the positive class in the training set, and let m represent the number of words that occur in the resampled points.

For the case where there are no resampled points in the training set, the maximum likelihood estimator for the probability the kth word will occur is $\hat{\theta}_k = F_k/n$.

Typically, the maximum likelihood estimator is not used because Laplace smoothing is often introduced (a standard practice when using multinomials for problems like text mining). With Laplace smoothing, the estimator becomes

$$\tilde{\theta}_k = \frac{F_k + 1}{\sum_{k'=1}^{d}(F_{k'} + 1)} = \frac{n\hat{\theta}_k + 1}{n + d}. \tag{8.13}$$

Note that we will use the notation $\hat{\theta}$ to describe parameter estimates when Laplace smoothing has not been used and $\tilde{\theta}$ to indicate parameter estimates when Laplace smoothing has been used.

After random oversampling, we find that the parameter estimates learned from the resampled points and original training set if Laplace smoothing is used are:

$$E\left[\tilde{\theta}_k^{(r)}\right] = \frac{(n+m)\theta_k + 1}{n + m + d} \tag{8.14}$$

and

$$\text{Var}\left[\tilde{\theta}_k^{(r)}\right] = \left[1 + \frac{m(n - 2d - 1) - d(2n + d)}{(n + m + d)^2}\right]\frac{\theta_k(1 - \theta_k)}{n}. \tag{8.15}$$

Thus, provided $m < d(2n+d)/(n-2d-1)$, the variance of the pooled, smoothed estimates will be smaller than that of the original sample.

In generative oversampling, one generates points based on an assumed probability distribution. If one uses a multinomial model to generate the points, the parameters one uses in the initial estimation can be either $\hat{\theta}$ (i.e., without Laplace smoothing) or $\tilde{\theta}$ (i.e., with Laplace smoothing). When using generative oversampling with multinomial naive Bayes, there are two places where Laplace smoothing can possibly be used: when performing the initial parameter estimates for generative oversampling and when performing the parameter estimates for multinomial naive Bayes. In our experiments, we always use Laplace smoothing when estimating parameters for multinomial naive Bayes. The question of whether to use Laplace smoothing will therefore always refer to the initial parameter estimates in generative oversampling.

As shown in the Appendix, if one uses $\hat{\theta}$ for initial parameter estimates in generative oversampling, then $E\left[\hat{\theta}_k^{(g)}\right] = E\left[\tilde{\theta}_k^{(r)}\right]$ and $\text{Var}\left[\hat{\theta}_k^{(g)}\right] = \text{Var}\left[\tilde{\theta}_k^{(r)}\right]$. If one performs Laplace smoothing and uses $\tilde{\theta}$ in the initial parameter estimates used for generative oversampling, however, the parameter vector $\tilde{\theta}^{(g)}$ estimated after generative oversampling will be different from the parameter vector $\tilde{\theta}^{(r)}$ estimated after random oversampling or the parameter vector $\tilde{\theta}$ used in cost-sensitive learning. Our empirical results show that the relative performance of using either $\hat{\theta}$ or $\tilde{\theta}$ when estimating the initial parameters used in generative oversampling depends on which classifier is used.

Since a cost-sensitive naive Bayes classifier uses the parameter estimates $\tilde{\theta}$ from the original set of points, it is clear that there will be a difference, in expectation, between the parameters estimated via random oversampling, generative oversampling, and cost-sensitive learning. We will see empirically that resampling produces better classifiers than cost-sensitive learning. Thus, even though resampling incurs additional overhead in terms of time and memory, the improvement in classification may justify this additional effort.

8.6 Empirical Comparison of Resampling and Cost-Sensitive Learning

In this section, we will provide empirical support for our analysis in Section 8.5. We will show that, as predicted, there is minimal empirical difference between random oversampling, generative oversampling, and cost-sensitive learning when Gaussian naive Bayes is used as the classifier. In contrast, when dealing with high-dimensional text data sets where a multinomial model is more suitable, there is a difference between random oversampling, generative oversampling, and cost-sensitive learning. The magnitude of the difference with regard to generative oversampling is related to whether Laplace smoothing is used to build the model used to generate artificial points.

While the primary goal of this chapter is not to perform extensive empirical benchmarks of all possible resampling methods, for the sake of comparison, we have also included some common resampling techniques (namely SMOTE and random undersampling).

8.6.1 Explaining Empirical Differences Between Resampling and Cost-Sensitive Learning

Our experiments compare the results of classifiers learned after resampling against a cost-sensitive classifier that estimates its parameters from the original training set. In this section, we will describe why comparing naive Bayes after resampling with cost-sensitive naive Bayes can answer the question of whether the benefits of resampling are limited to merely evening out the imbalance of the class priors, or if additional effects (from changing the estimates of the likelihoods) are responsible.

Oversampling the positive class has two possible effects on a Bayesian classifier: (1) it changes the estimated priors $\hat{P}(Y = y_+)$ and $\hat{P}(Y = y_-)$ to $\hat{P}^{(rs)}(Y = y_+)$ and $\hat{P}^{(rs)}(Y = y_-)$ and (2) it may or may not change the parameter estimate $\hat{P}^{(rs)}(\mathbf{X} = \mathbf{x}|Y = y_+, \hat{\theta}_+^{(rs)})$ such that $\hat{P}(\mathbf{X} = \mathbf{x}|Y = y_+, \hat{\theta}_+) \neq \hat{P}^{(rs)}(\mathbf{X} = \mathbf{x}|Y = y_+, \hat{\theta}_+^{(rs)})$.

The decision rule of the Bayesian classifier after resampling is to assign a point x to the positive class if

$$\frac{\hat{P}^{(rs)}(Y = y_+|\mathbf{X} = \mathbf{x})}{\hat{P}^{(rs)}(Y = y_-|\mathbf{X} = \mathbf{x})} = \frac{\hat{P}^{(rs)}(\mathbf{X} = \mathbf{x}|Y = y_+, \hat{\theta}_+^{(rs)})\hat{P}^{(rs)}(Y = y_+)}{\hat{P}^{(rs)}(\mathbf{X} = \mathbf{x}|Y = y_-, \hat{\theta}_-^{(rs)})\hat{P}^{(rs)}(Y = y_-)} \geq 1. \quad (8.16)$$

As described in the previous section, if $\hat{P}(\mathbf{X} = \mathbf{x}|Y = y_+, \hat{\theta}_+) = \hat{P}^{(rs)}(\mathbf{X} = \mathbf{x}|Y = y_+, \hat{\theta}_+^{(rs)})$ and $\hat{P}(\mathbf{X} = \mathbf{x}|Y = y_-, \hat{\theta}_-) = \hat{P}^{(rs)}(\mathbf{X} = \mathbf{x}|Y = y_-, \hat{\theta}_-^{(rs)})$, then the only effect of resampling is to change the learned class priors. Under this (possibly incorrect) assumption,

$$\frac{\hat{P}^{(rs)}(\mathbf{X} = \mathbf{x}|Y = y_+, \hat{\theta}_+^{(rs)})\hat{P}^{(rs)}(Y = y_+)}{\hat{P}^{(rs)}(\mathbf{X} = \mathbf{x}|Y = y_-, \hat{\theta}_-^{(rs)})\hat{P}^{(rs)}(Y = y_-)} = \frac{\hat{P}(\mathbf{X} = \mathbf{x}|Y = y_+, \hat{\theta}_+)\hat{P}^{(rs)}(Y = y_+)}{\hat{P}(\mathbf{X} = \mathbf{x}|Y = y_-, \hat{\theta}_-)\hat{P}^{(rs)}(Y = y_-)}.$$
$$(8.17)$$

This is the same as adjusting the decision rule learned on the original training set by setting $\alpha = \frac{\hat{P}^{(rs)}(Y=y_-)\hat{P}(Y=y_+)}{\hat{P}^{(rs)}(Y=y_+)\hat{P}(Y=y_-)}$ and assigning $\mathbf{x}$ to the positive class if $\frac{\hat{P}(Y=y_+|\mathbf{X}=\mathbf{x})}{\hat{P}(Y=y_-|\mathbf{X}=\mathbf{x})} \geq \alpha$. One can do this by training a cost-sensitive naive Bayes classifier with $\alpha = \frac{c_-}{c_+} = \frac{\hat{P}^{(rs)}(Y=y_-)\hat{P}(Y=y_+)}{\hat{P}^{(rs)}(Y=y_+)\hat{P}(Y=y_-)}$.

Thus, one can duplicate the beneficial effect of evening out the class priors via resampling if $\hat{P}(\mathbf{X} = \mathbf{x}|Y = y_+, \hat{\theta}_+) = \hat{P}^{(rs)}(\mathbf{X} = \mathbf{x}|Y = y_+, \hat{\theta}_+^{(rs)})$ and $\hat{P}(\mathbf{X} = \mathbf{x}|Y = y_-, \hat{\theta}_-) = \hat{P}^{(rs)}(\mathbf{X} = \mathbf{x}|Y = y_-, \hat{\theta}_-^{(rs)})$ by using a cost-sensitive naive Bayes classifier where $\frac{c_-}{c_+} = \frac{\hat{P}^{(rs)}(Y=y_-)\hat{P}(Y=y_+)}{\hat{P}^{(rs)}(Y=y_+)\hat{P}(Y=y_-)}$. Any empirical difference observed between a naive Bayes classifier after resampling and a cost-sensitive naive Bayes classifier with the appropriate values of c_- and c_+ is therefore attributable to the fact it is incorrect to assume that the estimated parameters modeling our probability distributions are equal before and after resampling.

Therefore, we can examine whether $\hat{P}(\mathbf{X} = \mathbf{x}|Y = y_+, \hat{\theta}_+) = \hat{P}^{(rs)}(\mathbf{X} = \mathbf{x}|Y = y_+, \hat{\theta}_+^{(rs)})$ by comparing a naive Bayes classifier that uses resampling and an equivalent cost-sensitive naive Bayes classifier where $\frac{c_-}{c_+} = \frac{\hat{P}^{(rs)}(Y=y_-)\hat{P}(Y=y_+)}{\hat{P}^{(rs)}(Y=y_+)\hat{P}(Y=y_-)}$. Such a comparison allows us to isolate and study only the part of resampling that could cause $\hat{P}(\mathbf{X} = \mathbf{x}|Y = y_+, \hat{\theta}_+) \neq \hat{P}^{(rs)}(\mathbf{X} = \mathbf{x}|Y = y_+, \hat{\theta}_+^{(rs)})$. We perform this comparison in Sections 8.6.2 and 8.6.3.

8.6.2 Naive Bayes Comparisons on Low-Dimensional Gaussian Data

In this section, we will provide some simple examples of classifying low-dimensional data with a Gaussian naive Bayes classifier to support the theory presented in Section 8.5.3.

We use f1-measure, a natural evaluation metric in information retrieval for high-dimensional data sets, as our evaluation metric. F1-measure is the harmonic mean of two other evaluation metrics, precision and recall. Precision $= n_{tp}/(n_{tp} + n_{fp})$ and recall $= n_{tp}/(n_{tp} + n_{fn})$, where n_{tp} is the number of true positives, n_{fp} is the number of false positives, and n_{fn} is the number of false negatives. F1-measure ranges from 0 to 1, with 1 being the best possible f1-measure achievable on the test set. F1-measure has some additional advantages over traditional ROC and AUC metrics when interpreting our experiments, as discussed in Section 8.6.2.2. In order to keep our results consistent, we use f1-measure for both the low-dimensional and the high-dimensional experiments.

To illustrate that there is minimal benefit in using either random oversampling, generative oversampling, or cost-sensitive learning when a Gaussian naive Bayes classifier is used, we present two sets of experiments.

8.6.2.1 Gaussian Naive Bayes on Artificial, Low-Dimensional Data

The first set of experiments utilizes an artificially generated data set consisting of two classes drawn from two one-dimensional Gaussians with true variance equal to 1. The location of the means of the two Gaussians is controlled such that a specific optimal Bayes error rate could potentially be achieved if the points in the test data were sampled equally from the two distributions (the Bayes error rate is defined as the lowest theoretical error rate achievable if we knew the true values of the parameters in our mixture of Gaussians [5]). We vary the optimal Bayes error rate (denoted as BER in Fig. 8.1) between 0.1 and 0.3. In order to introduce imbalance, 90% of the training set consists of points in the negative class and 10% of the training set

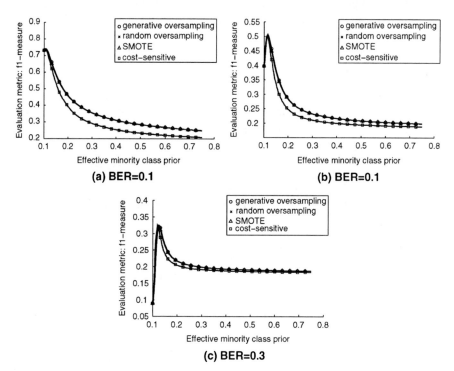

Fig. 8.1 Results on artificial Gaussian data for naive Bayes

consists of points in the positive class. In our experiments, we varied the amount of training data between 100 and 300 points, but found that the results were consistent regardless of how much training data was used; here, we present results where there are 100 training points. The test set consists of 1000 points with the same priors as the training set. We average our results over 100 trials, where each trial includes creating a completely new data set.

When resampling, one has control over the value of the estimated priors $\hat{P}^{(rs)}(Y = y_+)$ and $\hat{P}^{(rs)}(Y = y_-)$. Since $\hat{P}^{(rs)}(Y = y_+) + \hat{P}^{(rs)}(Y = y_-) = 1$, controlling $\hat{P}^{(rs)}(Y = y_+)$ is sufficient to control both $\hat{P}^{(rs)}(Y = y_+)$ and $\hat{P}^{(rs)}(Y = y_-)$. In our experiments, we vary $\hat{P}^{(rs)}(Y = y_+)$ between the prior estimated without resampling $\hat{P}(Y = y_+) = 10\%$ and a maximum possible value of 75%. Note that when $\hat{P}^{(rs)}(Y = y_+) = \hat{P}(Y = y_+)$, no actual resampling has been performed (this corresponds to the left-most point on each graph plotting f1-measure where all performance curves converge). When running cost-sensitive learning, we control the misclassification costs c_- and c_+. In order to directly compare cost-sensitive learning against results for resampling, we use the term "effective minority class prior" in our graphs. That is, a particular value of the effective minority class prior means that (1) when resampling is used, the resampled prior $\hat{P}^{(rs)}(Y = y_+)$ is equal to the effective minority class prior and (2) when cost-sensitive learning is used, $\frac{c_-}{c_+} = \frac{\hat{P}^{(rs)}(Y=y_-)\hat{P}(Y=y_+)}{\hat{P}^{(rs)}(Y=y_+)\hat{P}(Y=y_-)}$, where $\hat{P}^{(rs)}(Y = y_+)$ is equal to the effective minority class prior. In interpreting our results, we simply look at our results on the test set across a range of resampling rates. Choosing a resampling rate that yields optimal performance is an unsolved problem. That is, there is no closed-form solution for determining the appropriate effective minority class prior to maximize a particular evaluation metric, so this becomes a model selection problem.

In Fig. 8.1, we plot the f1-measure versus different effective minority class priors for Gaussian naive Bayes after random oversampling, Gaussian naive Bayes after generative oversampling, Gaussian naive Bayes after SMOTE, and cost-sensitive naive Bayes. Interestingly, regardless of the separability of the two Gaussian distributions, the curves have similar characteristics. In particular, the best possible f1-measure obtained by each is almost exactly the same. That is, in practice, random oversampling, generative oversampling, and SMOTE seem to have little effect on Gaussian naive Bayes that cannot be accomplished via cost-sensitive learning. This supports the theory presented in Section 8.5.3, which shows that, in expectation, the value of the sample mean estimated after random oversampling, generative oversampling, or cost-sensitive learning (which uses the original set of points) is equivalent.

8.6.2.2 A Note on ROC and AUC

Figure 8.2 contains ROC curves and values for AUC for the same set of experiments presented in Fig. 8.1 where BER = 0.3. When evaluating results on imbalanced data sets, ROC and AUC are often very useful. However, ROC and AUC can hide the

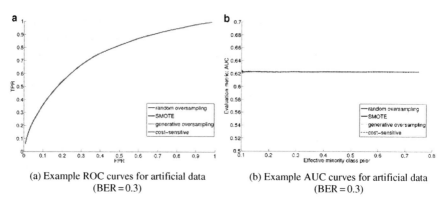

(a) Example ROC curves for artificial data
(BER = 0.3)

(b) Example AUC curves for artificial data
(BER = 0.3)

Fig. 8.2 Example results using AUC and ROC as evaluation metrics; note that using these evaluation metrics, it is difficult to determine whether the effective minority class prior has any effect on how well the classifier performs

effect that different resampling rates have on the classifier. To fully examine the effect of resampling rate using ROC curves would require an unwieldy number of curves per graph, since each resampling rate for each method being used would require a separate ROC curve. Figure 8.2, which seems to contain only a single ROC curve, is an exception since, as theory predicts, each ROC curve produced by each resampling method and cost-sensitive Gaussian naive Bayes is essentially the same (thus retraced multiple times in the figure). Another problem is that to create a ROC curve, one uses several different thresholds for each point on the curve; cost-sensitive naive Bayes also uses different thresholds to produce results using different costs. Thus, the differences in performance across different costs are hidden on a single ROC curve for this type of classifier. AUC, which is based on ROC, aggregates results over several thresholds. Thus, the AUC for cost-sensitive naive Bayes will always be about the same (sans statistical variation in the training/test sets) regardless of the cost used and is not particularly interesting. In fact, this is exactly what we see in Fig. 8.2, where the AUC remains essentially constant regardless of the effective minority class prior. The results in Fig. 8.2 can be extremely misleading, because it may lead one to conclude that different cost parameters or resampling rates have no effect on how well a classifier performs. In comparison, using f1-measure to plot unintegrated results corresponding to specific resampling rates and costs clearly shows the importance of choosing an appropriate resampling rate or cost.

8.6.2.3 Gaussian Naive Bayes on Real, Low-Dimensional Data

To complement the experiments on Gaussian naive Bayes on artificial data, we also present some results on low-dimensional data sets from the UCI data set repository to verify generalizations of the findings on real data. We have selected six data sets: pima indian, wine, breast cancer Wisconsin diagnostic (wdbc), breast cancer Wisconsin prognostic (wpbc),[1] page-blocks (using "non-text" versus "rest" as our binary class problem), and ionosphere. The features of all of the dimensions for wine and breast cancer Wisconsin diagnostic (wdbc) pass the Kolmogorov–Smirnov test for normality for a p value of 0.05; most of the features of the breast cancer Wisconsin prognostic (wpbc) data set also pass the Kolmogorov–Smirnov test for normality. In contrast, the majority of the features of the remaining data sets did not. We use both features that passed and did not pass the Kolmogorov–Smirnov test in our experiments, so the assumption that Gaussians can be used to model the various data sets does not hold very well in some cases.

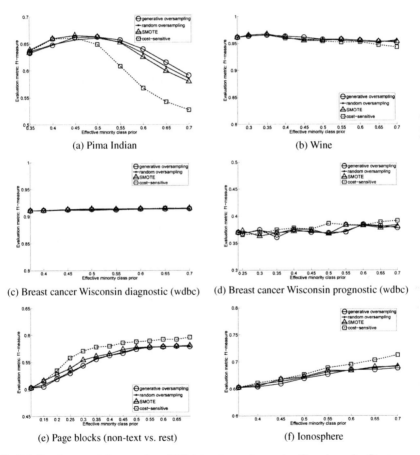

Fig. 8.3 Results on real data sets from UCI data set repository using Gaussian naive Bayes

---

[1] Note that the two breast cancer data sets are separate data sets in the UCI data set repository and not the same data set used for two different tasks in our experiments.

Our results are shown in Fig. 8.3. The results of these experiments support the same conclusion as before: there is little advantage to using either random oversampling, generative oversampling, or cost-sensitive learning when using Gaussian naive Bayes. For the sake of comparison, SMOTE is included in these datasets as well. While SMOTE has been shown to work well with other classifiers [4], it performs similarly to the other techniques when using Gaussian naive Bayes. Thus, given that it is much easier to use cost-sensitive learning instead of resampling and that there is no empirical advantage of using resampling, it is preferable to simply use a cost-sensitive version of naive Bayes when Gaussian distributions are used to model the data.

8.6.3 Multinomial Naive Bayes

In this section and the next, we examine the empirical effect of resampling on high-dimensional data. In particular, we will use text classification as an example domain. We first examine the effect of random and generative oversampling on multinomial naive Bayes [17], a classifier often used for text classification. Our experiments on text classification are more extensive than the experiments on low-dimensional data for two primary reasons: (1) additional results more fully illustrate the empirical differences between the different resampling methods and (2) empirical studies comparing resampling methods and/or cost-sensitive learning typically focus on low-dimensional data, so there are less published results available for high-dimensional data.

We compare the effect of random oversampling, generative oversampling, and cost-sensitive learning on multinomial naive Bayes using six text data sets drawn from different sources. The text data sets come from several past studies on information retrieval including TREC (http://trec.nist.gov), OHSUMED [8], and WebAce [7]. The data sets from TREC are all newspaper stories from either the LA Times (la12 data set) or the San Jose Mercury (hitech, reviews, sports data sets) classified into different topics. The ohscal data set contains text related to medicine, while the k1b data set contains documents from the Yahoo! subject hierarchy. All six of these data sets were included in the CLUTO toolkit [11].[2]

All text data sets were converted into a standard bag of words model. In addition, we used TFIDF weighting and normalized each document vector with the L2 norm after resampling[3] (see [22] for a discussion of TFIDF weighting and other common preprocessing tasks in text classification). Finally, we created several two-class problems based on our text data sets. For each data set, we chose the smallest class

---

[2] We use the data sets available at http://www.ideal.ece.utexas.edu/data/docdata.tar.gz, which have some additional preprocessing as described in [28].

[3] Note that the order in which one applies TFIDF weighting, normalization, and resampling appears to be important in terms of generating good classification performance; further analysis is required to determine the reasons for this.

in each data set as the minority class and aggregated all other classes to form the majority class.

For each data set, we create ten different training and test splits by randomly selecting 50% of the data using stratified sampling as the training set and the rest as the test set. We again use f1-measure as our evaluation metric and results are averaged over the ten training/test splits.

As in the experiments with Gaussian data, we control the effective minority class prior either through resampling or through cost-sensitive learning. Again, we vary the effective minority class prior between the prior estimated without resampling $\hat{P}(Y = y_+)$ and 70% (note that the prior before resampling varies for each data set but is always less than 10%).

Details about these data sets are given in Table 8.1, including the number of minority class points in the data and the "natural" value of the minority class prior in the data set.

Table 8.1 Data set characteristics

Data set	Num min class pts	Min class prior
hitech	116	0.0504
k1b	60	0.0256
la12	521	0.0830
ohscal	709	0.0635
reviews	137	0.0337
sports	122	0.0142

The results of our experiments on multinomial naive Bayes are shown in Fig. 8.4. The results indicate that resampling improves the resulting f1-measure when compared to classification without resampling (i.e., the left-most point in each graph). The results also indicate that there is a significant difference between the best possible f1-measure obtained from resampling (across all resampling rates) and the best possible f1-measure obtained through cost-sensitive learning (across all tested costs). In particular, the best possible f1-measure obtained after oversampling (regardless of which oversampling method we tested) is always better than cost-sensitive learning. In some cases, this value is much higher than the best possible f1-measure obtained via cost-sensitive learning.

Random oversampling, generative oversampling,[4] and SMOTE produce comparable f1-measure curves. These results indicate that, in practice, one can produce a classifier with better performance by oversampling instead of adjusting the decision boundary using cost-sensitive learning.

---

[4] Here, we use generative oversampling as described in the Appendix where $\hat{\theta}$ (i.e., without smoothing during parameter estimation) is used instead of $\tilde{\theta}$ to generate points; we will see in Section 8.6.4 why this distinction is important.

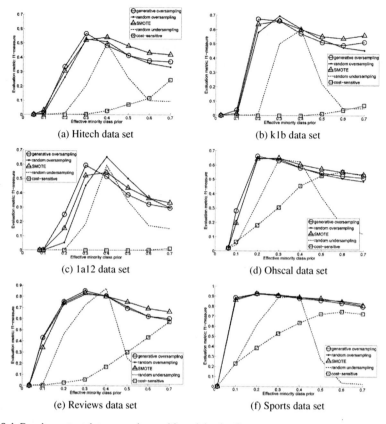

Fig. 8.4 Results on text data sets using multinomial naive Bayes

8.6.4 SVMs

In this section, we empirically test the effect of resampling on linear SVMs in the domain of text classification and compare performance against cost-sensitive SVMs.[5] In our experiments, we use an SVM with a linear kernel, which has been shown to work well in text classification [25]. Specifically, we use the SVM light implementation by Joachims [10]. Note that, for SVMs, there is an additional parameter used to control the trade-off between the margin and training errors. In SVM light this is

_____

[5] All SVM results presented use generative oversampling for multinomials with smoothing during parameter estimation ($\tilde{\theta}$) except for Fig. 8.7, where we present results for generative oversampling with both $\tilde{\theta}$ and $\hat{\theta}$. Generative oversampling for multinomials without smoothing (i.e., when $\hat{\theta}$ is used when estimating the parameters for generative oversampling) performs poorly for SVMs as shown at the end of this section.

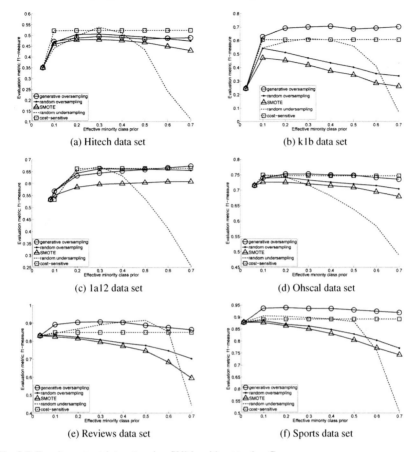

Fig. 8.5 Results on text data sets using SVMs without tuning C

the parameter C.[6] Adjusting C can affect the location of the separating hyperplane. In SVM light, one can either specify C or use a default value of C estimated from the data. We run experiments using both settings and the results are presented separately in Fig. 8.5 and Fig. 8.6.

Figure 8.5 plots our results comparing generative oversampling, random oversampling, SMOTE, random undersampling, and cost-sensitive SVMs on each of the six data sets when all default parameters for linear SVMs are used. The plots show f1-measure as a function of effective minority class prior, as presented for the naive Bayes results.

Figure 8.6 shows the results when the trade-off C between margin size and number of training errors is tuned via a validation set. When specifying C, we perform a search for the best value of C by using a validation set; we further split each training set into a smaller training set (70% of the initial training set) and validation set (the

---

[6] Not to be confused with a cost matrix **C**.

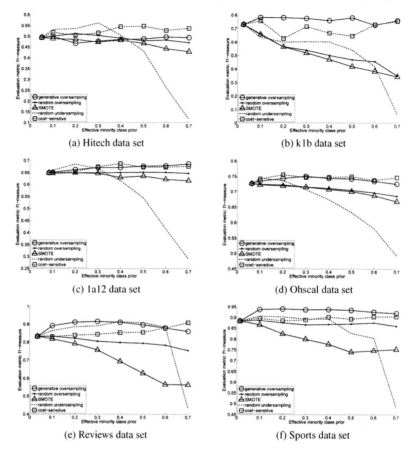

Fig. 8.6 Results on text data sets using SVMs while tuning C

remaining 30% of the training set) and search for the best value of C between 2^{-6} and 2^6. Note that this tuning is done separately for every experiment (i.e., once for every combination of data set, training set split, resampled minority class prior, and resampling method).

In both sets of experiments, generative oversampling performs well compared to random oversampling, SMOTE, random undersampling, and cost-sensitive SVMs. Generative resampling also shows robustness to the choice of the minority class prior (i.e., its performance does not vary significantly across resampling rates), which may be otherwise difficult to optimize, in practice. In contrast, SMOTE and random oversampling often cannot improve over using the natural prior (i.e., no resampling), particularly when C is tuned. Results with random undersampling depend heavily on choosing the minority class training prior. In all six of our data sets, there is a very clear degradation in f1-measure for random undersampling when the minority class training prior is either too low or too high regardless of whether C is tuned. Cost-sensitive SVMs perform quite well and are as robust to choice of cost parameter as generative oversampling is to choosing how much to resample, but on average, generative oversampling produces a higher f1-measure.

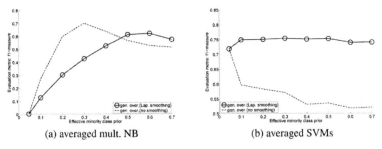

(a) averaged mult. NB   (b) averaged SVMs

Fig. 8.7 Average results on text data sets for multinomial naive Bayes and SVMs (where C is tuned) after running generative oversampling with different levels of smoothing

If one runs SVMs with resampled points created via generative oversampling with Laplace smoothing during parameter estimation (i.e., if one uses $\tilde{\theta}$), then generative oversampling potentially increases the size of the convex hull surrounding the minority class by producing artificial data points that occur both inside and outside of the original convex hull inscribing the minority class points in the training set. Figure 8.7 contains averaged results where we compare generative oversampling on SVMs using either $\tilde{\theta}$ or $\hat{\theta}$ (i.e., generative oversampling with and without Laplace smoothing). As one can see, the results when using $\hat{\theta}$ are much worse. When $\hat{\theta}$ is used, generative oversampling no longer effectively creates points outside of the original convex hull. Thus, generative oversampling with $\tilde{\theta}$ complements the SVMs well by increasing the size of the minority class convex hull.

Note that, in our multinomial naive Bayes experiments, we found that using $\hat{\theta}$ was always empirically superior to using $\tilde{\theta}$, while for SVMs, we found that using $\tilde{\theta}$ resulted in better empirical performance (see Fig. 8.7 for graphs of each case). That is, even the same resampling method in the same problem domain interacts very differently with different classifiers.

Finally, we observe that all of the results that work well with SVMs move the separating hyperplane in some fashion, either by (1) changing the shape of the convex hulls inscribing either the minority class (generative oversampling) or the majority class (random undersampling) or (2) changing the location of the separating hyperplane by controlling the trade-off between margin and number of empirical mistakes in the training set (tuning the parameter C or using cost-sensitive SVMs). Our analysis supports the conclusion that random oversampling and SMOTE, which work well for the multinomial naive Bayes classifier, have minimal effect on SVMs since neither are effective at changing the shape of the minority class convex hull. In fact, if one tunes the parameter C, then neither random oversampling nor SMOTE is useful.

8.6.5 Discussion

In summary, we have presented experiments which can be divided into two cases: in the first case, there is no advantage to performing resampling as opposed to simply performing cost-sensitive learning. Examples of this were presented for artificial and real low-dimensional data sets for a Gaussian naive Bayes classifier. In

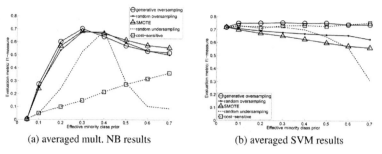

(a) averaged mult. NB results   (b) averaged SVM results

Fig. 8.8 Average results on text data sets for multinomial naive Bayes and SVMs (where C is tuned)

the second case, there is a clear difference in performing resampling as opposed to cost-sensitive learning due to various effects caused by the resampling methods. Several examples of this effect were presented using multinomial naive Bayes and linear SVMs on high-dimensional text data sets. Empirically, we showed that, for both naive Bayes and SVMs on text classification, resampling resulted in a better classifier than cost-sensitive learning.

Averaged results for the text data sets used in this chapter are presented in Fig. 8.8. When trying to achieve the optimal f1-measure on these text data sets, Fig. 8.8 shows that the best approach is to use linear SVMs with generative over-sampling.

Our results support the conclusion that the best resampling method (or whether cost-sensitive learning outperforms resampling) is dependent on the data set and classifier being used. This has been seen empirically in other papers such as [9]. Our analysis provides some insight for why these differences occur.

8.7 Conclusion

In this chapter, we have analyzed the relationship between resampling and cost-sensitive learning. In particular, we examine the effect of random and generative oversampling versus cost-sensitive learning from a theoretical perspective using Bayesian classifiers. The theoretical analysis is supported by empirical results, where we briefly examined the effect of resampling on low-dimensional data and included a much more extensive treatment of resampling versus cost-sensitive learning on high-dimensional text data sets using multinomial naive Bayes and linear SVMs.

Results vary depending on the dataset and the classifier used. In particular, for low-dimensional data sets where a Gaussian distribution is appropriate to model the classes, there seems to be no advantage to using resampling. Theoretically, resampling results in the same expected sample mean but with greater variance. Empirically, there is no benefit to resampling over cost-sensitive learning when used with a Gaussian naive Bayes classifier. In practice, this means that resampling is unnecessary if Gaussian naive Bayes is used; a cost-sensitive classifier that performs just as well can easily be trained without the overhead of resampling. When applying multinomial naive Bayes to text data sets, resampling results in changed class priors as well as different estimates of the parameters of the multinomial distribution modeling the resampled class. In this case, any of the oversampling methods tested result empirically in better classification of the minority class. Finally, when classifying imbalanced text data sets using an SVM classifier, we see that using generative oversampling, which helps to expand the convex hull of the minority class, can lead to consistently good performance. In particular, the best overall performance when classifying text data sets, regardless of classifier or method of resampling/incorporating costs, is generative oversampling coupled with SVMs.

Two of the most important results described in this chapter are as follows. First, while there is a theoretical equivalence between cost-sensitive learning and resampling under certain assumptions, these assumptions are often broken in practice. For example, all of the experiments on high-dimensional text data sets (for both naive Bayes and SVMs) break these assumptions, leading to an observed empirical difference between cost-sensitive and resampling methods. We also show that there is no resampling method that is always best. We give analytical and empirical results supporting why different resampling methods interact differently with certain classifiers on certain types of data.

Both of these results help explain why there are often differences between cost-sensitive learning and resampling methods in empirical studies such as [24]. Several areas of future work remain to explore other differences and further explain effects observed herein.

Appendix

In this section, we will derive some of the equations found in this chapter, particularly those in Sections 8.5.3 and 8.5.4.

Gaussian Random Oversampling

Let the set of n points in the positive class be denoted as $X_1, \ldots, X_n$. These points are a random sample from a normal distribution with true mean μ and true variance σ^2. Let us now consider a random sample $X_1^{(r)}, \ldots, X_m^{(r)}$ produced through random oversampling. Thus, each $X_j^{(r)}$ equals some X_{K_j}, where $K_1, \ldots, K_m$ are i.i.d. uniformly distributed random variables over the discrete set $\{1, \ldots, n\}$. The sample mean, $\bar{X}^{(r)}$, of only the randomly resampled data is an unbiased estimator of the true mean, since

$$E[\bar{X}^{(r)}] = \frac{1}{m}\sum_{j=1}^{m}E[X_j^{(r)}] = \frac{1}{m}\sum_{j=1}^{m}\sum_{i=1}^{n}E[X_{K_j}|K_j = i]P[K_j = i] = \frac{1}{m}\sum_{j=1}^{m}\sum_{i=1}^{n}\mu\frac{1}{n} = \mu,$$

(8.18)

noting that $E[X_{K_j}|K_j = i] = E[X_i] = \mu$ for all i and j. It follows that the sample mean, $\bar{X}_*^{(r)}$, of the pooled sample, $X_1,\ldots,X_n,X_1^{(r)},\ldots,X_m^{(r)}$, is also unbiased, since

$$E[\bar{X}_*^{(r)}] = \frac{E[n\bar{X} + m\bar{X}^{(r)}]}{n+m} = \frac{nE[\bar{X}] + mE[\bar{X}^{(r)}]}{n+m} = \mu.$$

(8.19)

The variance of the sample mean for only the resampled data is determined as follows. First, note that

$$\text{Var}[\bar{X}^{(r)}] = \frac{1}{m^2}\text{Var}\left[\sum_{j=1}^{m}X_j^{(r)}\right] = \frac{1}{m^2}\left[\sum_{j=1}^{m}\text{Var}[X_j^{(r)}] + \sum_{j\neq j'}\text{Cov}[X_j^{(r)},X_{j'}^{(r)}]\right].$$

(8.20)

Now, the variance of an individual sample is

$$\text{Var}[X_j^{(r)}] = \sum_{i=1}^{n}\text{Var}[X_{K_j}|K_j = i]P[K_j = i] = \sum_{i=1}^{n}\text{Var}[X_i]\frac{1}{n} = \sigma^2,$$

(8.21)

while the covariance between different samples is

$$\text{Cov}[X_j^{(r)},X_{j'}^{(r)}] = \sum_{i,i'}\text{Cov}[X_{K_j},X_{K_{j'}}|K_j = i, K_{j'} = i']P[K_j = i, K_{j'} = i']$$

$$= \sum_{i,i'}\frac{\sigma^2\,\delta_{i,i'}}{n^2} = \frac{\sigma^2}{n}.$$

(8.22)

Thus,

$$\text{Var}[\bar{X}^{(r)}] = \frac{1}{m^2}\left[m\sigma^2 + m(m-1)\frac{\sigma^2}{n}\right] = \left(1 + \frac{n-1}{m}\right)\frac{\sigma^2}{n},$$

(8.23)

and we note that the variance is larger than that for the original sample (i.e., $\text{Var}[\bar{X}] = \sigma^2/n$).

If the data is pooled, the variance of the sample mean becomes

$$\text{Var}[\bar{X}_*^{(r)}] = \frac{\text{Var}\left[n\bar{X} + m\bar{X}^{(r)}\right]}{(n+m)^2}$$

$$= \frac{n^2\text{Var}[\bar{X}] + m^2\text{Var}[\bar{X}] + 2\text{Cov}[n\bar{X}, m\bar{X}^{(r)}]}{(n+m)^2},$$

(8.24)

where

$$\mathrm{Cov}[n\bar{X}_n, m\bar{X}_m^{(r)}] = \sum_{i=1}^{n}\sum_{j=1}^{m}\sum_{i'=1}^{n}\mathrm{Cov}[X_i, X_{K_j}|K_j = i']P[K_j = i']$$

$$= \sum_{i=1}^{n}\sum_{j=1}^{m}\sum_{i'=1}^{n}\frac{\sigma^2 \delta_{i,i'}}{n} = m\sigma^2. \qquad (8.25)$$

Thus, we find

$$\mathrm{Var}[\bar{X}_*^{(r)}] = \left[1 + \frac{m(n-1)}{(n+m)^2}\right]\frac{\sigma^2}{n}, \qquad (8.26)$$

which is larger than $\mathrm{Var}[\bar{X}]$ but smaller than $\mathrm{Var}[\bar{X}^{(r)}]$.

Gaussian Generative Oversampling

Now consider points obtained by generative oversampling, wherein

$$X_j^{(g)} = \bar{X} + sZ_j, \qquad (8.27)$$

where $X_j^{(g)}$ is the jth point created by generative oversampling, s is the original sample standard deviation, and $Z_1, \ldots, Z_m$ are i.i.d. $N(0,1)$ and independent of $X_1, \ldots, X_n$ as well. For each resampled data point, the expected value is

$$E[X_j^{(g)}] = E[\bar{X}] + E[s]E[Z_j] = \mu + \sigma\sqrt{\frac{2}{n-1}}\frac{\Gamma(n/2)}{\Gamma((n-1)/2)}\cdot 0 = \mu, \qquad (8.28)$$

while the variance is given by

$$\begin{aligned}
\mathrm{Var}[X_j^{(g)}] &= \mathrm{Var}[\bar{X}] + \mathrm{Var}[sZ_j] + 2\mathrm{Cov}[\bar{X}, sZ_j] \\
&= \frac{\sigma^2}{n} + E[(sZ_j)^2] + 2E[(\bar{X}-\mu)sZ_j] \\
&= \frac{\sigma^2}{n} + E[s^2]E[Z_j^2] + 2E[(\bar{X}-\mu)s]E[Z_j] \\
&= \left(1 + \frac{1}{n}\right)\sigma^2.
\end{aligned} \qquad (8.29)$$

Furthermore, the pair-wise covariance is

$$\begin{aligned}
\mathrm{Cov}[X_j^{(g)}, X_{j'}^{(g)}] &= E[(\bar{X} + sZ_j - \mu)(\bar{X} + sZ_{j'} - \mu)] \\
&= E[(\bar{X}-\mu)^2] + E[(\bar{X}-\mu)s]\left(E[Z_j] + E[Z_{j'}]\right) + E[s^2]E[Z_j Z_{j'}] \\
&= \left(\delta_{j,j'} + \frac{1}{n}\right)\sigma^2.
\end{aligned}$$

$$(8.30)$$

The sample mean, $\bar{X}^{(g)}$, of the generatively resampled data is therefore unbiased, since

$$E[\bar{X}^{(g)}] = \frac{1}{m} \sum_{j=1}^{m} E[X_j^{(g)}] = \mu. \tag{8.31}$$

The variance of the sample mean may be computed as follows:

$$\mathrm{Var}[\bar{X}^{(g)}] = \frac{1}{m^2} \left[\sum_j \mathrm{Var}[X_j^{(g)}] + \sum_{j \neq j'} \mathrm{Cov}[X_j^{(g)}, X_{j'}^{(g)}] \right]$$

$$= \left(1 + \frac{n}{m} \right) \frac{\sigma^2}{n}. \tag{8.32}$$

Like the randomly resampled case, the variance of the sample mean is larger than that for the original sample.

Pooling the data leaves the sample mean, $\bar{X}_*^{(g)}$, unbiased. The variance of this estimator is determined as follows. First, note that

$$\mathrm{Var}[\bar{X}_*^{(g)}] = \frac{n^2 \mathrm{Var}[\bar{X}] + m^2 \mathrm{Var}[\bar{X}^{(g)}] + 2\mathrm{Cov}[n\bar{X}, m\bar{X}^{(g)}]}{(n+m)^2}, \tag{8.33}$$

where

$$\mathrm{Cov}[n\bar{X}, m\bar{X}^{(g)}] = \sum_{i=1}^{n} \sum_{j=1}^{m} \mathrm{Cov}[X_i, X_j^{(g)}] \tag{8.34}$$

and

$$\mathrm{Cov}[X_i, X_j^{(g)}] = \mathrm{Cov}[X_i, \bar{X} + sZ_j]$$

$$= \mathrm{Cov}[X_i, \bar{X}] + \mathrm{Cov}[X_i, sZ_j]. \tag{8.35}$$

Now, note that

$$\mathrm{Cov}[X_i, \bar{X}] = \frac{1}{n} \sum_{k=1}^{n} \mathrm{Cov}[X_i, X_k] = \frac{1}{n} \sum_{k=1}^{n} \sigma^2 \delta_{i,k} = \frac{\sigma^2}{n} \tag{8.36}$$

and

$$\mathrm{Cov}[X_i, sZ_j] = E[(X_i - \mu)sZ_j] = E[(X_i - \mu)s]E[Z_j] = 0. \tag{8.37}$$

Thus, $\mathrm{Cov}[X_i, X_j^{(g)}] = \sigma^2/n$, which implies $\mathrm{Cov}[n\bar{X}_n, m\bar{X}_m^{(g)}] = m\sigma^2$. This, in turn, implies

$$\mathrm{Var}[\bar{X}_*^{(g)}] = \left[1 + \frac{mn}{(n+m)^2} \right] \frac{\sigma^2}{n}. \tag{8.38}$$

As in the randomly resampled case, the variance of the sample mean is larger than that of the original sample.

Multinomial Random Oversampling

Let $X_1, \ldots, X_n$ be a random sample from a discrete distribution with values $1, \ldots, d$ and corresponding probabilities $\theta_1, \ldots, \theta_d$.[7] Let F_k denote the number of samples attaining the value $k \in \{1, \ldots, d\}$, i.e.,

$$F_k = \sum_{i=1}^{n} \delta_k(X_i), \qquad (8.39)$$

where $\delta_k(x) = 1$ if $x = k$ and is zero otherwise. The random variables $F_1, \ldots, F_d$ then follow a multi-nomial distribution. The maximum likelihood estimator (MLE) $\hat{\theta}_k = F_k/n$ is unbiased for θ_k and has variance $\theta_k(1 - \theta_k)/n$. With Laplace smoothing (a standard practice when using multinomials for problems like text mining), the estimator becomes

$$\tilde{\theta}_k = \frac{F_k + 1}{\sum_{k'=1}^{d}(F_{k'} + 1)} = \frac{n\hat{\theta}_k + 1}{n + d}, \qquad (8.40)$$

where we have used the fact that $F_1 + \cdots + F_d = n$ almost surely. The smoothed estimator is biased, having $E[\tilde{\theta}_k] = (n\theta_k + 1)/(n + d)$, but has a smaller variance, $\mathrm{Var}[\tilde{\theta}] = n\theta_k(1 - \theta_k)/(n + d)^2$.

Now consider a randomly resampled data set $X_1^{(r)}, \ldots, X_m^{(r)}$, for which each $X_j^{(r)}$ is drawn uniformly and independently from the original sample. The number of resampled data points with value k will correspondingly be denoted $F_k^{(r)}$. For the estimator $\hat{\theta}_k^{(r)} = F_k^{(r)}/m$ we find

$$E[F_k^{(r)}] = \sum_{j=1}^{m} E[\delta_k(X_j^{(r)})] = \sum_{j=1}^{m}\sum_{i=1}^{n} P[X_j^{(r)} = k | X_j^{(r)} = x_i] P[X_j^{(r)} = x_i] = m\theta_k, \quad (8.41)$$

since $P[X_j^{(r)} = k | X_j^{(r)} = x_i] = \theta_k$ and $P[X_j^{(r)} = x_i] = 1/n$. For the variance, first note that

$$\mathrm{Var}[F_k^{(r)}] = \sum_{j=1}^{m} \mathrm{Var}[\delta_k(X_j^{(r)})] + \sum_{j \neq j'} \mathrm{Cov}[\delta_k(X_j^{(r)}), \delta_k(X_{j'}^{(r)})]. \qquad (8.42)$$

Now,

$$\mathrm{Var}[\delta_k(X_j^{(r)})] = \sum_{i=1}^{n} \mathrm{Var}[\delta_k(X_j^{(r)}) | X_j^{(r)} = x_i] \frac{1}{n} = \theta_k(1 - \theta_k) \qquad (8.43)$$

and

---

[7] In the case of text mining, each X_i would correspond to a single word in the positive class and not a single document; in our analysis, the notation is much more straightforward if one looks at individual features instead of "blocks" of features that make up a datapoint (e.g., "blocks" of words making up a document); in addition, from the standpoint of parameter estimation, which document the word falls into does not matter.

$$\text{Cov}[\delta_k(X_j^{(r)}), \delta_k(X_{j'}^{(r)})] = \sum_{i=1}^n \sum_{i'=1}^n \text{Cov}[\delta_k(X_j^{(r)}), \delta_k(X_{j'}^{(r)})]|X_j^{(r)} = x_i, X_{j'}^{(r)} = x_{i'}]\frac{1}{n^2}$$

$$= \frac{1}{n^2} \sum_i \text{Cov}[\delta_k(X_i), \delta_k(X_{i'})] = \frac{\theta_k(1-\theta_k)}{n}.$$

$$(8.44)$$

Combining these results, we find

$$\text{Var}[\hat{\theta}_k^{(r)}] = \left(1 + \frac{n-1}{m}\right)\frac{\theta_k(1-\theta_k)}{n}. \tag{8.45}$$

For the pooled sample $X_1, \ldots, X_n, X_1^{(r)}, \ldots, X_m^{(r)}$, the MLE of θ is $(F_k + F_k^{(r)})/(n+m)$. Clearly, this estimator is unbiased. For the variance, note that

$$\text{Var}\left[F_k + F_k^{(r)}\right] = \text{Var}[F_k] + \text{Var}[F_k^{(r)}] + 2\text{Cov}[F_k, F_k^{(r)}]. \tag{8.46}$$

Now, the covariance is

$$\text{Cov}[F_k, F_k^{(r)}] = \sum_{i=1}^n \sum_{j=1}^m \text{Cov}[\delta_k(X_i), \delta_k(X_j^{(r)})]$$

$$= \sum_{i,j} \sum_{i'=1}^n \text{Cov}[\delta_k(X_i), \delta_k(X_j^{(r)})|X_j^{(r)} = x_{i'}]\frac{1}{n} \tag{8.47}$$

$$= \frac{1}{n} \sum_{i,i',j} \text{Cov}[\delta_k(X_i), \delta_k(X_{i'})] = m\theta_k(1-\theta_k).$$

After some simplification, we find that

$$\text{Var}\left[\frac{F_k + F_k^{(r)}}{n+m}\right] = \left[1 + \frac{m(n-1)}{(n+m)^2}\right]\frac{\theta_k(1-\theta_k)}{n}. \tag{8.48}$$

With Laplace smoothing, we have

$$E\left[\frac{F_k + F_k^{(r)} + 1}{n+m+d}\right] = \frac{(n+m)\theta_k + 1}{n+m+d} \tag{8.49}$$

and

$$\text{Var}\left[\frac{F_k + F_k^{(r)} + 1}{n+m+d}\right] = \left[1 + \frac{m(n-2d-1) - d(2n+d)}{(n+m+d)^2}\right]\frac{\theta_k(1-\theta_k)}{n}. \tag{8.50}$$

Thus, provided $m < d(2n+d)/(n-2d-1)$, the variance of the pooled, smoothed estimates will be smaller than that of the original sample, although the estimates themselves will be biased.

Multinomial Generative Oversampling

In generative oversampling, we use the probabilities estimated from $X_1, \ldots, X_n$ to generate a random sample $X_1^{(g)}, \ldots, X_m^{(g)}$ using the estimated parameters $\hat{\theta}_1, \ldots, \hat{\theta}_d$. The resampled data give unbiased estimates of the true parameters, since

$$E[F_k^{(g)}] = \sum_{j=1}^{m} E[\delta_k(X_j^{(g)})] = \sum_{j=1}^{m} P[X_j^{(g)} = k] \tag{8.51}$$

but

$$P[X_j^{(g)} = k] = \sum_{X_1, \ldots, X_n} P[X_j^{(g)} = k | X_1 = x_1, \ldots, X_n = x_n] P[X_1 = x_1, \ldots, X_n = x_n]$$

$$= \sum_{X_1, \ldots, X_n} \hat{\theta}_k P[X_1 = x_1, \ldots, X_n = x_n] = \theta_k. \tag{8.52}$$

For the variance, note that

$$\mathrm{Var}[F^{(g)}] = \sum_{j=1}^{m} \mathrm{Var}[\delta_k(X_j^{(g)})] + \sum_{j \neq j'} \mathrm{Cov}[\delta_k(X_j^{(g)}), \delta(X_{j'}^{(g)})]. \tag{8.53}$$

Now,

$$\mathrm{Var}[\delta_k(X_j^{(g)})] = E[\delta_k(X_j^{(g)})^2] - E[\delta_k(X_j^{(g)})]^2 = P[X_j^{(g)} = k] - P[X_j^{(g)} = k]^2 = \theta_k(1 - \theta_k) \tag{8.54}$$

and

$$\mathrm{Cov}[\delta_k(X_j^{(g)}), \delta(X_{j'}^{(g)})] = P[X_j^{(g)} = k, X_{j'}^{(g)} = k] - \theta_k^2. \tag{8.55}$$

The joint probability, for $j \neq j'$, is

$$P[X_j^{(g)} = k, X_{j'}^{(g)} = k] = \sum_{X_1, \ldots, X_n} \hat{\theta}_k^2 P[X_1 = x_1, \ldots, X_n = x_n] = \frac{\theta_k(1 - \theta_k)}{n} + \theta_k^2. \tag{8.56}$$

Thus,

$$\mathrm{Var}[f^{(g)}] = m\theta_k(1 - \theta_k) + m(m - 1)\frac{\theta_k(1 - \theta_k)}{n}, \tag{8.57}$$

from which we readily deduce that

$$\mathrm{Var}[\hat{\theta}_k^{(g)}] = \left(1 + \frac{n-1}{m}\right)\frac{\theta_k(1 - \theta_k)}{n}. \tag{8.58}$$

For the pooled sample $X_1, \ldots, X_n, X_1^{(g)}, \ldots, X_m^{(g)}$ we again have an unbiased estimator. The variance is given by

$$\mathrm{Var}[F_k + F_k^{(g)}] = \mathrm{Var}[F_k] + \mathrm{Var}[F_k^{(g)}] + 2\mathrm{Cov}[F_k, F_k^{(g)}], \tag{8.59}$$

where

$$\mathrm{Cov}[F_k, F_k^{(g)}] = \sum_{i=1}^{n}\sum_{j=1}^{m} \mathrm{Cov}[\delta_k(X_i), \delta_k(X_j^{(g)})]$$

$$= \sum_{i=1}^{n}\sum_{j=1}^{m} \left\{ P[X_i = k, X_j^{(g)} = k] - P[X_i = k]P[X_j^{(g)} = k] \right\}. \tag{8.60}$$

Now, the joint probability is found to be

$$P[X_i = k, X_j^{(g)} = k] = \sum_{X_1,\dots,X_n} P[X_i = k, X_j^{(g)} = k | X_1 = x_1,\dots,X_n = x_n]$$

$$P[X_1 = x_1,\dots,X_n = x_n] = \sum_{X_1,\dots,X_n} \delta_k(X_i)\hat{\theta}_k P[X_1 = x_1,\dots,X_n = x_n]. \tag{8.61}$$

But

$$\sum_{i=1}^{n} P[X_i = k, X_j^{(g)} = k] = \sum_{X_1,\dots,X_n} n\hat{\theta}_k\hat{\theta}_k P[X_1 = x_1,\dots,X_n = x_n] = \theta_k(1-\theta_k) + n\theta_k^2, \tag{8.62}$$

so

$$\mathrm{Cov}[F_k, F_k^{(g)}] = m\theta_k(1-\theta_k). \tag{8.63}$$

Combining these results, we find

$$\mathrm{Var}[F_k + F_k^{(g)}] = n\theta_k(1-\theta_k) + m^2 \left(1 + \frac{n-1}{m}\right)\frac{\theta_k(1-\theta_k)}{n} + 2m\theta_k(1-\theta_k), \tag{8.64}$$

and, thus,

$$\mathrm{Var}\left[\frac{F_k + F_k^{(g)}}{n+m}\right] = \left[1 + \frac{m(n-1)}{(n+m)^2}\right]\frac{\theta_k(1-\theta_k)}{n}. \tag{8.65}$$

These results are identical to those of the random oversampling case.

References

1. Batista, G.E.A.P.A., Prati, R.C., Monard, M.C.: A study of the behavior of several methods for balancing machine learning training data. SIGKDD Explorations 6(1), 20–29 (2004)
2. Bucila, C., Caruana, R., Niculescu-Mizil, A.: Model compression. KDD pp. 535–541 (2006)
3. Chan, P.K., Stolfo, S.J.: Toward scalable learning with non-uniform class and cost distributions: A case study in credit card fraud detection. Knowledge Discovery and Data Mining pp. 164–168 (1998)

4. Chawla, N., Bowyer, K.W., Hall, L.O., Kegelmeyer, W.P.: SMOTE: Synthetic Minority Over-sampling TEchnique. Journal of Artificial Intelligence Research 16, 321–357 (2002)
5. Duda, R., Hart, P., Stork, D.: Pattern Classification. John Wiley & Sons, New York (2001)
6. Elkan, C.: The foundations of cost-sensitive learning. Proceedings of the Seventeenth International Joint Conference on Artificial Intelligence pp. 973–978 (2001)
7. Han, E.H., Boley, D., Gini, M., Gross, R., Hastings, K., Karypis, G., Kumar, V., Mobasher, B., Moore, J.: WebACE: A web agent for document categorization and exploration. Proceedings of the Second International Conference on Autonomous Agents pp. 408–415 (1998)
8. Hersh, W., Buckley, C., Leone, T.J., Hickam, D.: Ohsumed: An interactive retrieval evaluation and new large test collection for research. Proceedings of ACM SIGIR pp. 192–201 (1994)
9. Hulse, J.V., Khoshgoftaar, T.M., Napolitano, A.: Experimental perspectives on learning from imbalanced data. In: ICML '07: Proceedings of the 24th international conference on Machine learning, pp. 935–942 (2007)
10. Joachims, T.: Text categorization with support vector machines: Learning with many relevant features. Proceedings of ECML-98, 10th European Conference on Machine Learning (1398), 137–142 (1998)
11. Karypis, G.: CLUTO – a clustering toolkit. University of Minnesota technical report 02-017 (2002)
12. Kubat, M., Holte, R.C., Matwin, S.: Machine learning for the detection of oil spills in satellite radar images. Machine Learning 30(2–3), 195–215 (1998)
13. Lewis, D., Gale, W.: Training text classifiers by uncertainty sampling. Proceedings of the Seventh Annual International ACM SIGIR Conference on Research and Development in Information Retrieval (1994)
14. Ling, C.X., Li, C.: Data mining for direct marketing: Problems and solutions. Knowledge Discovery and Data Mining pp. 73–79 (1998)
15. Liu, A., Ghosh, J., Martin, C.: Generative oversampling for mining imbalanced datasets. DMIN '07: International Conference on Data Mining (2007)
16. Maloof, M.: Learning when data sets are imbalanced and when costs are unequal and unknown. ICML-2003 Workshop on Learning from Imbalanced Data Sets II (2003)
17. McCallum, A., Nigam, K.: A comparison of event models for naive bayes text classification. AAAI-98 Workshop on Learning for Text Categorization (1998)
18. McCarthy, K., Zabar, B., Weiss, G.: Does cost-sensitive learning beat sampling for classifying rare classes? UBDM '05: Proceedings of the 1st international workshop on Utility-based data mining pp. 69–77 (2005)
19. Melville, P., Mooney, R.J.: Diverse ensembles for active learning. ICML '04: Proceedings of the twenty-first international conference on Machine learning pp. 584–591 (2004)
20. Morik, K., Brockhausen, P., Joachims, T.: Combining statistical learning with a knowledge-based approach – a case study in intensive care monitoring. Proceedings of the 16th International Conference on Machine Learning (ICML-99) (1999)
21. Phua, C., Alahakoon, D., Lee, V.: Minority report in fraud detection: Classification of skewed data. SIGKDD Explor. Newsl. 6(1), 50–59 (2004)
22. Sebastiani, F.: Machine learning in automated text categorization. ACM Computing Surveys 34(1), 1–47 (2002)
23. Turney, P.D.: Learning algorithms for keyphrase extraction. Information Retrieval 2(4), 303–336 (2000)
24. Weiss, G., McCarthy, K., Zabar, B.: Cost-sensitive learning vs. sampling: Which is best for handling class imbalance? DMIN '07: International Conference on Data Mining (2007)
25. Yang, Y.: An evaluation of statistical approaches to text categorization. Information Retrieval 1(1/2), 69–90 (1999)
26. Zadrozny, B., Langford, J., Abe, N.: Cost-sensitive learning by cost-proportionate example weighting. ICDM '03: Proceedings of the Third IEEE International Conference on Data Mining (2003)

27. Zhang, Mani: kNN approach to unbalanced data distributions: A case study involving information extraction. ICML '03: Proceedings of the twentieth international conference on Machine learning (2003)
28. Zhong, S., Ghosh, J.: A comparative study of generative models for document clustering. SDM Workshop on Clustering High Dimensional Data and Its Applications (2003)

Chapter 9
The Impact of Small Disjuncts on Classifier Learning

Gary M. Weiss

Abstract Many classifier induction systems express the induced classifier in terms of a disjunctive description. Small disjuncts are those that classify few training examples. These disjuncts are interesting because they are known to have a much higher error rate than large disjuncts and are responsible for many, if not most, of all classification errors. Previous research has investigated this phenomenon by performing ad hoc analyses of a small number of data sets. In this chapter we provide a much more systematic study of small disjuncts and analyze how they affect classifiers induced from 30 real-world data sets. A new metric, error concentration, is used to show that for these 30 data sets classification errors are often heavily concentrated toward the smaller disjuncts. Various factors, including pruning, training set size, noise, and class imbalance are then analyzed to determine how they affect small disjuncts and the distribution of errors across disjuncts. This analysis provides many insights into why some data sets are difficult to learn from and also provides a better understanding of classifier learning in general. We believe that such an understanding is critical to the development of improved classifier induction algorithms.

9.1 Introduction

It has long been observed that certain classification problems are quite difficult and that high levels of classification performance are not achievable in these cases. In certain circumstances entire classes of problems tend to be difficult, such as classification problems that deal with class imbalance [18]. These problems have often been studied in detail and sometimes methods have even been proposed for improving classification performance, but generally there is little explanation for why these techniques work and the research instead relies on empirical evaluations of the methods. As just one example, most of the research aimed at improving the performance

Gary M. Weiss
Fordham University, Bronx, NY 10458, USA, e-mail: gweiss@cis.fordham.edu

R. Stahlbock et al. (eds.), *Data Mining*, Annals of Information Systems 8,
DOI 10.1007/978-1-4419-1280-0_9, © Springer Science+Business Media, LLC 2010

of classifiers induced from imbalanced data sets provides little or no justification for the methods. In this chapter we focus on the role of small disjuncts in classifier learning and in so doing provide the terms and concepts necessary to provide these justifications. Additionally, we provide a number of conclusions about what makes classifier learning hard and under what circumstances.

Classifier induction programs often express the learned classifier as a disjunction. For example, such systems often express the classifier as a decision tree or a rule set, in which case each leaf in the decision tree or rule in the rule set corresponds to a disjunct. The *size* of a disjunct is defined as the number of training examples that the disjunct correctly classifies [9]. A number of empirical studies have shown that learned concepts include disjuncts that span a wide range of disjunct sizes and that small disjuncts – those disjuncts that correctly classify only a few training examples – collectively cover a significant percentage of the total test examples. These studies also show that small disjuncts have a much higher error rate than large disjuncts, a phenomenon sometimes referred to as the "problem with small disjuncts" and that these small disjuncts collectively contribute a significant portion of the total test errors.

One problem with past studies is that each study analyzes classifiers induced from only a few data sets. In particular, Holte et al. [9] analyze two data sets, Ali and Pazzani [1] one data set, Danyluk and Provost [8] one data set, Weiss [17] two data sets, Weiss and Hirsh [19] two data sets, and Carvalho and Freitas [3] two data sets. Because of the small number of data sets analyzed, and because there was no established way to measure the degree to which errors were concentrated toward the small disjuncts, these studies were not able to quantify the problem with small disjuncts. This chapter addresses these concerns. First, a new metric, error concentration, is introduced which quantifies, in a single number, the extent to which errors are concentrated toward the smaller disjuncts. This metric is then used to measure the error concentration of the classifiers induced from 30 data sets. Because we analyze a large number of data sets, we are able to draw general conclusions about the role that small disjuncts play in classifier learning.

Small disjuncts are of interest because they are responsible for many – if not most – of the errors that result when the induced classifier is applied to new (test) data. This in turn leads to two reasons for studying small disjuncts. First, we hope that what we learn about small disjuncts may enable us to build more effective classifier induction programs by addressing the problem with small disjuncts. Specifically, such learners would improve the classification performance of the examples covered by the small disjuncts without excessively degrading the accuracy of the examples covered by the larger disjuncts, such that the *overall* performance of the classifier is improved. Existing efforts to do just this, which are described in Section 9.9, have produced, at best, only marginal improvements. A better understanding of small disjuncts and their role in learning may be necessary before further advances are possible.

The second reason for studying small disjuncts is to provide a better understanding of small disjuncts and, by extension, of classifier learning in general. Most of the research on small disjuncts has not focused on this, which is the main focus of this chapter. Essentially, small disjuncts are used as a lens through which to examine factors that are important to classifier learning, which is perhaps the most common data mining method. Pruning, training set size, noise, and class imbalance are each analyzed to see how they affect small disjuncts and the distribution of errors throughout the disjuncts – and, more generally, how this impacts classifier learning.

This chapter is an expanded version of an earlier paper [20]. It is organized as follows. In Section 9.2 we analyze the role of small disjuncts in classifier learning and introduce relevant metrics and terminology. Section 9.3 then describes the methodology used to conduct our experiments. Our experimental results and the analysis of these results are then presented in the next five sections. We provide a general analysis of the impact that small disjuncts have on learning in Section 9.4 and then, over the next four sections, we then analyze how each of the following factors interact with small disjuncts during the learning process: pruning (Section 9.5), training set size (Section 9.6), noise (Section 9.7), and class imbalance (Section 9.8). Related work is covered in Section 9.9 and our conclusions and future work are discussed in Section 9.10.

9.2 An Example: The Vote Data Set

In order to illustrate the problem with small disjuncts, the performance of a classifier induced by C4.5 [14] from the vote data set is shown in Fig. 9.1. This figure shows how the correctly and incorrectly classified test examples are distributed across the disjuncts in the induced classifier. The overall test set error rate for the classifier is 6.9%.

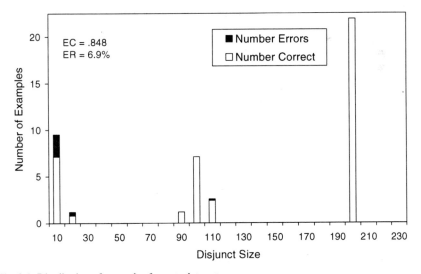

Fig. 9.1 Distribution of examples for vote data set

Each bar in the histogram in Fig. 9.1 covers 10 sizes of disjuncts. The leftmost bin shows that those disjuncts that correctly classify 0–9 training examples cover 9.5 test examples, of which 7.1 are classified correctly and 2.4 classified incorrectly (fractional values occur because the results are averaged over 10 cross-validated runs). Figure 9.1 clearly shows that the errors are concentrated toward the smaller disjuncts. Analysis at a finer level of granularity shows that the errors are skewed even more toward the small disjuncts –75% of the errors in the leftmost bin come from disjuncts of size 0 and 1. One may also be interested in the distribution of disjuncts by its size. The classifier associated with Fig. 9.1 is made up of 50 disjuncts, of which 45 are associated with the leftmost bin (i.e., have a disjunct size less than 10). Note that disjuncts of size 0 were formed because when the decision tree learner used to generate the classifier splits a node N using a feature f, the split will branch on all possible values of f – even if a feature value does not occur within the training data at N.

In order to more effectively show the extent to which errors are concentrated toward the small disjuncts, we plot the percentage of total test errors vs. the percentage of correctly classified test examples contributed by a set of disjuncts. The curve in Fig. 9.2 is generated by starting with the smallest disjunct from the classifier induced from the vote data set and then progressively adding larger disjuncts. This curve shows, for example, that disjuncts with size 0–4 cover 5.1% of the correctly classified test examples but 73% of the total test errors. The line $Y = X$ represents a classifier in which classification errors are distributed uniformly across the disjuncts, independent of the size of the disjunct. Since the "error concentration" curve in Fig. 9.2 falls above the line $Y = X$, the errors produced by this classifier are more concentrated toward the smaller disjuncts than to the larger disjuncts.

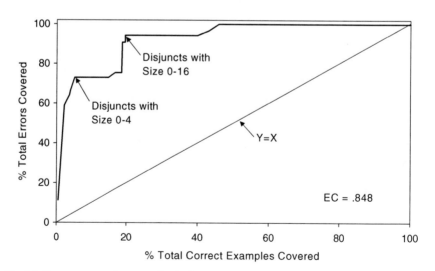

Fig. 9.2 Error concentration curve for the Vote data set

To make it easy to compare the degree to which errors are concentrated toward the smaller disjuncts for different classifiers, we introduce the *error concentration* (EC) metric. The error concentration of a classifier is defined as the fraction of the total area *above* the line $Y = X$ that falls below its error concentration curve. Using this scheme, the higher the error concentration, the more concentrated the errors are toward the smaller disjuncts. Error concentration may range from a value of +1, which indicates that all test errors are contributed by the smallest disjuncts, before even a single correctly classified test example is covered, to a value of −1, which indicates that all test errors are contributed by the largest disjuncts, after all correctly classified test examples are covered. Based on previous research, which indicates that small disjuncts have higher error rates than large disjuncts, one would expect the error concentration of most classifiers to be greater than 0. The error concentration for the classifier described in Fig. 9.2 is 0.848, indicating that the errors are highly concentrated toward the small disjuncts.

9.3 Description of Experiments

The majority of results presented in this chapter are based on an analysis of 30 data sets, of which 19 were obtained from the UCI repository [2] and 11, identified with a "+," were obtained from researchers at AT&T [6, 7]. These data sets are summarized in Table 9.1.

Table 9.1 Description of 30 data sets

No. Data set	Size	No. Data set	Size
1 adult	21,280	16 market1+	3180
2 bands	538	17 market2+	11,000
3 blackjack+	15,000	18 move+	3028
4 breast-wisc	699	19 network1+	3577
5 bridges	101	20 network2+	3826
6 coding	20,000	21 ocr+	2688
7 crx	690	22 promoters	106
8 german	1000	23 sonar	208
9 heart-hungarian	293	24 soybean-large	682
10 hepatitis	155	25 splice-junction	3175
11 horse-colic	300	26 ticket1+	556
12 hypothyroid	3771	27 ticket2+	556
13 kr-vs-kp	3196	28 ticket3+	556
14 labor	57	29 vote	435
15 liver	345	30 weather+	5597

Numerous experiments are run on these data sets to assess the impact that small disjuncts have on learning. The majority of the experimental results presented in this chapter are based on C4.5 [14], a popular program for inducing decision trees. C4.5 was modified by the author to collect a variety of information related to disjunct size. Note that disjunct size is defined based on the number of examples covered by the training data but, as is typical in data mining, the classification results are measured based on the performance on the test data. Many experiments were repeated using Ripper [6], a program for inducing rule sets, to ensure the generality of our results. Because Ripper exports detailed information about the performance of individual rules, internal modifications to the program were not required in order to track the statistics related to disjunct size. All experiments for both learners employ 10-fold cross-validation and all results are based on the averages over these 10 runs. Pruning tends to eliminate most small disjuncts and, for this reason, research on small disjuncts generally disables pruning [8, 9, 17, 19]. If this were not done, then pruning would mask the problem with small disjuncts. While this means that the analyzed classifiers are not the same as the ones that would be generated using the learners in their standard configurations, these results are nonetheless important, since the performance of the unpruned classifiers constrains the performance of the pruned classifiers. However, in this chapter both unpruned and pruned classifiers are analyzed, for both C4.5 and Ripper. This makes it possible to analyze the effect that pruning has on small disjuncts and to evaluate pruning as a strategy for addressing the problem with small disjuncts. As the results for pruning in Section 9.5 will show, the problem with small disjuncts is still evident after pruning, although to a lesser extent.

All results, other than those described in Section 9.5, are based on the use of C4.5 and Ripper with their pruning strategies disabled. For C4.5, when pruning is disabled the −m 1 option is also used, to ensure that C4.5 does not stop splitting a node before the node contains examples belonging to a single class (the default is −m 2). Ripper is configured to produce unordered rules so that it does not produce a single default rule to cover the majority class.

9.4 The Problem with Small Disjuncts

Previous research claims that errors tend to be concentrated most heavily in the smaller disjuncts [1, 3, 8, 9, 15, 17, 19]. In this section we provide the most comprehensive analysis of this claim to date, by measuring the degree to which errors are concentrated toward the smaller disjuncts for the 30 data sets listed in Table 9.1, for classifiers induced by C4.5 and Ripper.

The experimental results for C4.5 and Ripper, in order of decreasing error concentration, are displayed in Tables 9.2 and 9.3, respectively. In addition to specifying the error concentration, these tables also list the error rate of the induced classifier, the size of the data set, and the size of the largest disjunct in the induced classifier. They also specify the percentage of the total test errors that are contributed by the smallest disjuncts that collectively cover 10% of the correctly classified test examples and then the percentage of the total correctly classified examples that are covered by the smallest disjuncts that collectively cover half of the total errors.

Table 9.2 Error concentration results for C4.5

EC rank	Data set name	Error rate	Data set size	Largest disjunct	% Errs at 10% correct	% Correct at 50% errors	Error conc.
1	kr-vs-kp	0.3	3196	669	75.0	1.1	0.874
2	hypothyroid	0.5	3771	2697	85.2	0.8	0.852
3	vote	6.9	435	197	73.0	1.9	0.848
4	splice-junction	5.8	3175	287	76.5	4.0	0.818
5	ticket2	5.8	556	319	76.1	2.7	0.758
6	ticket1	2.2	556	366	54.8	4.4	0.752
7	ticket3	3.6	556	339	60.5	4.6	0.744
8	soybean-large	9.1	682	56	53.8	9.3	0.742
9	breast-wisc	5.0	699	332	47.3	10.7	0.662
10	ocr	2.2	2688	1186	52.1	8.9	0.558
11	hepatitis	22.1	155	49	30.1	17.2	0.508
12	horse-colic	16.3	300	75	31.5	18.2	0.504
13	crx	19.0	690	58	32.4	14.3	0.502
14	bridges	15.8	101	33	15.0	23.2	0.452
15	heart-hungar.	24.5	293	69	31.7	21.9	0.450
16	market1	23.6	3180	181	29.7	21.1	0.440
17	adult	16.3	21,280	1441	28.7	21.8	0.424
18	weather	33.2	5597	151	25.6	22.4	0.416
19	network2	23.9	3826	618	31.2	24.2	0.384
20	promoters	24.3	106	20	32.8	20.6	0.376
21	network1	24.1	3577	528	26.1	24.1	0.358
22	german	31.7	1000	56	17.8	29.4	0.356
23	coding	25.5	20,000	195	22.5	30.9	0.294
24	move	23.5	3028	35	17.0	30.8	0.284
25	sonar	28.4	208	50	15.9	32.9	0.226
26	bands	29.0	538	50	65.2	54.1	0.178
27	liver	34.5	345	44	13.7	40.3	0.120
28	blackjack	27.8	15,000	1989	18.6	39.3	0.108
29	labor	20.7	57	19	33.7	49.1	0.102
30	market2	46.3	11,000	264	10.3	45.5	0.040

As an example of how to interpret the results in these tables, consider the entry for the kr-vs-kp data set in Table 9.2. The error concentration for the classifier induced from this data set is 0.874. Furthermore, the smallest disjuncts that collectively cover 10% of the correctly classified test examples contribute 75% of the total test errors, while the smallest disjuncts that contribute half of the total errors cover only 1.1% of the total correctly classified examples. These measurements provide a concrete indication of just how concentrated the errors are toward the smaller disjuncts.

The results for C4.5 and Ripper show that although the error concentration values are, as expected, almost always positive, the values vary widely, indicating that the induced classifiers suffer from the problem of small disjuncts to varying degrees. The classifiers induced using Ripper have a slightly smaller average error concentration than those induced using C4.5 (0.445 vs. 0.471), indicating that the classifiers induced by Ripper have the errors spread slightly more uniformly across the disjuncts. Overall, Ripper and C4.5 tend to generate classifiers with similar error concentration values. This can be seen by comparing the EC rank in Table 9.3

Table 9.3 Error concentration results for Ripper

EC rank	C4.5 rank	Data set name	Error rate	Data set size	Largest disjunct	% Errors 10% correct	% Correct 50% Errs	Error conc.
1	2	hypothyroid	1.2	3771	2696	96.0	0.1	0.898
2	1	kr-vs-kp	0.8	3196	669	92.9	2.2	0.840
3	6	ticket1	3.5	556	367	69.4	1.6	0.802
4	7	ticket3	4.5	556	333	61.4	5.6	0.790
5	5	ticket2	6.8	556	261	71.0	3.2	0.782
6	3	vote	6.0	435	197	75.8	3.0	0.756
7	4	splice-junction	6.1	3175	422	62.3	7.9	0.678
8	9	breast-wisc	5.3	699	355	68.0	3.6	0.660
9	8	soybean-large	11.3	682	61	69.3	4.8	0.638
10	10	ocr	2.6	2688	804	50.5	10.0	0.560
11	17	adult	19.7	21,280	1488	36.9	15.0	0.516
12	16	market1	25.0	3180	243	32.2	16.9	0.470
13	12	horse-colic	22.0	300	73	20.7	23.9	0.444
14	13	crx	17.0	690	120	32.5	19.7	0.424
15	15	heart-hungar.	23.9	293	67	25.8	24.8	0.390
16	26	bands	21.9	538	62	25.6	29.2	0.380
17	25	sonar	31.0	208	47	32.6	23.9	0.376
18	23	coding	28.2	20,000	206	22.6	29.2	0.374
19	18	weather	30.2	5597	201	23.8	24.8	0.356
20	24	move	32.1	3028	45	25.9	25.6	0.342
21	14	bridges	14.5	101	39	41.7	35.5	0.334
22	20	promoters	19.8	106	24	20.0	20.0	0.326
23	11	hepatitis	20.3	155	60	19.3	20.8	0.302
24	22	german	30.8	1000	99	12.1	35.0	0.300
25	19	network2	23.1	3826	77	25.6	22.9	0.242
26	27	liver	34.0	345	28	28.2	32.0	0.198
27	28	blackjack	30.2	15,000	1427	12.3	42.3	0.0108
28	21	network1	23.4	3577	79	18.9	46.0	0.090
29	29	labor	24.5	57	21	0.0	18.3	−0.006
30	30	market2	48.8	11,000	55	10.4	49.8	0.018

for Ripper (column 1) with the EC rank for C4.5 (column 2), which is displayed graphically in the scatter plot in Fig. 9.3, where each point represents the error concentration for a single data set. Since the points in Fig. 9.3 are clustered around the line $Y = X$, both learners tend to produce classifiers with similar error concentrations and hence tend to suffer from the problem with small disjuncts to similar degrees. The agreement is especially close for the most interesting cases, where the error concentrations are large – the largest 10 error concentration values in Fig. 9.3, for both C4.5 and Ripper, are generated by the same 10 data sets.

With respect to classification accuracy, the two learners perform similarly, although C4.5 performs slightly better (it outperforms Ripper on 18 of the 30 data sets, with an average error rate of 18.4% vs. 19.0%). However, as will be shown in the next section, when pruning is used Ripper slightly outperforms C4.5.

The results in Tables 9.2 and 9.3 indicate that, for both C4.5 and Ripper, there is a relationship between the error rate and error concentration of the induced classifiers.

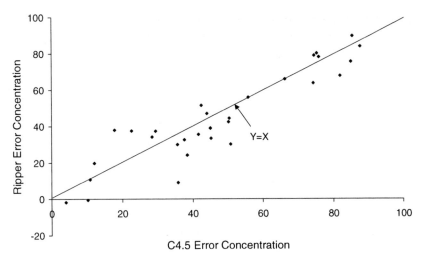

Fig. 9.3 Comparison of C4.5 and Ripper error concentration values

These results show that, for the 30 data sets, when the induced classifier has an error rate less than 12%, then the error concentration is always greater than 0.50. Based on the error rate and error concentration values, the induced classifiers seem to fit naturally into the following three categories:

1. High EC/moderate ER   data sets 1–10 for C4.5 and Ripper
2. Medium EC/high ER   data sets 11–22 for C4.5; 11–24 for Ripper
3. Low EC/high ER   data sets 23–30 for C4.5; 25–30 for Ripper

It is interesting to note that for those data sets in the high-EC/moderate-ER category, the largest disjunct generally covers a very large portion of the total training examples. As an example, consider the hypothyroid data set. Of the 3394 examples (90% of the total data) used for training, nearly 2700 of these examples, or 79%, are covered by the largest disjunct induced by C4.5 and Ripper. To see that these large disjuncts are extremely accurate, consider the vote data set, which falls within the same category. The distribution of errors for the vote data set was shown previously in Fig. 9.1. The data used to generate this figure indicates that the largest disjunct, which covers 23% of the total training examples, does not contribute a single error when used to classify the test data. These observations lead us to speculate that concepts that can be learned well (i.e., have low error rates) are often made up of very general cases that lead to highly accurate large disjunct – and therefore to classifiers with very high error concentrations. Concepts that are difficult to learn, on the other hand, either are not made up of very general cases or, due to limitations with the expressive power of the learner, these general cases cannot be represented using large disjuncts. This leads to classifiers without very large, highly accurate disjuncts and with many small disjuncts. These classifiers tend to have much smaller error concentrations.

9.5 The Effect of Pruning on Small Disjuncts

The results in the previous section, consistent with previous research on small disjuncts, were generated using C4.5 and Ripper with their pruning strategies disabled. Pruning is generally not used when studying small disjuncts because of the belief that it disproportionately eliminates small disjuncts from the induced classifier and thereby obscures the very phenomenon we wish to study. However, because pruning is employed by many learning systems, it is worthwhile to understand how it affects small disjuncts and the distribution of errors across disjuncts – as well as how effective it is at addressing the problem with small disjuncts. In this section we investigate the effect of pruning on the distribution of errors across the disjuncts in the induced classifier. We begin with an illustrative example. Figure 9.4 shows the distribution of errors for the classifier induced from the vote data set using C4.5 with pruning. This distribution can be compared to the corresponding distribution in Fig. 9.1 that was generated using C4.5 without pruning, to show the effect that pruning has on the distribution of errors.

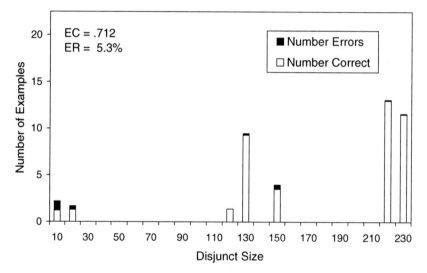

Fig. 9.4 Distribution of examples with pruning for the vote data set

A comparison of Figs. 9.4 with 9.1 shows that with pruning the errors are less concentrated in the small disjuncts. This is also confirmed by the error concentration value, which is reduced from 0.848 to 0.712. It is also apparent that with pruning far fewer examples are classified by disjuncts with size 0–9 and 10–19. The underlying data indicates that without pruning the induced classifiers typically (i.e., over the 10 runs) contain 48 disjuncts, of which 45 are of size 10 or less, while with pruning only 10 disjuncts remain, of which 7 have size 10 or less. So, in this case pruning eliminates 38 of the 45 disjuncts with size 10 or less. This confirms the assumption that pruning eliminates many, if not most, small disjuncts. The emancipated examples – those that would have been classified by the eliminated disjuncts – are now classified by larger disjuncts. It should be noted, however, that even with pruning the error concentration is still quite positive (0.712), indicating that the errors still tend to be concentrated toward the small disjuncts. In this case pruning also causes the overall error rate of the classifier to decrease from 6.9 to 5.3%.

The performance of the classifiers induced from the 30 data sets, using C4.5 and Ripper with their default pruning strategies, is presented in Tables 9.4 and 9.5, respectively. The induced classifiers are again placed into three categories, although in this case the patterns that were previously observed are not nearly as evident. In particular, with pruning some classifiers continue to have low error rates but no longer have large error concentrations (e.g., ocr, soybean-lg, and ticket3 for C4.5 only). In these cases pruning has caused the rarely occurring classification errors to be distributed much more uniformly throughout the disjuncts.

Table 9.4 Error concentration results for C4.5 with pruning

EC rank	Data set	Error rate	Data set size	Largest disjunct	% Errors 10% correct	% Correct 50% errors	Error conc.
1	hypothyroid	0.5	3771	2732	90.7	0.7	0.818
2	ticket1	1.6	556	410	46.7	10.3	0.730
3	vote	5.3	435	221	68.7	2.9	0.712
4	breast-wisc	4.9	699	345	49.6	10.0	0.688
5	kr-vs-kp	0.6	3196	669	35.4	15.6	0.658
6	splice-junction	4.2	3175	479	41.6	25.9	0.566
7	crx	15.1	690	267	45.2	11.5	0.516
8	ticket2	4.9	556	442	48.1	12.8	0.474
9	weather	31.1	5597	573	26.2	22.2	0.442
10	adult	14.1	21,280	5018	36.6	17.6	0.424
11	german	28.4	1000	313	29.6	21.9	0.404
12	soybean-large	8.2	682	61	48.0	14.4	0.394
13	network2	22.2	3826	1685	30.8	21.2	0.362
14	ocr	2.7	2688	1350	40.4	34.3	0.348
15	market1	20.9	3180	830	28.4	23.6	0.336
16	network1	22.4	3577	1470	24.4	27.2	0.318
17	ticket3	2.7	556	431	37.0	20.9	0.310
18	horse-colic	14.7	300	137	35.8	19.3	0.272
19	coding	27.7	20,000	415	17.2	34.9	0.216
20	sonar	28.4	208	50	15.1	34.6	0.202
21	heart-hung.	21.4	293	132	19.9	31.8	0.198
22	hepatitis	18.2	155	89	24.2	26.3	0.168
23	liver	35.4	345	59	17.6	34.8	0.162
24	promoters	24.4	106	26	17.2	37.0	0.128
25	move	23.9	3028	216	14.4	42.9	0.094
26	blackjack	27.6	15,000	3053	16.9	44.7	0.092
27	labor	22.3	57	24	14.3	40.5	0.082
28	bridges	15.8	101	67	14.9	50.1	0.064
29	market2	45.1	11,000	426	12.2	44.7	0.060
30	bands	30.1	538	279	0.8	58.3	0.184

Table 9.5 Error concentration results for Ripper with pruning

EC rank	C4.5 rank	Data set	Error rate	Data set size	Largest disjunct	% Errors 10% correct	% Correct 50% errs	Error conc.
1	1	hypothyroid	0.9	3771	2732	97.2	0.6	0.930
2	5	kr-vs-kp	0.8	3196	669	56.8	5.4	0.746
3	2	ticket1	1.6	556	410	41.5	11.9	0.740
4	6	splice-junction	5.8	3175	552	46.9	10.7	0.690
5	3	vote	4.1	435	221	62.5	2.8	0.648
6	8	ticket2	4.5	556	405	73.3	7.8	0.574
7	17	ticket3	4.0	556	412	71.3	9.0	0.516
8	14	ocr	2.7	2688	854	29.4	24.5	0.306
9	20	sonar	29.7	208	59	23.1	25.4	0.282
10	30	bands	26.0	538	118	22.1	24.0	0.218
11	9	weather	26.9	5597	1148	18.8	35.4	0.198
12	23	liver	32.1	345	69	13.6	34.7	0.146
13	12	soybean-large	9.8	682	66	17.8	47.4	0.128
14	11	german	29.4	1000	390	14.7	32.4	0.128
15	4	breast-wisc	4.4	699	370	14.4	31.4	0.124
16	15	market1	21.3	3180	998	19.0	43.4	0.114
17	7	crx	15.1	690	272	16.4	39.1	0.108
18	13	network2	22.6	3826	1861	15.3	39.5	0.090
19	16	network1	23.3	3577	1765	16.0	42.0	0.090
20	18	horse-colic	15.7	300	141	13.8	36.6	0.086
21	21	hungar-heart	18.8	293	138	17.9	42.6	0.072
22	19	coding	28.3	20,000	894	12.7	46.5	0.052
23	26	blackjack	28.1	15,000	4893	16.8	45.3	0.040
24	22	hepatitis	22.3	155	93	25.5	57.2	0.004
25	29	market2	40.9	11,000	2457	7.7	50.2	0.016
26	28	bridges	18.3	101	71	19.1	55.0	0.024
27	25	move	24.1	3028	320	10.9	63.1	0.094
28	10	adult	15.2	21,280	9293	9.8	67.9	0.146
29	27	labor	18.2	57	25	0.0	70.9	0.228
30	24	promoters	11.9	106	32	0.0	54.1	0.324

The results in Tables 9.4 and 9.5, when compared to the results in Tables 9.2 and 9.3, show that pruning tends to reduce the error concentration of most classifiers. This is shown graphically by the scatter plot in Fig. 9.5. Since most of the points fall below the line $Y = X$, we conclude that for both C4.5 and Ripper, pruning, as expected, tends to reduce error concentration. However, Fig. 9.5 makes it clear that pruning has a more dramatic impact on the error concentration for classifiers induced using Ripper than those induced using C4.5. Pruning causes the error concentration to decrease for 23 of the 30 data sets for C4.5 and for 26 of the 30 data sets for Ripper. More significant, however, is the magnitude of the changes in error concentration. On average, pruning causes the error concentration for classifiers induced using C4.5 to drop from 0.471 to 0.375, while the corresponding drop when using Ripper is from 0.445 to 0.206. These results indicate that the pruned classifiers produced by Ripper have the errors much less concentrated toward the small disjuncts than those produced by C4.5. Given that Ripper is generally known

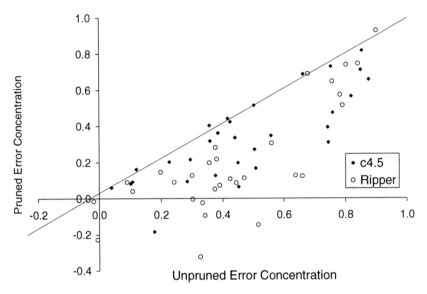

Fig. 9.5 Effect of pruning on error concentration

to produce very simple rule sets, this larger decrease in error concentration is likely due to the fact that Ripper has a more aggressive pruning strategy than C4.5.

The results in Tables 9.4 and 9.5 and in Fig. 9.5 indicate that, even with pruning, the "problem with small disjuncts" is still quite evident for both C4.5 and Ripper. For both learners the error concentration, averaged over the 30 data sets, is still decidedly positive. Furthermore, even with pruning both learners produce many classifiers with error concentrations greater than 0.50. However, it is certainly worth noting that with pruning, seven of the classifiers induced by Ripper have *negative* error concentrations. Comparing the error concentration values for Ripper with and without pruning reveals one particularly interesting example. For the adult data set, pruning causes the error concentration to drop from 0.516 to –0.146. This large change likely indicates that many error-prone small disjuncts are eliminated. This is supported by the fact that the size of the largest disjunct in the induced classifier changes from 1488 without pruning to 9293 with pruning. Thus, pruning seems to have an enormous effect on this Ripper classifier.

The effect that pruning has on error rate is shown graphically in Fig. 9.6 for both C4.5 and Ripper. Because most of the points in Fig. 9.6 fall below the line $Y = X$, we conclude that pruning tends to reduce the error rate for both C4.5 and Ripper. However, the figure also makes it clear that pruning improves the performance of Ripper more than it improves the performance of C4.5. In particular, for C4.5 pruning causes the error rate to drop for 19 of the 30 data sets while for Ripper pruning causes the error rate to drop for 24 of the 30 data sets. Over the 30 data sets pruning causes C4.5's error rate to drop from 18.4 to 17.5% and Ripper's error rate to drop from 19.0 to 16.9%.

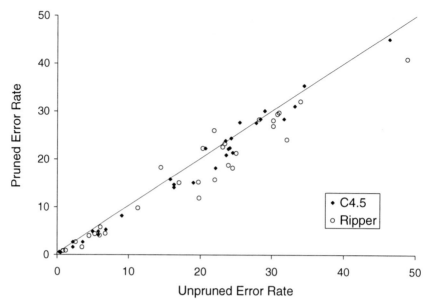

Fig. 9.6 Effect of pruning on error rate

Given that pruning tends to affect small disjuncts more than large disjuncts, an interesting question is whether pruning is more effective at reducing error rate when the errors in the unpruned classifier are most highly concentrated in the small disjuncts. Figure 9.7 addresses this by plotting the absolute reduction in error rate due to pruning vs. the error concentration rank of the unpruned classifier. The data sets with high and medium error concentrations show a fairly consistent reduction in error rate.[1] Finally, the classifiers in the low-EC/high-ER category show a net *increase* in error rate. These results suggest that pruning is most beneficial when the errors are most highly concentrated in the small disjuncts – and may actually hurt when this is not the case. The results for Ripper show a somewhat similar pattern, although the unpruned classifiers with low error concentrations do consistently show some reduction in error rate when pruning is used.

The results in this section show that pruned classifiers generally have lower error rates and lower error concentrations than their unpruned counterparts. Our analysis shows us that for the vote data set this change is due to the fact that pruning eliminates most small disjuncts. A similar analysis, performed for other data sets in this study, shows a similar pattern – pruning eliminates mostsmall disjuncts. In

---

[1] Note that although the classifiers in the medium-EC/high-ER category show a greater absolute reduction in error rate than those in the high-EC/moderate-ER group, this corresponds to a smaller relative reduction in error rate, due to the differences in the error rate of the unpruned classifiers.

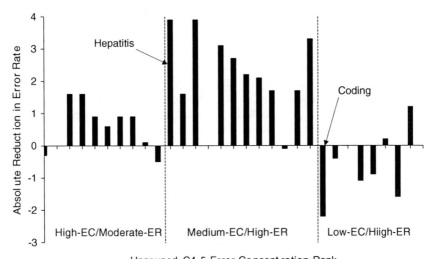

Fig. 9.7 Improvement in error rate versus error concentration rank

summary, pruning is a strategy for dealing with the "problem of small disjuncts." Pruning eliminates many small disjuncts and the emancipated examples that would have been classified by the eliminated disjuncts are then classified by other, typically much larger, disjuncts. The result of pruning is that there is a decrease in the average error rate of the induced classifiers and the remaining errors are more uniformly distributed across the disjuncts.

One can gauge the effectiveness of pruning as a strategy for addressing the problem with small disjuncts by comparing it to an "ideal" strategy that causes the error rate of the small disjuncts to equal the error rate of the larger disjuncts. Table 9.6 shows the average error rates of the classifiers induced by C4.5 for the 30 data sets, without pruning, with pruning, and with two variants of this idealized strategy. The error rates for the idealized strategies are determined by first identifying the smallest disjuncts that collectively cover 10% (20%) of the training examples and then calculating the error rate of the classifier as if the error rate of these small disjuncts equaled the error rate of the examples classified by all of the other disjuncts.

Table 9.6 Comparison of pruning to idealized strategy

	Strategy			
	No Pruning (%)	Pruning (%)	Idealized (10%)	Idealized (20%)
Average error rate	18.4	17.5	15.2%	13.5%
Relative improvement		4.9	17.4%	26.6%

The results in Table 9.6 show that the idealized strategy yields much more dramatic improvements in error rate than pruning, even when it is only applied to the disjuncts that cover 10% of the training examples. This indicates that pruning is not very effective at addressing the problem with small disjuncts and provides a strong motivation for finding better strategies for handling small disjuncts (several such strategies are discussed in Section 9.9). Note, however, that we are not suggesting that the performance of the idealized strategies can necessarily ever be realized.

For many real-world problems, it is more important to classify a reduced set of examples with high precision than in finding the classifier with the best overall accuracy. For example, if the task is to identify customers likely to buy a product in response to a direct marketing campaign, it may be impossible to utilize all classifications – budgetary concerns may permit one to only contact the 10,000 people most likely to make a purchase. Given that our results indicate that pruning *decreases* the precision of the larger, more precise disjuncts (compare Figs. 9.1 and 9.4), this suggests that pruning may be harmful in such cases – even though pruning leads to an overall increase in the accuracy of the induced classifier. To investigate this further, classifiers were generated by starting with the largest disjunct and then progressively adding smaller disjuncts. A classification decision is made only if an example is covered by one of the added disjuncts; otherwise no classification is made. The error rate (i.e., precision) of the resulting classifiers, generated with and without pruning, is shown in Table 9.7, as is the difference in error rates. A negative difference indicates that pruning leads to an improvement (i.e., a reduction) in error rate, while a positive difference indicates that pruning leads to an increase in error rate. Results are reported for classifiers with disjuncts that collectively cover 10, 30, 50, 70, and 100% of the training examples.

The last row in Table 9.7 shows the error rates averaged over the 30 data sets. These results clearly show that, over the 30 data sets, pruning only helps for the last column – when all disjuncts are included in the evaluated classifier. Note that these results, which correspond to the accuracy results presented earlier, are typically the only results that are described. This leads to an overly optimistic view of pruning, since in other cases pruning results in a *higher* overall error rate. As a concrete example, consider the case where we use only the disjuncts that collectively cover 50% of the training examples. In this case C4.5 with pruning generates classifiers with an average error rate of 12.9% whereas C4.5 without pruning generates classifiers with an average error rate of 11.4%. Looking at the individual results for this situation, pruning does worse for 17 of the data sets, better for 9 of the data sets, and the same for 4 of the data sets. However, the magnitude of the differences is much greater in the cases where pruning performs worse.

The results from the last row of Table 9.7 are displayed graphically in Fig. 9.8, which plots the error rates, with and without pruning, averaged over the 30 data sets. Note, however, that unlike the results in Table 9.7, Fig. 9.8 shows classifier performance at each 10% increment.

Figure 9.8 clearly demonstrates that under most circumstances pruning does *not* produce the best results. While it produces marginally better results when predictive

Table 9.7 Effect of pruning when classification based only on largest disjuncts

| Data set name | Error rate with pruning (Yes) and without pruning (No) | | | | | | | | | | | |
| | 10% covered | | | 30% covered | | | 70% covered | | | 100% covered | | |
Pruning used:	Yes	No	Δ	Yes	No	Δ	Yes	No	Δ	Yes	No	Δ
kr-vs-kp	0.0	0.0	0.0	0.0	0.0	0.0	0.1	0.0	0.1	0.6	0.3	0.3
hypothyroid	0.1	0.3	−0.2	0.2	0.1	0.1	0.1	0.0	0.0	0.5	0.5	0.0
vote	3.1	0.0	3.1	1.0	0.0	1.0	2.3	0.7	1.6	5.3	6.9	−1.6
splice-junction	0.3	0.9	−0.6	0.2	0.3	−0.1	2.4	0.6	1.8	4.2	5.8	−1.6
ticket2	0.3	0.0	0.3	2.7	0.8	1.9	2.5	1.0	1.5	4.9	5.8	−0.9
ticket1	0.1	2.1	−1.9	0.3	0.6	−0.3	0.3	0.3	0.0	1.6	2.2	−0.5
ticket3	2.1	2.0	0.1	1.7	1.2	0.5	1.5	0.5	1.0	2.7	3.6	−0.9
soybean-large	1.5	0.0	1.5	5.4	1.0	4.4	4.7	1.3	3.5	8.2	9.1	−0.9
breast-wisc	1.5	1.1	0.4	1.0	1.0	0.0	1.0	1.4	−0.4	4.9	5.0	−0.1
ocr	1.5	1.8	−0.3	1.9	0.8	1.1	1.9	1.0	0.9	2.7	2.2	0.5
hepatitis	5.4	6.7	−1.3	15.0	2.2	12.9	12.8	12.1	0.6	18.2	22.1	−3.9
horse-colic	20.2	1.8	18.4	14.6	4.6	10.0	10.7	10.6	0.1	14.7	16.3	−1.7
crx	7.0	7.3	−0.3	7.9	6.5	1.4	7.8	9.3	−1.6	15.1	19.0	−3.9
bridges	10.0	0.0	10.0	17.5	0.0	17.5	14.9	9.4	5.4	15.8	15.8	0.0
heart-hung.	15.4	6.2	9.2	18.4	11.4	7.0	16.0	16.4	−0.4	21.4	24.5	−3.1
market1	16.6	2.2	14.4	12.2	7.8	4.4	14.5	15.9	−1.4	20.9	23.6	−2.6
adult	3.9	0.5	3.4	3.6	4.9	−1.3	8.3	10.6	−2.3	14.1	16.3	−2.2
weather	5.4	8.6	−3.2	10.6	14.0	−3.4	22.7	24.6	−1.9	31.1	33.2	−2.1
network2	10.8	9.1	1.7	12.5	10.7	1.8	15.1	17.2	−2.1	22.2	23.9	−1.8
promoters	10.2	19.3	−9.1	10.9	10.4	0.4	19.6	16.8	2.8	24.4	24.3	0.1
network1	15.3	7.4	7.9	13.1	11.8	1.3	16.7	17.3	−0.6	22.4	24.1	−1.7
german	10.0	4.9	5.1	11.1	12.5	−1.4	20.4	25.7	−5.3	28.4	31.7	−3.3
coding	19.8	8.5	11.3	18.7	14.3	4.4	23.6	20.6	3.1	27.7	25.5	2.2
move	24.6	9.0	15.6	19.2	12.1	7.1	22.6	18.7	3.8	23.9	23.5	0.3
sonar	27.6	27.6	0.0	23.7	23.7	0.0	24.4	24.3	0.1	28.4	28.4	0.0
bands	13.1	0.0	13.1	34.3	16.3	18.0	33.8	26.6	7.2	30.1	29.0	1.1
liver	27.5	36.2	−8.8	32.4	28.1	4.3	30.7	31.8	−1.2	35.4	34.5	0.9
blackjack	25.3	26.1	−0.8	25.1	25.8	−0.8	26.1	24.4	1.7	27.6	27.8	−0.2
labor	25.0	25.0	0.0	17.5	24.8	−7.3	24.4	17.5	6.9	22.3	20.7	1.6
market2	44.1	45.5	−1.4	43.1	44.3	−1.2	43.3	45.3	−2.0	45.1	46.3	−1.2
Average	11.6	8.7	2.9	12.5	9.7	2.8	14.2	13.4	0.8	17.5	18.4	−0.9

accuracy is the evaluation metric (i.e., all examples must be classified), it produces much poorer results when one can be very selective about the classification "rules" that are used. These results confirm the hypothesis that when pruning eliminates some small disjuncts, the emancipated examples cause the error rate of the more accurate large disjuncts to decrease. The overall error rate is reduced only because the error rate for the emancipated examples is lower than their original error rate. Thus, pruning redistributes the errors such that the errors are more uniformly distributed than without pruning. This is exactly what one does not want to happen when one can be selective about which examples to classify (or which classifications to act upon). We find the fact that pruning improves only classifier performance when disjuncts covering more than 80% of the training examples are used to be quite compelling.

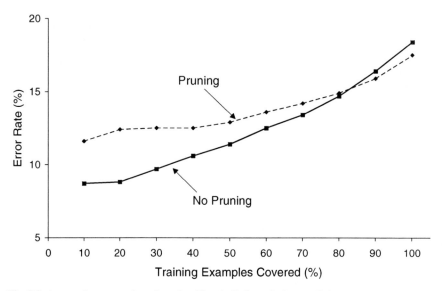

Fig. 9.8 Averaged error rate based on classifiers built from the largest disjuncts

9.6 The Effect of Training Set Size on Small Disjuncts

The amount of training data available for learning has several well-known effects. Namely, increasing the amount of training data will tend to increase the accuracy of the classifier and increase the number of "rules," as additional training data permits the existing rules to be refined. In this section we analyze the effect that training set size has on small disjuncts and error concentration.

Figure 9.9 returns to the vote data set example, but this time shows the distribution of examples and errors when the training set is limited to use only 10% of the total data. These results can be compared with those in Fig. 9.1, which are based upon 90% of the data being used for training. Thus, the results in Fig. 9.9 are based on 1/9th the training data used in Fig. 9.1. Note that the size of the bins, and consequently the scale of the x-axis, has been reduced in Fig. 9.9.

A comparison of the relative distribution of errors between Figs. 9.9 and 9.1 shows that errors are more concentrated toward the smaller disjuncts in Fig. 9.1, which has a higher error concentration (0.848 vs. 0.628). This indicates that increasing the amount of training data increases the degree to which the errors are concentrated toward the small disjuncts. Like the results in Fig. 9.1, the results in Fig. 9.9 show that there are three groupings of disjuncts, which one might be tempted to refer to as small, medium, and large disjuncts. The size of the disjuncts within each group differs between the two figures, due to the different number of training examples used to generate each classifier (note the change in scale of the x-axis). It is informative to compare the error concentrations for classifiers induced

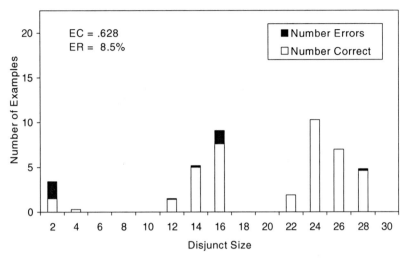

Fig. 9.9 Distribution of examples for the vote data set (using 1/9 of the normal training data)

using different training set sizes because error concentration is a relative measure – it measures the distribution of errors within the classifier relative to the disjuncts within the classifier and relative to the total number of errors produced by the classifier (which will be less when more training data is available). Summary statistics for all 30 data sets are shown in Table 9.8.

Table 9.8 shows the error rate and error concentration for the classifiers induced from each of the 30 data sets using three different training set sizes. The last two columns highlight the impact of training set size, by showing the change in error concentration and error rate that occurs when the training set size is increased by a factor of 9. As expected, the error rate tends to decrease with additional training data while the error concentration, consistent with the results associated with the vote data set, shows a consistent increase – for 27 of the 30 data sets the error concentration increases when the amount of training data is increased by a factor of 9.

The observation that an increase in training data leads to an increase in error concentration can be explained by analyzing how an increase in training data affects the classifier that is learned. As more training data becomes available, the induced classifier is better able to sample, and learn, the general cases that exist within the concept. This causes the classifier to form highly accurate large disjuncts. As an example, note that the largest disjunct in Fig. 9.1 does not cover a single error and that the medium-sized disjuncts, with sizes between 80 and 109, cover only a few errors. Their counterparts in Fig. 9.9, with size between 20 and 27 and 10 and 15, have a higher error rate. Thus, an increase in training data leads to more accurate large disjuncts and a higher error concentration. The small disjuncts that are formed using the increased amount of training data may correspond to rare cases within the concept that previously were not sampled sufficiently to be learned.

Table 9.8 The effect of training set size on error concentration

Data Set	Amount of total data used for training						Δ from 10 to 90%	
	10%		50%		90%			
	ER	EC	ER	EC	ER	EC	ER	EC
kr-vs-kp	3.9	0.742	0.7	0.884	0.3	0.874	−3.6	0.132
hypothyroid	1.3	0.910	0.6	0.838	0.5	0.852	−0.8	−0.058
vote	9.0	0.626	6.7	0.762	6.9	0.848	−2.1	0.222
splice-junction	8.5	0.760	6.3	0.806	5.8	0.818	−2.7	0.058
ticket2	7.0	0.364	5.7	0.788	5.8	0.758	−1.2	0.394
ticket1	2.9	0.476	3.2	0.852	2.2	0.752	−0.7	0.276
ticket3	9.5	0.672	4.1	0.512	3.6	0.744	−5.9	0.072
soybean-large	31.9	0.484	13.8	0.660	9.1	0.742	−22.8	0.258
breast-wisc	9.2	0.366	5.4	0.650	5.0	0.662	−4.2	0.296
ocr	8.9	0.506	2.9	0.502	2.2	0.558	−6.7	0.052
hepatitis	22.2	0.318	22.5	0.526	22.1	0.508	−0.1	0.190
horse-colic	23.3	0.452	18.7	0.534	16.3	0.504	−7.0	0.052
crx	20.6	0.460	19.1	0.426	19.0	0.502	−1.6	0.042
bridges	16.8	0.100	14.6	0.270	15.8	0.452	−1.0	0.352
heart-hungarian	23.7	0.216	22.1	0.416	24.5	0.450	0.8	0.234
market1	26.9	0.322	23.9	0.422	23.6	0.440	−3.3	0.118
adult	18.6	0.486	17.2	0.452	16.3	0.424	−2.3	−0.062
weather	34.0	0.340	32.7	0.380	33.2	0.416	−0.8	0.076
network2	27.8	0.354	24.9	0.342	23.9	0.384	−3.9	0.030
promoters	36.0	0.108	22.4	0.206	24.3	0.376	−11.7	0.268
network1	28.6	0.314	25.1	0.354	24.1	0.358	−4.5	0.044
german	34.3	0.248	33.3	0.334	31.7	0.356	−2.6	0.108
coding	38.4	0.214	30.6	0.280	25.5	0.294	−12.9	0.080
move	33.7	0.158	25.9	0.268	23.5	0.284	−10.2	0.126
sonar	40.4	0.028	27.3	0.292	28.4	0.226	−12.0	0.198
bands	36.8	0.100	30.7	0.152	29.0	0.178	−7.8	0.078
liver	40.5	0.030	36.4	0.054	34.5	0.120	−6.0	0.090
blackjack	29.4	0.100	27.9	0.094	27.8	0.108	−1.6	0.008
labor	30.3	0.114	17.0	0.044	20.7	0.102	−9.6	−0.012
market2	47.3	0.032	45.7	0.028	46.3	0.040	−1.0	0.008
Average	23.4	0.347	18.9	0.438	18.4	0.471	−5.0	0.124

In this section we noted that additional training data reduces the error rate of the induced classifier and increases its error concentration. These results help to explain the pattern, described in Section 9.4, that classifiers with low error rates tend to have higher error concentrations than those with high error rates. That is, if we imagine that additional training data were made available to those data sets where the associated classifier has a high error rate, we would expect the error rate to decline and the error concentration to increase. This would tend to move classifiers into the high-EC/moderate-ER category. Thus, to a large extent, the pattern that was established in Section 9.4 between error rate and error concentration reflects the degree to which a concept has been learned – concepts that have been well-learned tend to have very large disjuncts which are extremely accurate and hence have low error concentrations.

9.7 The Effect of Noise on Small Disjuncts

Noise plays an important role in classifier learning. Both the structure and performance of a classifier will be affected by noisy data. In particular, noisy data may cause many erroneous small disjuncts to be induced. Danyluk and Provost [8] speculated that the classifiers they induced from (systematic) noisy data performed poorly because of an inability to distinguish between these erroneous consistencies and correct ones. Weiss [17] and Weiss and Hirsh [19] explored this hypothesis using, respectively, two artificial data sets and two real-world data sets and showed that noise can make rare cases (i.e., true exceptions) in the true, unknown, concept difficult to learn. The research presented in this section further investigates the role of noise in learning, and, in particular, shows how noisy data affects induced classifiers and the distribution of the errors across the disjuncts within these classifiers.

The experiments described in this section involve applying random class noise and random attribute noise to the data. The following experimental scenarios are explored:

Scenario 1: Random class noise applied to the training data
Scenario 2: Random attribute noise applied to the training data
Scenario 3: Random attribute noise applied to both training and test data

Class noise is applied only to the training set since the uncorrupted class label in the test set is required to properly measure classifier performance [12]. The second scenario, in which random attribute noise is applied only to the training set, permits us to measure the sensitivity of the learner to noise (if attribute noise were applied to the test set then even if the correct concept were learned there would be classification errors). The third scenario, in which attribute noise is applied to both the training and test sets, corresponds to the real-world situation where errors in measurement affect all examples. A level of $n\%$ random class noise means that for $n\%$ of the examples the class label is replaced by a randomly selected class value, including possibly the original value. Attribute noise is defined similarly, except that for numerical attributes a random value is selected between the minimum and maximum values that occur within the data set. Note that only when the noise level reaches 100% is all information contained within the original data lost.

The vote data set is used to illustrate the effect that noise has on the distribution of examples, by disjunct size. The results are shown in Fig. 9.10a–f, with the graphs in the left column corresponding to the case when there is no pruning and the graphs in the right column corresponding to the case when pruning is employed. Figure 9.10a, which is an exact copy of Fig. 9.1, and Fig. 9.10b, which is an exact copy of Fig. 9.4, show the results without any noise and are provided for comparison purposes. Figures 9.10c and 9.10d correspond to the case where 10% attribute noise is applied to the training data and Figs. 9.10e and 9.10f to the case where 10% class noise is applied to the train-ing data.

A comparison of Fig. 9.10a,c and e shows that both attribute and class noise cause more test examples to be covered by small disjuncts, although this shift is

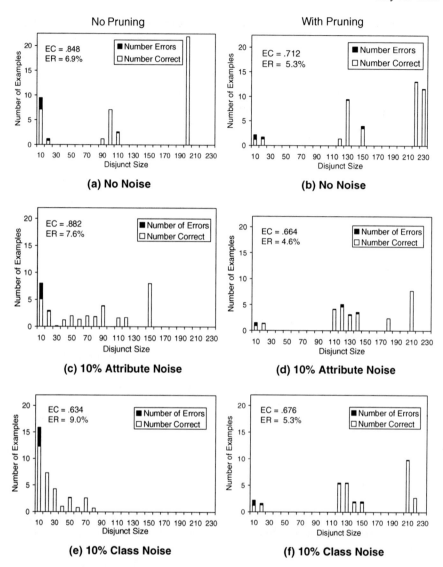

Fig. 9.10 The effect that noise has on the distribution of examples, by disjunct size

more dramatic for class noise than for attribute noise. The underlying data indicates that this shift occurs because noisy data causes more small disjuncts to be formed. This comparison also shows that the error concentration remains fairly stable when attribute noise is added but decreases significantly when class noise is added.

A careful examination of Fig. 9.10 makes it clear that pruning reduces the shift in distribution of (correctly and incorrectly) examples that is observed when pruning is not used. A comparison of the error rates for classifiers with and without pruning also shows that pruning is able to combat the effect of noise on the ability of the classifier to learn the concept. Surprisingly, when pruning is used, classifier accuracy for the vote data set actually improves when 10% attribute noise is added – the error rate decreases from 5.3 to 4.6%. This phenomenon, which is discussed in more detail shortly, is actually observed for many of the 30 data sets, but only when low (e.g., 10%) levels of attribute noise are added. The error concentration results also indicate that even with pruning, noise causes the errors to be distributed more uniformly throughout the disjuncts than when no noise is applied.

The results presented in the remainder of this section are based on averages over 27 of the 30 data sets listed in Table 9.1 (the coding, ocr, and bands data sets were omitted due to difficulties applying our noise model to these data sets). The next three figures show, respectively, how noise affects the number of leaves, the error rate, and the error concentration of the induced classifiers. Measurements are taken at the following noise levels: 0, 5, 10, 20, 30, 40, and 50%. The curves in these figures are labeled to identify the type of noise that is applied, whether it is applied to the training set or training and test sets, and whether pruning is used. The labels are interpreted as follows: the "Class" and "Attribute" prefix indicate the type of noise, the "-Both" term, if included, indicates that the noise is applied to the training and test sets rather than to just the training set, and the "-Prune" suffix is used to indicate that the results are with pruning.

Figure 9.11 shows that without pruning the number of leaves in the induced decision tree increases dramatically with increasing levels of noise, but that pruning effectively eliminates this increase. The effect that noise has on error rate is shown in Fig. 9.12. Error rate increases with increasing levels of noise, with one exception. When attribute noise is applied to only the training data and pruning is used, the error rate decreases slightly from 17.7% with 5% noise to 17.5% with 10% noise.

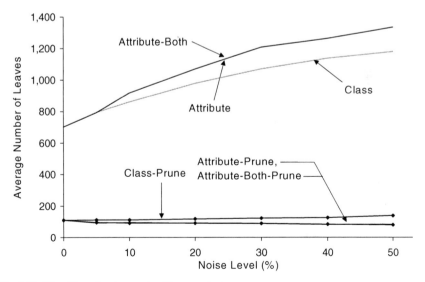

Fig. 9.11 The effect of noise on classifier complexity

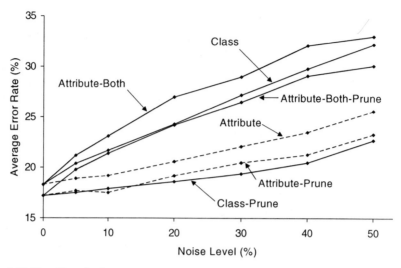

Fig. 9.12 The effect of noise on error rate

This decrease is no anomaly, since it occurs for many of the data sets analyzed. We believe the decrease in error rate may be due to the fact that attribute noise leads to more aggressive pruning (most of the data sets that show the decrease in error rate have high overall error rates, which perhaps are more likely to benefit from aggressive pruning). Figure 9.12 also shows that pruning is far more effective at handling class noise than attribute noise.

Figure 9.13 shows the effect of noise on error concentration. When pruning is not employed, increasing levels of noise lead to decreases in error concentration,

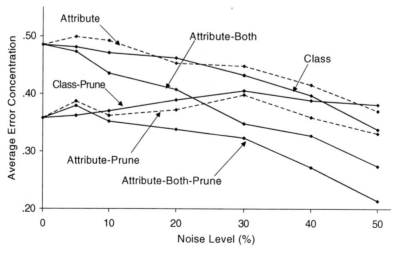

Fig. 9.13 The effect of noise on error concentration The effect of class distribution on error concentration

indicating that errors become more uniformly distributed based on disjunct size. This helps explain why we find a low-ER/high-EC group of classifiers and a high-ER/medium-EC group of classifiers: adding noise to classifiers in the former increases their error rate and decreases their error concentration, making them look more like classifiers in the latter group. The results in Fig. 9.13 also show, however, that when there is noise only in the training set, then pruning causes the error concentration to remain relatively constant (this is especially true for class noise).

The results in this section demonstrate that pruning enables the learner to combat noisy training data. Specifically, pruning removes many of the disjuncts that are caused by the noise (Fig. 9.11) and this yields a much smaller increase in error rate than if pruning were not employed (Fig. 9.12). Because pruning eliminates many of the erroneous small disjuncts, the errors are not nearly as concentrated in the small disjuncts (Fig. 9.13). We believe that the increase in error rate that comes from noisy training data when pruning is employed is at least partly due to the inability of the learner to distinguish between true exceptions and noise.

The detailed results associated with the individual data sets show that for class noise there is a trend for data sets with high error concentrations to experience a greater increase in error rate from class noise. What is much more apparent, however, is that many classifiers with low error concentrations are *extremely* tolerant of class noise, whereas none of the classifiers with high error concentrations exhibit this tolerance. For example, the blackjack and labor data sets, both of which have low error concentrations, are so tolerant of noise that when 50% random class noise is added to the training set, the error rate on the induced classifier on the test data increases by less than 1%. These results are consistent with the belief that noise makes learning difficult because it makes of an inability to distinguish between true exceptions and noise. Even without the addition of noise, none of the concepts can be induced perfectly (i.e., they have nonzero error rate). The classifiers with a high error concentration already show an inability to properly learn the rare cases in the concept (which show up as small disjuncts) – the addition of noise simply worsens the situation. Those concepts with very general cases that can be learned well without noise (leading to highly accurate large disjuncts and low error concentrations) are less susceptible to noise. For example, corrupting the class labels for a few examples belonging to a very large disjunct is unlikely to change the class label learned for that disjunct.

9.8 The Effect of Class Imbalance on Small Disjuncts

A data set exhibits class imbalance if the number of examples belonging to each class is unequal. A great deal of recent research, some of which is described in Section 9.9, has studied the problem of learning classifiers from imbalanced data, since this has long been recognized as commonly occurring and difficult data mining problem. However, with few exceptions [11, 22], this research has not examined the role of small disjuncts when learning from imbalanced data.

The study by Weiss and Provost [22] showed that examples truly belonging to the minority class are misclassified much more often than examples belonging to the majority class and that examples labeled by the classifier as belonging to the minority class (i.e., minority-class predictions) have much higher error rates than those labeled with the majority class. That study further showed that the minority-labeled disjuncts tend to cover fewer training examples than the majority-labeled disjuncts. This result is not surprising given that the minority class has, by definition, fewer training examples than the majority class.[2] The study concluded that part of the reason that minority-class predictions are more error prone than majority-class predictions is because the minority-class predictions have a lower average disjunct size and hence suffer more from the problem with small disjuncts. The work by Jo and Japkowicz [11] is discussed in Section 9.9.

In this section we extend the research by Weiss and Provost [22] to consider whether there is a causal link between class imbalance and the problem with small disjuncts in the opposite direction. That is, we consider whether class imbalance causes small disjuncts to have a higher error rate than large disjuncts, or, more generally, whether an increase in class imbalance will cause an increase in error concentration. Before evaluating this hypothesis empirically, it is useful to speculate why such a causal link might exist. Weiss and Provost suggested that one reason that minority-class predictions are more error prone than the majority-class predictions is because, by definition, there are more majority-class test examples than minority-class test examples. To see why this is so, imagine a data set for which there are nine majority-class examples for every one minority-class example. If one *randomly* generates a classifier and *randomly* labels each disjunct (e.g., leaf), then the minority-labeled disjuncts will have an expected error rate of 90% while the majority-labeled disjuncts will have an expected error rate of only 10%. Thus, this test-distribution effect favors majority-class predictions. Given that Weiss and Provost showed that small disjuncts are disproportionately likely to be labeled with the minority class, one would therefore expect this test-distribution effect to favor the larger disjuncts over the smaller disjuncts.

We evaluate this hypothesis by altering the class distribution of data sets and then measuring the error concentration associated with the induced classifiers. For simplicity, we look at only two class distributions for each data set: the naturally occurring class distribution and a perfectly balanced class distribution, in which each class is represented in equal proportions. By comparing the error concentrations for these two class distributions, we can also determine how much of the "problem with small disjuncts" is due to class imbalance in the data set.

We form data sets with the natural and balanced class distributions using the methodology described by Weiss and Provost [22]. This methodology employs stratified sampling, without replacement, to form the desired class distribution from the

---

[2] The detailed results show that the induced classifiers have more majority-labeled disjuncts than minority-labeled disjuncts, but the ratio of majority-labeled disjuncts to minority-labeled disjuncts is smaller than the ratio of majority-class examples to minority-class examples. Thus the majority-class disjuncts cover more examples than the minority-class examples.

original data set. The number of examples selected for training is the same for the natural and balanced versions of each data set, to ensure that any differences in performance are due solely to the difference in class distribution (the actual number of training examples that are used is reduced from what is available, to ensure that the balanced class distribution can be formed without duplicating any examples). Because this methodology reduces the number of training examples, we exclude the small data sets when studying class imbalance, so that all classifiers are induced from using a "reasonable" number of examples. The data sets employed in this section include the larger data sets from Table 9.1 plus some additional data sets. These data sets, listed in Table 9.9, are identical to the ones studied by Weiss and Provost [22]. They include 20 data sets from the UCI repository, 5 data sets, identified with a "+," from previously published work by researchers at AT&T [7] and one new data set, the phone data set, generated by the author. The data sets are listed in order of decreasing class imbalance (the percentage of minority-class examples in each data set is included). In order to simplify the presentation and analysis of the results, data sets with more than two classes were mapped into two classes by designating the least frequently occurring class as the minority class and mapping the remaining classes into a new, majority class. Each data set that originally started with more than two classes is identified with an asterisk (*).

Table 9.9 Description of data sets for class imbalance experiments

No.	Data set	Min. (%)	Size	No.	Data set	Min. (%)	Size
1	letter-a*	3.9	20,000	14	network2	27.9	3826
2	pendigits*	8.3	13,821	15	yeast*	28.9	1484
3	abalone*	8.7	4177	16	network1+	29.2	3577
4	sick-euthyroid	9.3	3163	17	car*	30.0	1728
5	connect-4*	9.5	11,258	18	german	30.0	1.000
6	optdigits*	9.9	5620	19	breast-wisc	34.5	699
7	covertype*	14.8	581,102	20	blackjack+	35.6	15,000
8	solar-flare*	15.7	1389	21	weather+	40.1	5597
9	phone	18.2	652,557	22	bands	42.2	538
10	letter-vowel*	19.4	20,000	23	market1+	43.0	3181
11	contraceptive*	22.6	1473	24	crx	44.5	690
12	adult	23.9	48,842	25	kr-vs-kp	47.8	3196
13	splice-junction*	24.1	3175	26	move+	49.4	3029

Figure 9.14 shows the error concentration for the classifiers induced by C4.5 from the natural and balanced versions of the data sets listed in Table 9.9. Since the error concentrations are all greater than zero when there is no class imbalance, we conclude that even with a balanced data set errors tend to be concentrated toward the smaller disjuncts. However, by comparing the error concentrations associated with the classifiers induced from the balanced and natural class distributions, we see that when there is class imbalance, with few exceptions, the error concentration increases. The differences tend to be larger when the data set has greater class imbalance (the leftmost data set has the most natural class imbalance and the class imbalance decreases from left to right).

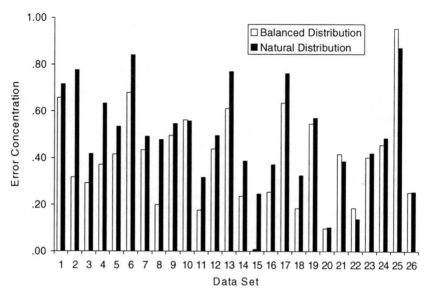

Fig. 9.14 The effect of class distribution on error concentration

If we look at the average error concentration for the classifiers induced from the natural and balanced versions of the 26 data sets, we see that the balanced versions have an average error concentration of 0.396 while the natural versions have an average error concentration of 0.496. This corresponds to a 20% reduction in error concentration when class imbalance is removed. If we restrict our attention to the first 18 data sets, which contain at most 30% minority-class examples, then the differences in error concentration are 28% (0.387 for the balanced data sets versus 0.537 for the data sets with the natural class distributions). We therefore conclude that for data sets with class imbalance, part of the reason why small disjuncts have a higher error rate than the large disjuncts is due to the fact that minority-class predictions are more likely to be erroneous due to the test-distribution effect described earlier. This is empirical evidence that class imbalance is partly responsible for the problem with small disjuncts. This also indicates that if one artificially modifies the class distribution of the training data to be more balanced, then the error concentration will decrease. This observation may help explain why, as noted by Weiss and Provost [22], classifiers built using balanced class distributions tend to be quite robust.

9.9 Related Work

Research on small disjuncts can be placed into the following three categories, which we use to organize our discussion of related work. These three categories are based on whether the purpose of the research is to:

1. characterize and/or measure the role of small disjuncts in learning,
2. provide a better understanding of small disjuncts (e.g., why they are more error prone than large disjuncts), or
3. design better classifiers that address the problem with small disjuncts.

Most previous research on small disjuncts only incidentally tried to characterize or measure the role of small disjuncts in learning and only analyzed one or two data sets [1, 3, 8, 9, 17, 19]. This made it impossible to form any general conclusions. We addressed this problem by analyzing 30 data sets.

Some research has focused on providing a better understanding of small disjuncts. Danyluk and Provost [8] observed that in the domain they were studying, when they trained using noisy data, classifier accuracy suffered severely. They speculated that this occurred because (1) it is difficult to distinguish between noise and true exceptions and (2) in their domain, errors in measurement and classification often occur systematically rather than randomly. Thus, they speculated that it was difficult to distinguish between erroneous consistencies and correct ones. This speculation formed the basis for the work by Weiss [17] and Weiss and Hirsh [19]. Weiss [17] investigates the interaction between noise, rare cases, and small disjuncts using synthetic data sets, for which the true "concept" is known and can be manipulated. Some synthetic data sets were constructed from concepts that included many rare, or exceptional cases, while others were constructed from concepts that mainly included general cases. The research showed that the rare cases tended to form small disjuncts in the induced classifier. It further showed that systematic attribute noise, class noise, and missing attributes can each cause the small disjuncts to have higher error rates than the large disjuncts, and also cause those test examples that correspond to rare cases to be misclassified more often than those test examples corresponding to common cases. That paper also provided an explanation for this behavior: it is asserted that attribute noise in the training data can cause the common cases to look like the rare cases, thus "overwhelming" the rare cases and causing the wrong subconcept to be learned.

The majority of research on small disjuncts focuses on ways to address the problem with small disjuncts. Holte et al. [9] evaluate several strategies for improving learning in the presence of small disjuncts. They show that the strategy of eliminating all small disjuncts is ineffective, because the emancipated examples are then even more likely to be misclassified. The authors focus on a strategy of making small disjuncts highly specific and argue that while a maximum generality bias, which is used by systems such as ID3, is appropriate for large disjuncts, it is not appropriate for small disjuncts. To test this claim, they ran experiments where a maximum generality bias is used for the large disjuncts and a maximum specificity bias is used for the small disjuncts (for a maximum specificity bias *all* conditions satisfied by the training examples covered by a disjunct are added to the disjunct). The experimental results show that with the maximum specificity bias, the resulting disjuncts cover fewer cases but have much lower error rates. Unfortunately, the emancipated examples increase the error rate of the large disjuncts to the extent that the overall error rates remain roughly the same. Although the authors also experiment with a more selective bias that produces interesting results, it does not demonstrably improve learning.

Ting [15] evaluates a method for improving the performance of small disjuncts that also uses a maximum specificity bias. However, unlike the method employed by Holte et al. [9], this method does not affect (and therefore cannot degrade) the performance of the large disjuncts. The basic approach is to use C4.5 to determine if an example is covered by a small or large disjunct. If it is covered by a large disjunct, then C4.5 is used to classify the example. However, if the example is covered by a small disjunct, then IB1, an instance-based learner, is used to classify the example. Instance-based learning is used in this case because it can be considered an extreme example of the maximum specificity bias. In order to use this hybrid learning method, there must be a specific criterion for determining what is a small disjunct. The paper empirically evaluates alternative criteria, based on a threshold value and (1) the absolute size of the disjunct, (2) the relative size of the disjunct, or (3) the error rate of the disjunct. For each criterion, only the best result, produced using the best threshold, is displayed. The results are therefore overly optimistic because the criteria/threshold values are selected using the test data rather than an independent holdout set. Thus, although the observed results are encouraging, it cannot be claimed that the composite learner is very successful in addressing the problem with small disjuncts.

Carvalho and Freitas [3] employ a hybrid method similar to that used by Ting [15]. They also use C4.5 to build a decision tree and then, for each training example, use the size of the leaf covering that example to determine if the example is covered by a small or large disjunct. The training examples that fall into each small disjunct are then fed together into a genetic algorithm-based learner that forms rules to specifically cover the examples that fall into that individual disjunct. Test examples that fall into leaves corresponding to large disjuncts are then assigned a class label based on the decision tree; test examples that fall into a small disjunct are classified by the rules learned by the genetic algorithm for that particular disjunct. Their results are also encouraging, but, because they are based on only a few data sets, and because, as with the results by Ting [15], the improvements in error rate are only seen for certain specific definitions of "small disjunct," it cannot be concluded that this research substantially addresses the problem with small disjuncts.

Several other approaches are advocated for addressing the problem with small disjuncts. Quinlan [13] tries to minimize the problem by improving the probability estimates used to assign a class label to a disjunct. A naive estimate of the error rate of a disjunct is the proportion of the training examples that it misclassifies. However, this estimate performs quite poorly for small disjuncts, due to the small number of examples used to form the estimate. Quinlan describes a method for improving the accuracy estimates of the small disjuncts by taking the class distribution into account. The motivation for this work is that for unbalanced class distributions one would expect the disjuncts that predict the majority class to have a lower error rate than those predicting the minority class (this is the test-distribution effect described in Section 9.8). Quinlan incorporates these *prior probabilities* into the error rate estimates. However, instead of using the overall class distribution as the prior probability, Quinlan generates a more representative measure by calculating the class distribution only on those training examples that are "close" to the small disjunct – that is, fail to satisfy at most one condition in the disjunct. The experimental results demonstrate that Quinlan's error rate estimation model outperforms the naive method, most significantly for skewed distributions.

Van den Bosch et al. [16] advocate the use of instance-based learning for domains with many small disjuncts. They are mainly interested in language learning tasks, which they claim result in many small disjuncts, or "pockets of exceptions." In particular, they focus on the problem of learning word pronunciations. Because instance-based learning does not form disjunctive concepts, rather than determining disjunct sizes, they instead compute cluster sizes, which they view as analogous to disjunct size. They determine cluster sizes by repeatedly selecting examples from the data, forming a ranked list of the 100 nearest neighbors, and then they determine the rank of the nearest neighbor with a different class value – this value minus 1 is considered to be the cluster size. This method, as well as the more conventional method of measuring disjunct size via a decision tree, shows that the word pronunciation domain has many small disjuncts. The authors also try an information-theoretic weighted similarity matching function, which effectively rescales the feature space so that "more important" features have greater weight. When this is done, the size of the average cluster is increased from 15 to 25. Unfortunately, error rates were not specified for the various clusters and hence one cannot measure how effective this strategy is for addressing the problem with small disjuncts.

The problem of learning from imbalanced data where the classes are represented in unequal proportions is a common problem that has received a great deal of attention [4, 5, 10, 21]. Our results in Section 9.8 provide a link between the problem of learning from imbalanced data and the small disjuncts problem. A similar link was provided by Jo and Japkowicz [11], who also showed that a method that deals with the problem of small disjuncts, cluster-based oversampling, can also improve the performance of classifiers that learn from imbalanced data. This supports the notion that a better understanding of small disjuncts can lead the design of better classification methods.

9.10 Conclusion

This chapter makes several contributions to the study of small disjuncts and, more generally, classifier learning. First, the degree to which small disjuncts affect learning is quantified using a new measure, error concentration. Because error concentration is measured for a large collection of data sets, for the first time it is possible to draw general conclusions about the impact that small disjuncts have on learning. The experimental results show that, as expected, for many classifiers errors are highly concentrated toward the smaller disjuncts – however the results also show that for a substantial number of classifiers this simply is not true. Our research also indicates that the error concentration for the classifiers induced using C4.5 and Ripper is highly correlated, indicating that error concentration measures some "real" aspect of the concept being learned and is not totally an artifact of the learner. Finally, our results indicate that classifiers with relatively low error rates almost always have high error concentrations while this is not true of classifiers with high error rates. Analysis indicates that this is due to the fact that classifiers with low error rates generally contain some very accurate large disjuncts. We conclude from this that concepts that can be learned well tend to contain very general cases and that C4.5 and Ripper generate classifiers with similar error concentrations because they are both able to form accurate large disjuncts to cover these general cases.

Another contribution of this chapter is that it takes an in-depth look at pruning. This is particularly important because previous research into small disjuncts largely ignores pruning. Our results indicate that pruning eliminates many of the small disjuncts in the induced classifier and that this leads to a reduction in error concentration. These results also show that pruning is more effective at reducing the error rate of a classifier when the unpruned classifier has a high error concentration. Pruning is evaluated as a method for addressing the problem with small disjuncts and is shown to be of limited effectiveness. Our analysis also shows that because pruning distributes the errors that were concentrated in small disjuncts to the more accurate, larger disjuncts, pruning can actually degrade classifier performance when one may be selective in applying the induced classification rules.

In this chapter we also show how factors such as training set size, noise, and class imbalance affect small disjuncts and error concentration. This provides a better understanding not only of small disjuncts, but also of how these important, real-world factors affect inductive learning. As an example, the results in Section 9.6 permit us to explain how increasing the amount of training data leads to an improvement in classifier accuracy. These results, which show that increasing the amount of training data leads to an increase in error concentration, suggest that the additional training data allows the general cases within the concept to be learned better than before, but that it also introduces many new small disjuncts. These small disjuncts, which correspond to rare cases in the concept, are formed because there is now sufficient training data to ensure that they are sampled. These small disjuncts are error prone, however, due to the small number of training examples used to determine the classification. The small disjuncts in the induced classifier may also be error prone because, as the results in Section 9.7 and previous research [17, 19] indicate, noisy data causes erroneous small disjuncts to be formed. Our results indicate that pruning is somewhat effective at combating the effect of noise on classifier accuracy because of its ability to handle small disjuncts. Finally, the results in this chapter also indicate that class imbalance can worsen the problem with noise and small disjuncts. This may help explain why a balanced class distribution often leads to classifiers that are more robust than those induced from the naturally occurring class distribution.

We believe that an understanding of small disjuncts is important in order to properly appreciate the difficulties associated with classifier learning, because, as this chapter clearly shows, it is often the small disjuncts that determine the overall performance of a classifier. We therefore hope that the metrics provided in this chapter can be used to better evaluate the performance of classifiers and will ultimately lead to the design of better classifiers. The research in this chapter also enables us to better understand how various real-world factors, like noise and class imbalance, impact classifier learning. This is especially important as data mining tackles more difficult problems.

References

1. Ali, K.M., Pazzani, M.J.: Reducing the small disjuncts problem by learning probabilistic concept Descriptions. In: Petsche, T. (ed.) Computational Learning Theory and Natural Learning Systems, Volume 3, MIT Press, Cambridge, MA (1992)
2. Asuncion, A., Newman, D.J.: UCI Machine Learning Repository. University of California, Irvine, School of Information and Computer Science. http://www.ics.uci.edu/~mlearn/MLRepository.html. Cited Sept 2008
3. Carvalho D.R., Freitas A.A.: A hybrid decision tree/genetic algorithm for coping with the problem of small disjuncts in data mining. In: Proceedings of the 2000 Genetic and Evolutionary Computation Conference, pp. 1061–1068 (2000)
4. Chawla N.V., Bowyer K.W., Hall L.O., Kegelmeyer W.P.: SMOTE: Synthetic minority oversampling technique. Journal of Artificial Intelligence Research, 16, 321–357 (2002)
5. Chawla N.V., Cieslak D.A., Hall L.O., Joshi A.: Automatically countering imbalance and its empirical relationship to cost. Data Mining and Knowledge Discovery, 17(2), 225–252 (2008)
6. Cohen W.: Fast effective rule induction. In: Proceedings of the Twelfth International Conference on Machine Learning, pp. 115–123 (1995)
7. Cohen W., Singer Y.: A simple, fast, and effective rule learner. In: Proceedings of the Sixteenth National Conference on Artificial Intelligence, pp. 335–342 (1999)
8. Danyluk A.P., Provost F.J.: Small disjuncts in action: learning to diagnose errors in the local loop of the telephone network. In: Proceedings of the Tenth International Conference on Machine Learning, pp. 81–88 (1993)
9. Holte R.C., Acker L.E., Porter B.W.: Concept learning and the problem of small disjuncts. In: Proceedings of the Eleventh International Joint Conference on Artificial Intelligence, pp. 813–818 (1989)
10. Japkowicz N., Stephen S.: The class imbalance problem: a systematic study. Intelligent Data Analysis 6(5), 429–450 (2002)
11. Jo T., Japkowicz, N. Class imbalances versus small disjuncts. SIGKDD Explorations 6(1), 40–49 (2004)
12. Quinlan J.R.: The effect of noise on concept learning. In: Michalski R.S., Carbonell J.G., Mitchell T.M. (eds.), Machine Learning, an Artificial Intelligence Approach, Volume II, Morgan Kaufmann, San Francisco, CA (1986)
13. Quinlan J.R.: Technical note: improved estimates for the accuracy of small disjuncts. Machine Learning, 6(1) (1991)
14. Quinlan J.R.: C4.5: Programs for Machine Learning. Morgan Kaufmann, San Mateo, CA (1993)
15. Ting K.M.: The problem of small disjuncts: its remedy in decision trees. In: Proceedings of the Tenth Canadian Conference on Artificial Intelligence, pp. 91–97 (1994)
16. Van den Bosch A., Weijters A., Van den Herik H.J., Daelemans W.: When small disjuncts abound, try lazy learning: A case study. In: Proceedings of the Seventh Belgian-Dutch Conference on Machine Learning, pp. 109–118 (1997)
17. Weiss G.M.: Learning with rare cases and small disjuncts. In: Proceedings of the Twelfth International Conference on Machine Learning, pp. 558–565 (1995)

18. Weiss G.M.: Mining with rarity: A unifying framework, SIGKDD Explorations 6(1), 7–19 (2004)
19. Weiss G.M., Hirsh H.: The problem with noise and small disjuncts. In: Proceedings of the Fifteenth International Conference on Machine Learning, pp. 574–578 (1998)
20. Weiss G.M., Hirsh H.: A quantitative study of small disjuncts. In: Proceedings of the Seventeenth National Conference on Artificial Intelligence, Austin, Texas, pp. 665–670 (2000)
21. Weiss G.M., McCarthy K., Zabar B.: Cost-Sensitive Learning vs. Sampling: Which is best for handling unbalanced classes with unequal error costs? In: Proceedings of the 2007 International Conference on Data Mining, pp. 35–41 (2007)
22. Weiss G.M., Provost F.: Learning when training data are costly: the effect of class distribution on tree induction. Journal of AI Research 19, 315–354 (2003)

Part IV
Hybrid Data Mining Procedures

Chapter 10
Predicting Customer Loyalty Labels in a Large Retail Database: A Case Study in Chile

Cristián J. Figueroa

Abstract Although loyalty information is a key part in customer relationship management, it is hardly available in industrial databases. In this chapter, a data mining approach for predicting customer loyalty labels in a large Chilean retail database is presented. First, unsupervised learning techniques are used for segmenting a representative sample of the database. Second, the multilayer perceptron neural network is used for classifying the remaining population. Results show that 19% of the customers can be considered loyal. Finally, a set of validation tasks using data about in-store minutes charges for prepaid cell phones and distribution of products is presented.

10.1 Introduction

It is necessary for companies to have a thorough understanding of their customer base [15]. In this sense, customer loyalty, which is one of the most important concepts in marketing [32], can really improve customer relationship management (CRM). The knowledge of a customer's loyalty and the evolution therein could be useful for evaluating the results of CRM-related investments [4]. Knowing customer loyalty may also improve a customer's perception about the benefits received from products and services offered by companies [21, 43].

When companies, such as the retail ones, plan advertisement campaigns, it is essential to have updated information available about customer loyalty through labels. A loyalty label is a mark in the database which allows the identification of different groups of customers in function of their different purchasing behaviors related to the recency, frequency, and monetary information.

Cristián J. Figueroa
Department of Electrical Engineering, University of Chile, Santiago, Chile,
e-mail: cristian.figueroa@gmail.com

R. Stahlbock et al. (eds.), *Data Mining*, Annals of Information Systems 8,
DOI 10.1007/978-1-4419-1280-0_10, © Springer Science+Business Media, LLC 2010

Several analytical applications such as cross-selling, up-selling, and churn models are constructed to strengthen CRM. The outputs of these models are sets of customers who are ranked in function of their respective probabilities. These ranked customers act as inputs to these campaigns. Typically, these models are built on the entire customer database. However, it could be interesting to build these models on loyal customers only, because only for these customers, their total product needs are known. The core of a valuable customer base consists of loyal customers [15]. In this context, it seems suboptimal to include non-loyal customers into the analysis [4]. In consequence, customer loyalty labels may help to select the most appropriate customers for each advertisement campaign.

Nevertheless, the customer loyalty concept has to be understood in this type of applications in a non-contractual setting. This suffers from the problem that customers have the opportunity to continuously change their purchase behavior without informing the company about it [3].

For this reason, obtaining updated customer loyalty labels for each single customer is not an easy task. Moreover, industrial databases may contain millions of records associated with a huge number of customers [4]. In practice many companies do not manage this relevant information. Knowledge is limited in providing insights to companies regarding the differences within their customer base [15].

Although companies are able to collect huge quantities of heterogenous data through Corporate Data Warehouses, marketers must find ways of working smarter if they are to coordinate disparate customer information, ensure customer loyalty, and have a high marketing campaign success rate in an increasingly fragmented and sophisticated market [29]. Data must be analyzed.

Novelty information can be extracted automatically from this data by using the iterative process called Knowledge Discovery in Databases (KDD). Data mining is an important part within KDD [12, 16]. If the data mining process is used properly, companies have an unbeatable opportunity to generate advanced business conclusions about customer preferences and loyalty.

In this chapter, a data mining approach for predicting loyalty labels in a large Chilean retail customer database is presented. This retail company deals with the customer loyalty concept in a non-contractual setting. Preliminary results of this work were presented in [13]. Before this study, the retail company did not have any available updated customer loyalty labels. This avoided targeting appropriate marketing actions, delivering low success rates.

The proposed neural mining approach uses first a set of unsupervised learning algorithms such as the self-organizing map (SOM) for exploring data and discovering customer loyalty labels, initially nonexistent. Using different unsupervised algorithms, it is possible to gain different insights of the multidimensional data, define loyalty labels which also make sense from a business perspective, and segment a representative sample of the retail customer database. As a second step, the multilayer perceptron (MLP) neural network [1] is used for classifying the remaining Chilean retail customer database by using the loyalty labels discovered in the first step.

In this case study, customer loyalty is defined by seven variables, which resume purchase behaviors of customers in terms of the recency, frequency, and monetary information. To distinguish loyal customers, the recency and frequency information is more important than the monetary information [3]. Different values for these seven variables yield four types of loyalty labels in the database: loyal, common, potentially loyal, and non-loyal customers.

Applying these clustering algorithms directly to the entire large Chilean retail database for exploratory data analysis and segmentation is totally unworkable because analysts must deal with a huge computational load. Furthermore, each time that a new customer appears, new training of the clustering algorithms has to be carried out. For these reasons, the proposed two-step approach is appropriate. Customer loyalty labels can be obtained periodically without additional work. It is only necessary to execute monthly the trained MLP neural network for obtaining updated customer loyalty labels in the retail database.

This chapter is organized as follows. Section 10.2 delivers a detailed revision of the literature. Section 10.3 presents the features of this case study based on data mining. Section 10.4 shows results of this work. Section 10.5 establishes a set of validation tasks to show that these customer loyalty labels make sense from a business perspective by using data which is not used as input for clustering such as in-store minutes charges for prepaid cell phones and distribution of products in the retail stores. Finally, Section 10.6 describes the conclusions of this work with some discussions.

10.2 Related Work

Within customer loyalty related work, Buckinx et al. [4] presented a multiple linear regression which is compared against random forests and automatic relevance determination neural networks for predicting customer's behavioral loyalty. Two studies which show differences among internal customer groups in a service industry, are examined in [15]. As a result, customers who have switched service providers because of dissatisfaction seem to differ significantly from other customer groups in their satisfaction and loyalty behaviors.

A specific loyalty program with data from an online merchant that specializes in grocery and drugstore items is evaluated in [27]. Through simulation and policy experiments, it is possible to evaluate and compare the long-term effects of the loyalty program and other marketing instruments (e.g., e-mail coupons, fulfillment rates, shipping fees) on customer retention. A two-stage approach for dynamically deriving behaviorally persistent segments and subsequent target marketing selection using retail purchase histories from loyalty program members is proposed in [38]. The underlying concept of behavioral persistence entails an in-depth analysis of complementary cross-category purchase interdependencies at a segment level.

In [43] a modeling framework to study consumer behavioral loyalty as evidenced by two types of loyalty is proposed. The first one is hard-core loyalty, when consumers exclusively repeat purchase on one product alternative, and the second is reinforcing loyalty, when consumers may switch among product alternatives, but predominantly repeat purchase on one or more product alternatives to a significant extent. The Diamond of Loyalty is showed as a new management tool for customer loyalty by categorizing customer purchasing styles according to their level of involvement and their purchasing portfolio across suppliers.

Associated to clustering and segmentation applications, the use of artificial neural networks is examined for segmenting retail databases. Specifically, the Hopfield–Kagmar clustering algorithm is used and empirically compared to K-means and mixture model clustering algorithms [2]. A determination of market segments by clustering households on the basis of their average choice elasticities across purchases and brands w.r.t. price, sales promotion, and brand loyalty is presented in [17].

In Kiang et al. [20] an extended version of the SOM network to a consumer data set from the American Telephone and Telegraph Company (AT&T) is applied. The results are compared against a two-step procedure that combines factor analysis and K-means cluster analysis in uncovering market segments. In [25] a comparison of a conventional two-stage method with proposed two-stage method through the simulated data is presented. The proposed two-stage method integrates artificial neural networks and multivariate analysis through the combination of SOM and K-means methods.

In [24] a method which combines the SOM and genetic K-means algorithms for segmenting a real-world problem of the freight transport industry is presented. Lee and Park [26] presented a multiagent-based system called the survey-based profitable customers segmentation system which executes the customer satisfaction survey and conducts the mining of customer satisfaction survey, socio-demographic, and accounting database through the integrated uses of Data Envelopment Analysis (DEA), SOM, and C4.5 Decision Tree for the profitable customers segmentation.

In Lingras et al. [28] changes in cluster characteristics of supermarket customer over a 24-week period by using temporal data mining based on the SOM network are studied. This approach is useful to understand the migrations of the customers from one group to another group. In [34] a mathematical programming-based clustering approach that is applied to a digital platform company's customer segmentation problem is presented. In [36] three clustering methods K-means, SOM, and fuzzy C-means are used to find graded stock market brokerage commission rates based on the 3-month long total trades of two different transaction modes.

As analytical prediction models to improve CRM, a model to predict partial defection by behaviorally loyal clients using three classification techniques, logistic regression, automatic relevance determination (ARD) neural networks, and random forests, is presented. Focusing on partial attrition of high-frequency shoppers who exhibit a regular visit pattern may overcome the problem of unidentifiability of total defection in non-contractual settings [3]. An LTV model considering past profit contribution, potential benefit, and defection probability of a customer is presented in [18]. This model also covers a framework for analyzing customer value and segmenting customers based on their value. Customer value is classified into three categories: current value, potential value, and customer loyalty. Customers are segmented according to three types of customer value. A case study on a wireless telecommunication company is also illustrated.

In [31] an investigation of RFM (recency, frequency, and monetary), CHAID (Chi-squared automatic interaction detector), and logistic regression as analytical methods for direct marketing segmentation using two data sets is enunciated. As results, CHAID tends to be superior to RFM when the response rate to mailing of a relatively small portion of the database is low. On the other hand, RFM is an acceptable procedure in other circumstances. In [37] it is shown that customer management decisions can be biased and misleading. Indeed, it presented a modeling approach that estimated the length of a customer's lifetime and made adjustments for this bias. Using the model, the financial impact of not accounting for the effect of acquisition on customer retention is also presented.

Within work related to marketing campaigns and strategy, Luxton [29] published the development of a dynamic marketing strategy and details of key steps, technologies, and strategies required to achieve this. This work illustrated the importance of marketing campaign systems and analysis in the battle to understand customer data and ensure lifelong customer loyalty. A customer case study illustrating the real-life effectiveness of technology and strategy in harnessing the value of customer data is also presented. A unified strategic framework that enables competing marketing strategy options to be traded off on the basis of projected financial return, which is operationalized as the change in a firm's customer equity relative to the incremental expenditure necessary to produce the change is presented in [33].

10.3 Objectives of the Study

The Chilean retail company under study had had several attempts at obtaining cross-selling and churn analysis by using the entire retail customer base. However, these models delivered very low success rates. There were no updated labels which can improve the success rates of the predictive models, by splitting the customer base into different groups according to different behavioral loyalty patterns.

In consequence, the business objective of this work is to incorporate updated loyalty labels in a column of the database such that the next predictive models that are constructed can be targeted more appropriately.

To obtain updated customer loyalty labels for each customer in the large Chilean retail database, the following objectives must be fulfilled:

1. The creation of a set of variables in function of the recency, frequency, and monetary information which will help to discriminate among the different types of purchasing behaviors of the customers. These behaviors are defined under the concept of loyalty in a non-contractual setting.
2. The use of a set of unsupervised learning algorithms for exploring the multidimensional data and discovering customer loyalty labels, initially nonexistent in the retail database.

3. The definition of a set of customer loyalty labels which make sense from a business perspective and the segmentation of a representative sample of the Chilean retail customer database given the new information.
4. The use of the MLP neural network to classify the remaining Chilean retail customer database by using the loyalty labels discovered in the exploratory analysis.
5. The generation of an additional column in the retail database which contains, for each customer, the updated loyalty label for each month. The trained MLP can be executed monthly to update the loyalty labels.

It is important to note that this work does not propose any methodological innovation nor new learning algorithms, but employs well-known techniques to solve a complex real-world decision problem.

10.3.1 Supervised and Unsupervised Learning

Learning from data comes in two types: supervised learning and unsupervised learning. In supervised learning the variables under investigation can be split into two groups: explanatory variables and one (or more) dependent variables. The target of the analysis is to specify a relationship between the explanatory variables and the dependent variables [42].

In unsupervised learning situations all variables are treated in the same way, there is no distinction between explanatory and dependent variables. Supervised learning requires that the target variable is well defined and that a sufficient number of its values are given. For unsupervised learning the target variable is unknown [42].

10.3.2 Unsupervised Algorithms

10.3.2.1 Self-Organizing Map

The self-organizing feature map (SOM) neural network was introduced by Kohonen [22] for mapping input patterns of a high dimensionality onto an output lattice of a lower dimensionality. The SOM has been widely utilized for vector quantization (VQ), data projection, and multidimensional visualization. The VQ techniques encode a manifold of data by using a finite set of codebook vectors. In the SOM, output units are arranged on a fixed grid of $N_x \times N_y$ units, i.e., the topological relations between the output nodes are specified a priori. These units have associated codebook vectors of the same dimension as the input vectors. The learning algorithm is an adaptive process, which has the ability to represent the data using the codebook vectors. This learning is carried out through a set of epochs of training. Typical training tasks consider a number of epochs higher than 400, although this number also depends on the size of the map and the complexity of the input vectors. The SOM performs VQ under the constraint of a predefined neighborhood between neurons in a discrete output grid. This mapping preserves the distance relationships between input and output spaces. The output grid is usually used for high-dimensional data clustering and visualization. Figure 10.1 shows the SOM architecture, both in the input space and in the output space. x_i represents each one of the input vectors, w_j represents each one of the codebook vectors, p_j represents each one of the output units. p_j is associated to w_j.

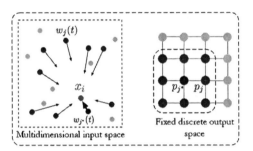

Fig. 10.1 The SOM architecture

In the SOM the distances between the codebook vectors are not directly represented on the map. A coloring scheme such as the U-matrix is required for visualizing the cluster boundaries [39, 40]. Figure 10.2 represents the output grid of a SOM trained with multidimensional data associated with technological attribute ratios of a set of countries. In the U-matrix, the darker the color between output units, the higher the distance between the respective codebook vectors.

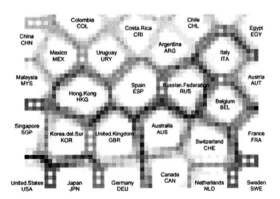

Fig. 10.2 The U-matrix applied to a SOM trained with technological data of countries

During the training of the SOM, each one of the codebook vectors w_j is compared with the input vector $x(t)$ through the Euclidean distance. Then, the best matching output unit (BMU) is found j^* using

$$j^* = \text{argmin}_{j=1...N}\|x_i(t) - w_j(t)\| \tag{10.1}$$

Each codebook vector $w_j(t)$ is updated using

$$\Delta w_j(t) = w_j(t+1) - w_j(t) = \alpha(t) \times h_{jj*}(t) \times (x_i(t) - w_j(t)) \tag{10.2}$$

where

$$\alpha(t) = \alpha_i \left(\frac{\alpha_f}{\alpha_i}\right)^{(t/t_{\max})} \tag{10.3}$$

corresponds to a monotonically decreasing learning rate. Typically, $\alpha_i > \alpha_f$. Typical values for α_i are between 0.9 and 0.5. Typical values for α_f are between 0.1 and 0.001.

$$h_{jj*}(t) = e^{-\dfrac{\|p_j - p_j^*\|^2}{L(t)}} \tag{10.4}$$

$h_{jj*}(t)$ represents a Gaussian neighborhood function measured in the output grid, where $L(t)$ is a monotonically decreasing function. Typically, $L_i > L_f$. Typical values for L_i are between 5.0 and 2.5. Typical values for L_f are between 0.5 and 0.01. It is important to note that these values also depend on the size of the map.

$$L(t) = L_0 \left(\frac{L_f}{L_i}\right)^{(t/t_{\max})} \tag{10.5}$$

10.3.2.2 Sammon Mapping

The Sammon's mapping (NLM) is a well-known distance preserving mapping technique which gives both insight in the presence and the structure of the clusters in the data, and each projection point corresponds with a data entry [35]. It is a nonlinear method for projecting high-dimensional vectors in a low-dimensional space, preserving the interpoint distances as close as possible. The purpose of the NLM is to provide a visual representation of the pattern of distances among a set of elements. On the basis of distances between a set of elements a set of points is returned so that the distances between the points are approximately equal to the original distances.

The Sammon stress, E, reflects the projection error of the input space on the output space. The lower the E value, the better the projection in the output space. The measure E is defined as

$$E = \frac{1}{\sum_{j=1}^{N}\sum_{i=1}^{j} d_{ijw}} \sum_{j=1}^{N}\sum_{i=1}^{j} \frac{[d_{ijw} - d_{ijp}]^2}{d_{ijw}} \tag{10.6}$$

where d_{ijw} denotes the distance between the codebook vectors i and j. Likewise, d_{ijp} denotes the distance between the position vectors i and j. Figure 10.3 shows the two-dimensional projection of a NLM using as input the well-known Iris data set [14], which contains 150 input vectors in a four-dimensional input space.

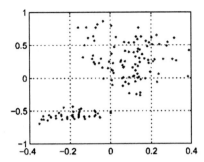

Fig. 10.3 The NLM applied to the well-known Iris data set

However, due to its complexity load $O(N^2)$, the NLM is typically used in combination with the SOM for large databases [23]. This means that the SOM and the Sammon mapping are combined (SOM/NLM) for projecting high-dimensional data. It is based on the vector quantization property of the SOM. Although computationally less expensive than the Sammon mapping, the SOM/NLM approach does not use the output lattice of the SOM. Figure 10.4 shows a schematic diagram of the SOM/NLM combination.

Fig. 10.4 The SOM/NLM combination

10.3.2.3 Curvilinear Component Analysis

The CCA (curvilinear component analysis) is an enhancement of the SOM which first performs VQ of the data manifold in the input space using the SOM [7]. Second, the CCA makes a nonlinear projection of the codebook vectors. The projection part of the CCA is similar to the NLM, since it minimizes a cost function based on the interpoint distances. The cost function of the CCA allows to unfold strongly nonlinear or closed structures being its complexity equal to $O(N)$. The lower the E value, the better the projection in the output space. The measure E is defined as

$$E = \frac{1}{2} \sum_{j=1}^{N} \sum_{k \neq j} (D_{j,k} - d_{j,k})^2 F(D_{j,k}, \lambda_D) = \frac{1}{2} \sum_{j=1}^{N} \sum_{k \neq j} E_{j,k} \qquad (10.7)$$

where $D_{j,k}$ represents the Euclidean distance between the output vectors p_j and p_k. $d_{j,k}$ represents the Euclidean distance between the codebook vectors w_j and w_k. F is a weighting function such as a step function which depends directly on the

Euclidean distance between p_j and p_k for emphasizing local topology conservation. λ_D represents a threshold for the step function.

In the CCA the output is not a fixed grid but a continuous space like in the NLM (Fig. 10.3) that is able to take the shape of the data manifold. The codebook vectors are projected as codebook positions in the output space, which are updated by a special adaptation rule.

Therefore, the CCA allows preserving local distances while the NLM tries to preserve global distances. Recently, the CCA has been compared with new visualization schemes [9, 10] which use the neural gas network as a clustering algorithm [30].

10.3.3 Variables for Segmentation

Several meetings were held with the retail experts of the Chilean company who explained the main aspects of their criteria for considering a customer loyal. Different degrees of loyalty which customers possess were also explained from their perspective in a non-contractual setting. In this sense, products which belong to loyalty-influencing departments and areas were identified. For disclosure contract with the retail store, the name of the company is confidential. A department of the retail store contains similar products to offer. An area contains more specific products within a department. A department may contain several areas.

This allowed to better understand the underlying business model, including discriminative behavioral patterns. During these meetings a set of 23 preliminary variables were designed to discriminate between loyalty and non-loyalty behaviors. These variables were analyzed along with retail experts to match with the data available in the database. Several tasks to assess the data quality were done. Checking for correlations among variables and discriminative power, as well as their consistency with empirical knowledge was also done. The time period used for the analysis was between August 2004 and September 2005. Finally, the following seven variables were selected for initiating the segmentation study:

- **STD**: Standard deviation of the amount of money spent in different departments of the retail store during the period.
- **TotalAmount**: Total amount of money spent during the period.
- **FromWhen**: Number of months from the last purchasing.
- **ProductsBuy**: Scoring of products bought during the period.
- **DepartmentsBuy**: Scoring of the number of loyalty-influencing departments where customer bought during the period.
- **AreasBuy**: Scoring of the number of loyalty-influencing areas within departments where customer bought during the period.
- **HOT**: Scoring of how much compulsive is a customer when he/she visits the retail store for payment reasons.

For further information about the creation of these variables, see Appendix.

10.3.4 Exploratory Data Analysis

The analysis started with a set of 944,253 customers which corresponds to the entire population under study. A single data set of 944,253 × 7 was calculated and stored into a unique repository for further data analysis. Due to the size of the database, the following step was to define a representative sample which can be used for clustering and segmentation tasks. A total of 25% (235,910) of the entire population was considered. In order to choose this sample, a simple random sampling method was used. It establishes as main requirement that distributions of all the variables in the sample must match with the ones of the entire population. In consequence, that sample that preserves empirically the distributions was selected.

To discover patterns from the sample obtained, the next activity was to do an exploratory data analysis. The first step was to cluster the sample of 25% by using the SOM. The configuration of the SOM was a grid of 4 × 3, considering a Gaussian neighborhood whose initial and final values were 3.0 and 0.5, respectively. 500 epochs of training were used. A learning rate whose initial and final values were 0.8 and 0.01 were used. From a business point of view, the reason why to use 12 neurons is for simplicity and an initial insight only needs to be obtained at this step.

Next, a U-matrix configuration which measures the distance in the input space between the codebook vectors through a coloring scheme was constructed. Figure 10.5 shows the results of the U-matrix showing clearly the nearness among the clusters 7, 8, and 11. Farther distances among the clusters 4, 5, 7, and 8 are evident too. Furthermore, it is possible to observe small distances among the clusters 6, 9, and 10.

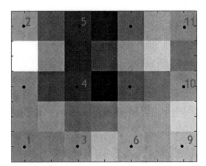

Fig. 10.5 The U-matrix of the Kohonen map of 4 × 3 by using the sample of 25%

In addition, the NLM was used for mapping the 11 codebook vectors generated by the SOM from the input space to a continuous output space preserving global distances. Figure 10.6 shows the resulting Sammon's mapping where it is possible to identify the formation of a group given by the clusters 7, 8, and 11. Moreover, a set of customers given by the clusters 1, 2, and 4 are located at the opposed side of the last group. At the middle of the map, three different groups can be identified. First, a group formed by clusters 3 and 6 is shown. Second, another group formed

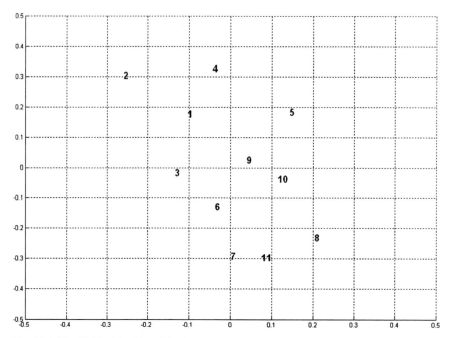

Fig. 10.6 The NLM of the 11 codebook vectors generated by the Kohonen map of 4 × 3 by using the sample of 25%

by the clusters 9 and 10 is located at the center. Finally, cluster 5 appears more separated at the right side.

The CCA was also used for mapping the 11 codebook vectors generated by the SOM. The difference is that the CCA's mapping allows preserving local distances rather global ones between the input space and the output space. Figure 10.7 shows the resulting CCA's mapping identifying the formation of the same relative groups shown by the NLM. Although in this map local distances are better preserved, the relative distances among all the codebook vectors are maintained too.

10.3.5 Results of the Segmentation

By taking into account all the results obtained by the different exploratory data analysis techniques, it is essential to say that clusters 7, 8, and 11 contain customers with profiles that are strongly loyal.

The quantitative analysis for all the clusters can be checked in Table 10.1 which shows details about average and standard deviation values for each codebook vector. Qualitative characteristics for each one of the 11 clusters can be checked in Table 10.2.

All this information allowed the segmentation of the sample under study. So, each one of the 11 clusters generated by the SOM was labeled according to the quantitative and qualitative characteristics that the retail experts carefully analyzed. Table 10.3 shows the class given by the retail experts to each one of the 11 clusters and the distribution of classes within the sample of 25%. From this table, it is possible to find 3 clusters as non-loyal, 1 cluster as potentially loyal, 4 clusters as common customers, and 3 clusters as loyal. The potentially loyal segment is very important because the amount of loyal customers may increase through appropriate marketing tasks for this segment. Each segment contains different characteristics which would allow differentiating even more each cluster. However, as the idea is to classify the remaining 75% of customers, these four categories will be maintained.

10.4 Results of the Classifier

The following step was to classify the remaining 75% of the entire population by using the sample of 25%. The entire population was equal to 708,343 customers. For classifying, a multilayer feedforward (MLP) neural network was used. The unlabeled 708,343 vectors were only used as a final testing data set, after training the MLP neural network.

The MLP was formed by the seven input units, three units in the hidden layer, and four output units. Sigmoidal activation functions were utilized. Clementine from

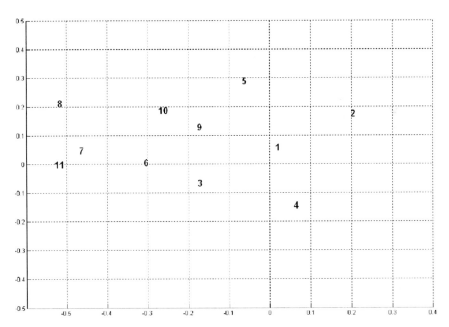

Fig. 10.7 The CCA of the 11 codebook vectors generated by the Kohonen map of 4×3 by using the sample of 25%

Table 10.1 Quantitative characteristics for each one of the 11 clusters obtained

Variables	1 (17%)	2 (32%)	3 (4%)
STD	126.92 ± 0.02	102.92 ± 0.01	170.01 ± 0.02
TotalAmount	198.43 ± 0.03	143.59 ± 0.02	320.66 ± 0.03
FromWhen	4.78 ± 0.39	4.88 ± 0.40	3.95 ± 0.33
ProductsBuy	0	0	0
DepartmentsBuy	4.00 ± 0.28	2.00 ± 0.14	6.00 ± 0.43
AreasBuy	1 ± 0.16	0	2.00 ± 0.33
HOT	0	0	0
Total	41,048	77,357	9268

Variables	4 (1%)	5 (3%)	6 (3%)
STD	150.03 ± 0.02	134.21 ± 0.02	205.50 ± 0.03
TotalAmount	210.82 ± 0.02	220.72 ± 0.02	410.72 ± 0.04
FromWhen	5.93 ± 0.49	3.51 ± 0.29	2.92 ± 0.24
ProductsBuy	0	0	1.00 ± 0.11
DepartmentsBuy	3.00 ± 0.21	6.00 ± 0.43	1.00 ± 0.11
AreasBuy	2.00 ± 0.33	0	2.00 ± 0.33
HOT	0	0	0
Total	2825	7655	8823

Variables	7 (3%)	8 (3%)	9 (13%)
STD	268.21 ± 0.04	297.97 ± 0.04	17.87 ± 0.01
TotalAmount	634.67 ± 0.07	742.52 ± 0.08	337.33 ± 0.04
FromWhen	2.37 ± 0.20	2.42 ± 0.20	3.25 ± 0.27
ProductsBuy	2.00 ± 0.22	3.00 ± 0.33	1.00 ± 0.11
DepartmentsBuy	9.00 ± 0.64	7.00 ± 0.50	6.00 ± 0.43
AreasBuy	2.00 ± 0.33	2.00 ± 0.33	0
HOT	0	8.00 ± 0.09	0
Total	82,726	7539	30,798

Variables	10 (4%)	11 (13%)
STD	198.40 ± 0.03	260.90 ± 0.04
TotalAmount	392.76 ± 0.04	608.04 ± 0.07
FromWhen	2.86 ± 0.23	2.19 ± 0.18
ProductsBuy	1.00 ± 0.11	2.00 ± 0.22
DepartmentsBuy	7.00 ± 0.50	9.00 ± 0.64
AreasBuy	1.00 ± 0.17	2.00 ± 0.33
HOT	8.00 ± 0.09	8.00 ± 0.09
Total	11,394	30,927

SPSS was used for obtaining this MLP model selection. The output unit indicates whether the sample corresponds to non-loyal, potentially loyal, common, or loyal behaviors. In order to build the multiclass classifier the labeled sample of 25% was divided into a training set, a validation set, and a testing set (Table 10.4). The training procedure minimized the error measured on the validation set while using the training set to adjust the networks' weights. This technique known as *early stopping*

Table 10.2 Qualitative characteristics for each one of the 11 clusters obtained

Id	Description
1	Non-loyal and non-compulsive customers who have a slight relationship with the retail company.
2	Non-loyal and non-compulsive customers who have no deep relationship with the retail company.
3	Two kinds of customers, they buy a lot of products in loyalty-influencing areas. They are not compulsive.
4	Infrequent customers who are compulsive and buy in loyalty-influencing areas.
5	From all the non-loyal clusters, this is the most loyal. Customers buy in loyalty-influencing departments. Almost the half of customers is compulsive. There are possibilities for loyalty.
6	Customers buy in loyalty-influencing areas. They have high purchasing power. Frequent visits to the retail store. They are non-compulsive.
7	Loyal customers. They buy in loyalty-influencing departments. They are non-compulsive. Non-compulsive housewives.
8	Loyal customers. They buy in loyalty-influencing departments. They are compulsive and have high purchasing power.
9	Non-loyal and non-compulsive customers who have no deep relationship with the retail company. High purchasing power.
10	Non-loyal, compulsive customers and high purchasing power.
11	Loyal customers. They buy in loyalty-influencing departments. Compulsive housewives.

Table 10.3 Segmentation of the sample of 25%

Cluster identifier	Label	Description	Distribution (%)
1, 2, 4	0	Non-loyal	51.39 (121,230)
5	1	Potentially loyal	3.24 (7655)
3, 6, 9, 10	2	Common customers	25.55 (60,283)
7, 8, 11	3	Loyal	19.82 (46,742)
Total			100.00 (235,910)

provides a principled method for selecting models that generalize well without sacrificing capacity, hence avoiding overfitting to the training data's noise while keeping the classifier's ability to learn nonlinear discriminant boundaries [5]. The testing set originated from the sample of 25% was used to estimate the generalization performance of the system. In this work, the final objective is to obtain a customer loyalty mark that allows labeling monthly each customer in the database. Later, this label may be used to distinguish customers based on different purchasing behaviors and therefore select the loyal customers to participating in each direct marketing application among other constraint-driven activities. This classifier does not consist in identifying a set of customers with the highest likelihood as in when cross-selling, up-selling, and churn models are created. This prediction of customer loyalty labels is a first step before creating predictive models to specific products and hence there is no product to offer to customers at this point yet.

For training purposes, the potentially loyal customers were multiplied 16 times randomly, the common customers were duplicated randomly, and the loyal customers were triplicated randomly to remove possible bias. Differences in prior class probabilities or class imbalances lower the performance of some standard classifiers, such as decision trees and MLP neural networks [19]. This problem has been studied extensively, being the random oversampling of minority classes an adequate strategy to deal with this in MLP neural networks [6, 19, 41, 44].

Table 10.4 Distribution of the labeled sample of 25%

Category	Set	Number of cases
Non-loyal	Training	51,029
Potentially loyal	Training	3079
Common customers	Training	26,282
Loyal	Training	18,600
Total	Training	98,990
Non-loyal	Validation	21,870
Potentially loyal	Validation	1319
Common customers	Validation	11,264
Loyal	Validation	7972
Total	Validation	42,424
Non-loyal	Testing	48,687
Potentially loyal	Testing	3066
Common customers	Testing	24,068
Loyal	Testing	18,675
Total	Testing	94,496

Such as in [8] and [11], Table 10.5 shows the percentage of correct classifications for each one of the customer loyalty labels. All these labels had a performance over 99% of correct classifications in the test set.

Next, the classification is applied over the remaining 75% of the entire unlabeled population. The distribution of labels generated for this final testing set is show in Table 10.6.

10.5 Business Validation

Two different tasks for validating the resulting segments from a business perspective are presented. Both tasks are only applied to customers who belong to the non-loyal and loyal segments.

Table 10.5 Percentage of correct classifications on the testing data set of the 25% of the population for four customer loyalty labels

Category	Test (%)
Non-loyal	99.97
Potentially loyal	99.64
Common customers	99.84
Loyal	99.52
Average	99.74

Table 10.6 Classification generated for the remaining 75% of the retail industrial database

Label description	Distribution (%)
Non-loyal	50.38 (356,895)
Potentially loyal	4.16 (29,478)
Common customers	25.48 (180,468)
Loyal	19.98 (141,502)
Total	100.00 (708,343)

The first task refers to analyzing in-store minutes charges done by the non-loyal and loyal customers for prepaid cell phones. The second one consists in constructing distribution of products bought by the non-loyal and loyal customers in the departments and areas of the store.

Both of these analyses deliver similar conclusions. Loyal customers have a higher degree of participation with the store than non-loyal customers, buying in departments and areas which are more profitable for the store.

These variables were not used as inputs for producing the customer loyalty segmentation.

10.5.1 In-Store Minutes Charges for Prepaid Cell Phones

Table 10.7 shows the number of customers per segment along with the number of customers who generate at least one in-store minutes charge during the analysis time window. The last column in Table 10.7 establishes the percentage of each segment with in-store minutes charges. A total of 15.1% of the non-loyal segment do at least one in-store minutes charge. The percentage is almost duplicated when referring to loyal customers. A total of 20.1% corresponds to the relative value of customers who do in-store minutes charges in relation with the total population.

Table 10.8 explains the difference between the non-loyal and loyal segments when the money spent in buying minutes charges for prepaid cell phones and the

Table 10.7 In-store minutes charges done by non-loyal and loyal customers for prepaid cell phones

Segment	Total population	Customers with in-store minutes charges	Percentage of customers with in-store minutes charges
Non-loyal	477,914	72,240	15.1
Potentially loyal	36,944	7160	19.4
Common customers	240,640	54,877	22.8
Loyal	188,705	55,644	29.5
Total	944,253	189,921	20.1

Table 10.8 Average spent money and average number of in-store minutes charges for non-loyal and loyal customers

Segment	Spent money	Average number of in-store minutes charges
Non-loyal	US $67.48	7.66
Loyal	US $93.12	10.58

number of in-store minutes charges are taken into account. As can be observed the loyal segment spend on average US $93.12 in buying minutes charges for pre-paid cell phones whereas the non-loyal segment only spends on average US $67.48. Similar situation occurs when comparing the number of in-store minutes charges between both segments. On average, 10.58 is the value of charges that a customer, who belongs to the loyal segment, does during the analysis time window. Non-loyal customers show on average a value of 7.66 to indicate the number of charges in the same period of time.

This information is valuable because the in-store minutes charges item corresponds to an external business from where the company wants to obtain more profitability in the future. All of these differences between segments are significative, showing that the discovered segments make sense from a business point of view.

10.5.2 Distribution of Products in the Store

Figure 10.8 shows the distribution of money spent by the total population in the different departments of the store during the analysis time window. The values are expressed in US dollars. It can be observed that the money spent in the Electronic equipment department is the highest reaching US $8,500,000. This is expected because this department contains products which are more expensive than those of other departments. On the other hand, the Women department has associated US $4,200,000 in purchases.

Figure 10.9 shows the distribution of money spent by the loyal segment in the different departments of the store during the analysis time window. The values are expressed in US dollars. It can be observed that the money spent in the Electronic equipment department is also the highest reaching US $290,000. On the other hand, the Women department has associated US $225,000 in purchases. In comparison with the Electronic equipment department the remaining departments maintain the same proportions in both cases.

Figure 10.10 shows the distribution of money spent by the non-loyal segment in the different departments of the store during the analysis time window. The values are expressed in US dollars. It can be observed that the money spent in the Electronic equipment department is also the highest reaching US $300,000. On the other hand, the Women department has associated US $78,000 in purchases.

In consequence, it is possible to verify that buying products in the Women department is a strong signal of how loyal a customer can become with the store. The subjective scores given by the retail experts for loyalty-influencing departments address appropriately the concept of loyalty in this case. The Women department is one of the most profitable departments in the store.

Figures 10.11 and 10.12 show the number of products bought by the loyal and non-loyal customers, respectively, in those areas that form the Women department: Lingerie and sleepwear, Casual clothing, Women's on trend clothing, Exclusive lin-

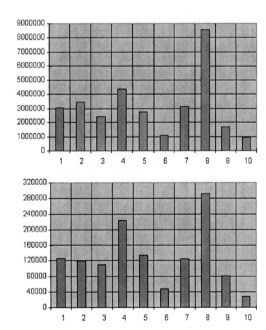

Fig. 10.8 Money spent by the total population per department in the store during the analysis time window. 1: Household, 2: Home appliances, 3: Men, 4: Women, 5: Kids, 6: Men's footwear, 7: Sports, 8: Electronic equipment, 9: Women's footwear, 10: Others

Fig. 10.9 Money spent by the loyal customers per department in the store during the analysis time window. 1: Household, 2: Home appliances, 3: Men, 4: Women, 5: Kids, 6: Men's footwear, 7: Sports, 8: Electronic equipment, 9: Women's footwear, 10: Others

Fig. 10.10 Money spent by
the non-loyal customers per
departments in the store during the analysis time window.
1: Household, 2: Home appliances, 3: Men, 4: Women, 5:
Kids, 6: Men's footwear, 7:
Sports, 8: Electronic equipment, 9: Women's footwear,
10: Others

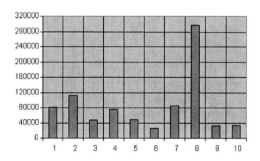

gerie, Women's underwear, Women's leathers, Teen female jeans, Female teenager,
Cosmetics and Fragrances, and Bathing suits.

Fig. 10.11 Total number
of products bought by the
loyal customers per area in
the store during the analysis
time window. 17: Lingerie
and sleepwear, 18: Casual
clothing, 19: Women's on
trend clothing, 28: Exclusive
lingerie, 33: Women's under-
wear, 35: Women's leathers,
37: Teen female jeans, 38: Female teenager, 45: Cosmetics
and fragrances, 53: Bathing
suits

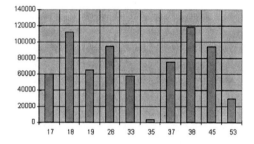

Figure 10.11 presents the distribution of products bought by the loyal customers
in the different areas of the Women department. Figure 10.12 presents the distribution of products bought by the non-loyal customers in the different areas of
the Women department. It can be observed that buying in the areas Exclusive lingerie, Women's underwear, among others is an important signal of loyalty in the
store. Consequently, the subjective scores given by the retail experts for loyalty-
influencing areas address also appropriately the concept of loyalty.

10.6 Conclusions and Discussion

A large database of customers who buy in a Chilean retail store has been segmented
by using the data mining process through the design of behavioral loyalty indicators
which measure the recency, frequency, and monetary information. Within the proposed neural mining approach, a number of seven variables were used for clustering
a representative sample of the entire population. Later, exploratory data analysis was

Fig. 10.12 Total number of products bought by the non-loyal customers per area in the store during the analysis time window. 17: Lingerie and sleepwear, 18: Casual clothing, 19: Women's on trend clothing, 28: Exclusive lingerie, 33: Women's under-wear, 35: Women's leathers, 37: Teen female jeans, 38: Fe-male teenager, 45: Cosmetics and fragrances, 53: Bathing suits

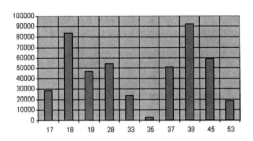

developed for discovering different insights of the sample of 25%. Specifically, the SOM, the NLM, and the CCA algorithms were used for clustering and segmentation. All of them generated analogue conclusions about the natural structure of the multidimensional data which allowed retail experts segment the sample into four different categories: non-loyal, potentially loyal, common customers, and loyal customers. Next, a MLP neural network was utilized to classify the remaining 75% of customers of the entire database.

Checking the results obtained, a 19% of the entire population was labeled as loyal, purchasing preferably and frequently products located in children and women clothes departments and spending a high amount of money monthly. This 19% is partitioned as follows. A 13% can be considered as compulsive loyal housewives segment because they can be easily tempted with some bargains when they visit the retail store. A 3% can be called loyal in a VIP sense because they possess a high purchasing power and are compulsive. Finally, 3% are called non-compulsive housewives who buy in loyalty-influencing departments. On the other hand, 4% of customers could become loyal in a strong sense if marketing strategies are well focused. A 50% of the population corresponds to non-loyal customers whereas 1% of the customers are infrequent in terms of purchasing behavior, buying products in non-important departments and areas according to loyalty. In addition, a few number of visits are registered during the year.

This segmentation produced robust results. The trained MLP can be executed monthly to update the loyalty labels. This is a critical point because the database is large. As a conclusion, the classification delivered similar percentages for each class in both the sample and the remaining vectors. This can be observed comparing Tables 10.3 and 10.6. The underlying motivations for using this neural framework through the clustering and classification stages were to know how the customer database was grouped naturally, to decrease computational load of the clustering algorithms when applied to a large database, and to establish a robust classification of the entire customer database. Generate an additional column in the retail database which contains, for each customer, the updated loyalty label for each month was the final objective of this study.

The results of this work have had a high impact within the retail store. First, although retail experts knew by intuition that the loyal customers percentage was about 20%, they did not have any objective measurement to confirm it. Second, this solution not only has allowed to obtain further and new information about loyal segment but also important figures and features (Tables 10.1 and 10.2, respectively) about other segments within the customer database have been reached. After applying this scheme, novelty knowledge has been generated through the data stored. Finally, many tasks for generating profitability such as cross-selling, churn analysis, advertisement campaigns, pricing analysis, and lifetime value analysis have become more successful. For instance, churn analysis was executed over the entire population before our solution without obtaining successful results. After our solution, churn analysis has been focused on each segment, predicting a higher churn rate each month. Furthermore, the capability of our solution has been emphasized by retail experts as a simple way of classifying which diminishes efforts considerably within the organization. Indeed, a few minutes are required each month to classify the entire customer database and fill the respective loyalty column in the database.

A set of business validation tasks to verify that customer loyalty labels make sense from a business perspective by using data about in-store minutes charges for prepaid cell phones and distribution of products per departments and areas of the retail stores was presented. As a conclusion of these validation tasks, it can be established that loyal customers have a higher degree of participation with the store than non-loyal customers, buying in departments and areas which are more profitable for the retail company. Furthermore, retail experts addressed appropriately the allocation of loyalty scores to departments and areas.

Acknowledgments Clementine was used under permission of SPPS Chile. Specifically, the author would like to thank Stephen Cressall who is with SPSS Chile for his valuable help and support on the Clementine's platform during the implementation of some models.

Furthermore, the author appreciates enormously the comments and suggestions about this business application given by Jacek M. Zurada during his visit made to Chile in January 2007. Jacek M. Zurada is a professor who belongs to the Department of Computer and Electrical Engineering, University of Louisville, USA.

Appendix

Table 10.9 shows the ProductsBuy variable defined along with the retail experts. The ProductsBuy variable is obtained by coding between 0 and 9 the number of different products bought during the analysis time. For instance, if a customer buys 34 different products during the period, the ProductsBuy variable scores 3 for that customer.

DepartmentsBuy and AreasBuy variables were defined according to subjective loyalty degrees established by retail experts. These degrees are 0, 1, 2, or 3 which are added up for a customer who buys in the respective departments and areas; 0

Table 10.9 ProductsBuy variable. It is obtained by coding between 0 and 9 the number of different products bought during the period

Number of products	ProductsBuy variable
0–9	0
10–19	1
20–29	2
30–39	3
40–49	4
50–59	5
60–69	6
70–79	7
80–89	8
90– ∞	9

corresponds to a department which does not generate loyalty in those customers who buy products from it. The analogue situation occurs with the AreasBuy variable.

The DepartmentBuy variable corresponds to the sum of department scores. The AreaBuy variable corresponds to the sum of areas scores.

HOT variable measures how much compulsive a customer is when he/she visits the retail store for payment reasons. This means that when a customer goes every month to the retail store for paying installments of products bought previously, some bargains in the store tempt customer to buy other products. This variable is given by retail experts supposing that the number of visits done by customer during the year is 12 (1 visit per month). The entire scoring of the HOT variable is described in Table 10.10.

Table 10.10 HOT variable. This variable is given by retail experts supposing that the number of visits done by customer during the year is 12 (1 visit per month)

Number of visits	HOT variable
0	0
1	8
2	16
3	25
4	33
5	41
6	50
7	58
8	66
9	75
10	83
11	91
12	100

References

1. Bishop CM (1995) Neural Networks for Pattern Recognition. Oxford University Press (ed), New York.
2. Boone DS, Roehm M (2002) Retail segmentation using artificial neural networks. International Journal of Research in Marketing 19:287–301.
3. Buckinx W, Van den Poel D (2005) Customer base analysis: Partial defection of behaviourally loyal clients in a non-contractual FMCG retail setting. European Journal of Operational Research 164(1):252–268.
4. Buckinx W, Verstraeten G, Van den Poel D (2007) Predicting customer loyalty using the internal transactional database. Expert Systems with Applications 32:125–134.
5. Caruana R, Lawrence S, Giles CL (2000) Overfitting in neural networks: Backpropagation, conjugate gradient, and early stopping. Neural Information Processing Systems, Denver, Colorado.
6. Chawla NV (2003) C4.5 and imbalanced datasets: Investigating the effect of sampling method, probabilistic estimate, and decision tree structure. In: ICML Workshop on Learning from Imbalanced Datasets. Washington, DC, USA.
7. Demartines P, Hérault J (1997) Curvilinear component analysis: A self-organizing neural network for nonlinear mapping of data sets. IEEE Transactions on Neural Networks 8:1: 148–154.
8. El Ayech H, Trabelsi and A (2006) Decomposition Method for Neural Multiclass Classification Problem. Proceeding of World Academy of Science, Engineering and Technology 15:150–153.
9. Estévez PA, Figueroa CJ (2006) Online data visualization using the neural gas network. Neural Networks 19(6–7):923–934.
10. Estévez PA, Figueroa CJ, Saito K (2005) Cross-entropy embedding of high-dimensional data using the neural gas model. Neural Networks 18(5–6):727–737.
11. Estévez PA, Perez C, Goles E (2003) Genetic Input Selection to a Neural Classifier for Defect Classification of Radiata Pine Boards. Forest Products Journal 53(7/8):87–94.
12. Fayyad UM (1996) Data mining and knowledge discovery: making sense out of data. IEEE Expert, Intelligent Systems and their Applications 20–25.
13. Figueroa CJ, Araya JA (2008) A Neural Mining Approach for Predicting Customer Loyalty in Large Retail Databases. Proceedings of the International Conference on Data Mining, DMIN'08, Las Vegas, Nevada, USA 1:26–34.
14. Fisher RA (1936) The use of multiple measurements in taxonomic problems. Annual Eugenics 7, Part II:179–188.
15. Ganesh J, Arnold MJ, Reynolds KE (2000) Understanding the customer base of service providers: An examination of the differences between switchers and stayers. Journal of Marketing 64(3):65–87.
16. Han J, Kamber M (2001) Data Mining: Concepts and Techniques. Morgan Kaufmann Publishers (ed), San Francisco.
17. Hruschka H, Fettes W, Probst M (2004) Market segmentation by maximum likelihood clustering using choice elasticities. European Journal of Operational Research 154:779–786.
18. Hwang H, Jung T, Suh E (2004) An LTV model and customer segmentation based on customer value: a case study on the wireless telecommunication industry. Expert Systems with Applications 26:181–188.
19. Japkovicz N, Stephen S (2002) The class imbalance problem: A systematic study. Intelligent Data Analysis 6(5):429–450.
20. Kiang MY, Hu MY, Fisher DM (2006) An extended self-organizing map network for market segmentation – a telecommunication example. Decision Support Systems 46:36–47.
21. Knox S (1998) Loyalty-based segmentation and the customer development process. European Management Journal 16:6:729–737.
22. Kohonen T (1995) Self-Organizing Maps. Springer-Verlag, Berlin.

23. Konig A (2000) Interactive visualization and analysis of hierarchical neural projections for data mining. IEEE Transactions on Neural Networks 11:3:615–624.
24. Kuo RJ, An YL, Wang HS, Chung WJ (2006) Integration of self-organizing feature maps neural network and genetic k-means algorithm for market segmentation. Expert Systems with Applications 30:313–324.
25. Kuo RJ, Ho LM, Hu CM (2002) Cluster analysis in industrial market segmentation through artificial neural network. Computers & Industrial Engineering 42:391–399.
26. Lee JH, Park SC (2005) Intelligent profitable customers segmentation system based on business intelligence tools. Expert Systems with Applications 29:145–152.
27. Lewis M (2004) The influence of loyalty programs and short-term promotions on customer retention. Journal of Marketing Research 41(3):281–292.
28. Lingras P, Hogo M, Snorek M, West C (2005) Temporal analysis of clusters of supermarket customers: conventional versus interval set approach. Information Sciences 172:215–240.
29. Luxton R (2002) Marketing campaign systems – the secret to life-long customer loyalty. Journal of Database Marketing 9(3):248–258.
30. Martinetz TM, Schulten KJ (1991) A neural gas network learns topologies. In T. Kohonen, K. Makisara, O. Simula, & J. Kangas (Eds.), Artificial Neural Networks, North-Holland, Amsterdam, pp. 397–402.
31. McCarty JA, Hastak M (2007) Segmentation approaches in data-mining: A comparison of RFM, CHAID, and logistic regression. Journal of Business Research 20:656–662.
32. Reichheld, FF, Sasser, WE Jr (1990). Zero defections: Quality comes to service. Harvard Business Review 68(5):105–111.
33. Rust RT, Lemon KN, Zeithaml VA (2004) Return of marketing: Using customer equity to focus marketing strategy. Journal of Marketing 68(1):109–127.
34. Sağlam B, Salman FS, Sayin S, Türkay M (2006) A mixed-integer programming approach to the clustering problem with an application in customer segmentation. European Journal of Operational Research 173:866–879.
35. Sammon JW (1969) A nonlinear mapping for data structure analysis. IEEE Transactions on Computers C-18:401–409.
36. Shin HW, Sohn SY (2004) Segmentation of stock trading customers according to potential value. Expert Systems with Applications 27:27–33.
37. Thomas JS (2001) A methodology for linking customer acquisition to customer retention. Journal of Marketing Research 38(2):262–268.
38. Thomas Reutterer, AMMNAT (2006) A dynamic segmentation approach for targeting and customizing direct marketing campaigns. Journal of Interactive Marketing 20(3–4):43–57.
39. Ultsch A (2003) Maps for the visualization of high-dimensional data spaces. Proceedings of the Workshop on Self-Organizing Maps (WSOM'05) 1:225–230.
40. Ultsch A (2005) Clustering with SOM: U*C. Proceedings of the Workshop on Self-Organizing Maps (WSOM'05) 1:75–82.
41. Weiss GM (2004) Mining with rarity: A unifying framework. ACM SIGKDD Explorations Newsletter 6(1):7–19.
42. Wilhelm AFX, Wegman EJ, Symanzik J (1999) Visual clustering and classification: The oronsay particle size data set revisited. Computational Statistics 14:109–146.
43. Yim CK, Kannan PK (1999) Consumer behavioral loyalty: A segmentation model and analysis. Journal of Business Research 44:75–92.
44. Zadrozny B, Elkan C (2001) Learning and making decisions when costs and probabilities are both unknown. Proceeding of the 7th ACM SIGKDD International Conference on Discovery and Data Mining. ACM Press, 204–213.

Chapter 11
PCA-Based Time Series Similarity Search

Leonidas Karamitopoulos, Georgios Evangelidis, and Dimitris Dervos

Abstract We propose a novel approach in multivariate time series similarity search for the purpose of improving the efficiency of data mining techniques without substantially affecting the quality of the obtained results. Our approach includes a representation based on principal component analysis (PCA) in order to reduce the intrinsically high dimensionality of time series and utilizes as a distance measure a variation of the squared prediction error (SPE), a well-known statistic in the Statistical Process Control community. Contrary to other PCA-based measures proposed in the literature, the proposed measure does not require applying the computationally expensive PCA technique on the query. In this chapter we investigate the usefulness of our approach in the context of query by content and 1-NN classification. More specifically, we consider the case where there are frequently arriving objects that need to be matched with the most similar objects in a database or that need to be classified into one of several pre-determined classes. We conduct experiments on four data sets used extensively in the literature, and we provide the results of the performance of our measure and other PCA-based measures with respect to classification accuracy and precision/recall. Experiments indicate that our approach is at least comparable to other PCA-based measures and a promising option for similarity search within the data mining context.

Leonidas Karamitopoulos · Georgios Evangelidis
Department of Applied Informatics, University of Macedonia, 156 Egnatia Str., GR-54006 Thessaloniki, Greece, e-mail: `lkaramit@uom.gr, gevan@uom.gr`

Dimitris Dervos
Information Technology Department, Alexander Technology Educational Institute of Thessaloniki, P.O. Box 141, GR-57400 Sindos, Greece, e-mail: `dad@it.teithe.gr`

R. Stahlbock et al. (eds.), *Data Mining*, Annals of Information Systems 8,
DOI 10.1007/978-1-4419-1280-0_11, © Springer Science+Business Media, LLC 2010

11.1 Introduction

Rapid advances in automated monitoring systems and storing devices have led to the generation of huge amounts of data in the form of time series, that is, series of measurements recorded through time. Inevitably, (most of) this volume of data remains unexploited, since the traditional methods of analyzing data do not adequately scale to the massive data sets frequently encountered. In the last decade, there has been an increasing interest in the data mining field, which involves techniques and algorithms capable of efficiently extracting patterns that can potentially constitute knowledge from very large databases.

The field of time series data mining mainly considers methods for the following tasks: clustering, classification, novelty detection, motif discovery, rule discovery, segmentation, and indexing [27]. At the core of these tasks lies the concept of similarity, since most of them require searching for similar patterns [18]. Two time series can be considered similar when they exhibit similar shape or pattern. However, the presence of high levels of noise demands the definition of a similarity/distance measure that allows imprecise matches among series [8]. In addition to that, the intrinsically high dimensionality of time series affects the efficiency of data mining techniques. Note that the dimensionality is defined by the length of the time series. In other words, each time point can be considered as a feature whose value is recorded. Thus, an appropriate representation of the time series is necessary in order to manipulate and efficiently analyze huge amounts of data. The main objective is to reduce the dimensionality of a time series by representing it in a lower dimension and analyze it in this dimension. There have been several time series representations proposed in the literature for the purpose of dealing with the problem of the "dimensionality curse" that appears frequently within real-world data mining applications [1, 23].

In this chapter, we consider the case of multivariate time series, that is, a set of time series recorded at the same time interval. Contrary to the univariate case, the values of more than one attribute are recorded through time. The objects under consideration can be expressed in the form of matrices, where columns correspond to attributes and rows correspond to time instances. Notice that a univariate time series can be expressed as a column (or row) vector that corresponds to the values of one attribute at consecutive time instances. Multivariate time series frequently appear in several diverse applications. Examples include human motion capture [24], geographical information systems [7], statistical process monitoring [17], or intelligent surveillance systems [31]. For instance, it is of interest to form clusters of objects that move similarly by analyzing data from surveillance systems or classify current operating conditions in a manufacturing process into one of several operational states.

As a motivating example, consider the task of automatically identifying people based on their gait. Suppose that data is generated using a motion capture system, which transmits the coordinates of 22 body joints every second (i.e., 66 values) for 2 min (i.e., 120 s). The resulting data set consists of 66 time series and 120 time instances and corresponds to a specific person. This data set can be expressed as a matrix $X_{120 \times 66}$. Also, suppose that we have obtained gait data for every known person under different conditions, for example, under varying gait speeds, and stored it in a database. Each record corresponds to one person and holds the gait data, which can be considered as a matrix, along with a label that indicates the identity of this person. Note that there is more than one record that corresponds to the same person, since we have obtained gait data under different conditions for every known person. Given this database, the objective is to identify a person under surveillance. In this case, we search the database for the most similar matrix to the one that is generated by this person. This task can be considered as a classification task. Each known person represents a class that consists of the gait data of this person generated under different conditions. The task is to classify (identify) a person under surveillance into a class.

This classification problem can be virtually handled by (other) classic classification techniques [32, 26], if each matrix is represented as a vector by concatenating its columns (i.e., the values of the corresponding attributes). However, we have to consider two issues with respect to this approach. The first issue is that the problem of high dimensionality deteriorates in the case of multivariate time series, since it is not only the length of the time series, but also the number of attributes that determines the dimensionality. In the previous example, the matrix $X_{120 \times 66}$ that corresponds to the gait data of one person constitutes an object of 7920 (120×66) dimensions. The second issue is that the correlations among attributes of the same multivariate time series are ignored. This loss of information may be of serious importance within a classification application.

We introduce a novel approach in identifying similar multivariate time series, which includes a PCA-based representation for the purpose of dimensionality reduction and a distance measure that is based on this representation. Principal component analysis (PCA) is a well-known statistical technique that can be used to reduce the dimensionality of a multivariate data set by condensing a large number of interrelated variables into a smaller set of variates, while retaining as much as possible of the variation present in the original data set [13]. In our case, the interrelated variables are in the form of time series. We provide a novel PCA-based measure that is a variation of the squared prediction error (SPE) or Q-statistic, which is broadly utilized in multivariate statistical process control [19]. Contrary to other PCA-based measures proposed in the literature, this measure does not require applying the computationally expensive PCA technique on the query. Moreover, we provide a method that further speeds up the calculations of the proposed measure by reducing the dimensionality of each one of the time series that form the query object during the pre-processing phase. Although our approach can be applied on other types of data, we concentrate on time series for two reasons. First, this type of data differs from other domains in that it exhibits high dimensionality, high feature correlation, and high levels of noise. Second, a large portion of data is generated in the form of time series in almost all real-world applications.

The objective of our approach is to provide a means for improving the efficiency of data mining techniques without substantially affecting the quality of the corresponding results. In particular, the dimensionality reduction of the original data improves the scalability of any data mining technique that will be applied subsequently, and the proposed measure aims at maintaining the quality of the results. In this chapter, we investigate the potential usefulness of our approach, mainly in the context of query by content and 1-NN classification. More specifically, we consider the case where there are frequently arriving objects that need to be matched with the most similar objects in a database or that need to be classified into one of several pre-determined classes.

In Section 11.2, we discuss PCA with respect to similarity search and we provide related work. Section 11.3 introduces our approach and provides a distance measure that is based on multivariate statistical process control. In Section 11.4, we describe the experimental settings with respect to the data sets, the methods, and the rival measures. The results of our experiments are presented and discussed in Section 11.5. Finally, conclusions and future work are provided in Section 11.6.

11.2 Background

We briefly review principal component analysis on multivariate data in Section 11.2.1. Similarity search is based on shapes, meaning that two time series are considered similar when their shapes are considered to be "close enough." Apparently, the notion of "close enough" depends heavily on the application itself, a fact that affects the decision of the pre-processing phase steps to be followed, the similarity measure to be utilized, and the representation to be applied on the raw data (Section 11.2.2). Finally, in Section 11.2.3, we review several PCA-based measures.

11.2.1 Review of PCA

PCA is applied on a multivariate data set, which can be represented as a matrix $X_{n \times p}$. In the case of time series, n represents their length (number of time instances), whereas p is the number of variables being measured (number of time series). Each row of X can be considered as a point in p-dimensional space. The objective of PCA is to determine a new set of orthogonal and uncorrelated composite variates $Y_{(j)}$, which are called principal components:

$$Y_{(j)} = a_{1j}X_1 + a_{2j}X_2 + \cdots + a_{pj}X_p, \quad j = 1, 2, \ldots, p \qquad (11.1)$$

The coefficients a_{ij} are called component weights and X_i denotes the ith variable. Each principal component is a linear combination of the original variables and is derived in such a manner that its successive component accounts for a smaller portion of variation in X. Therefore, the first principal component accounts for the largest portion of variance, the second one for the largest portion of the remaining variance, subject to being orthogonal to the first one, and so on. Hopefully, the first m components will retain most of the variation present in all of the original variables (p). Thus, an essential dimensionality reduction may be achieved by projecting the original data on the new m-dimensional space, as long as $m \ll p$.

The derivation of the new axes (components) is based on Σ, where Σ denotes the covariance matrix of X. Each eigenvector of Σ provides the component weights a_{ij} of the $Y_{(j)}$ component, while the corresponding eigenvalue, denoted λ_j, provides the variance of this component. Alternatively, the derivation of the new axes can be based on the correlation matrix, producing slightly different results. These two options are equivalent when the variables are standardized (i.e., they have mean equal to 0 and standard deviation equal to 1).

Intuitively, PCA transforms a data set X by rotating the original axes of a p-dimensional space and deriving a new set of axes (components), as in Fig. 11.1. The component weights represent the angles between the original and the new axes. In particular, the component weight a_{ij} is the cosine of the angle between the ith original axis and the jth component [10]. The values of $Y_{(j)}$ calculated from Equation 11.1 provide the coordinates of the original data in the new space.

Conclusively, the application of PCA on a multivariate data set $X_{n \times p}$ results in two matrices, in particular, the matrix of component weights $A_{p \times p}$ and the matrix of variances $\Lambda_{p \times 1}$. In addition to that, the matrix of the new coordinates $Y_{n \times p}$ of the original data can be calculated from A, since $Y = X \cdot A$.

11.2.2 Implications of PCA in Similarity Search

Regarding the pre-processing phase, there are four main distortions that may exist in raw data, namely offset translation, amplitude scaling, time warping, and noise. Distance measures may be seriously affected by the presence of any of these distortions, resulting most of the times in missing similar shapes. Offset translation refers to the case where there are differences in the magnitude of the values of two time series, while the general shape remains similar (Fig. 11.2). This distortion is inherently handled by PCA, since it is based on covariances, which are not affected by the magnitude of the values. This is a potential disadvantage of PCA,

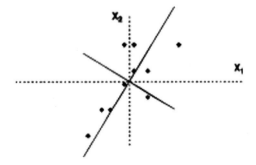

Fig. 11.1 A multivariate time series consisting of two variables (X_1 and X_2) and 10 time instances. *Dots* represent the time instances, while *solid lines* represent the principal components that have been derived by PCA. A dimensionality reduction can be achieved, if only the first component $Y\_(1)$ is retained and data is projected on it

if similarity search is to be based also on the magnitude of the values. Amplitude scaling refers to the case where there are differences in the magnitude of the fluctuations of two time series, while the general shape remains similar (Fig. 11.2). In this case, PCA representation can be based on the correlations among variables, instead of the covariances. This is an alternative way of deriving the principal components that produces slightly different results, but not essentially different in the context of dimensionality reduction. Time warping, which may be global or local, refers to the acceleration or deceleration of the evolvement of a time series through time. In the case of global time warping (i.e., two multivariate time series evolve in different rates), PCA representation is expected to be similar, since the shorter time series can be considered as a systematic random sample of the longer one, resulting to a similar covariance matrix. Intuitively, the existence of local time warping distortions may be captured by the covariances of the corresponding variables. Finally, noise is intrinsically handled by PCA, since the discarded principal components account mainly for variations due to noise.

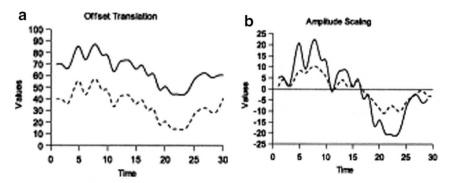

Fig. 11.2 Two of the distortions that may exist in raw time series data

Another issue in the pre-processing phase is the handling of time series of different lengths. PCA requires variables (time series) of equal length for the same object. For example, an object $X_{n \times p}$ consists of p time series that all have the same length n. Therefore, this is a limitation of this technique. However, similarity search is performed among objects, and thus, it is based on the produced matrices $A_{p \times p}$ and $\Lambda_{p \times 1}$, which are independent of the lengths of the series. For example, the comparison of two objects $X_{n \times p}$ and $Q_{m \times p}$ is feasible, since their PCA representations are independent of n and m, respectively.

Similarity search also requires a measure that quantifies the similarity or dissimilarity between two objects. Under PCA transformation, this measure should be based on at least one of the produced matrices, mentioned in the previous paragraph, $A_{p \times p}$, $\Lambda_{p \times 1}$, and $Y_{n \times p}$. The central concept is that, if two multivariate time series are similar, their PCA representations will be similar, that is, the produced matrices will be close enough. Searching similarity based on $A_{p \times p}$ means to compare the angles of principal components derived from two multivariate time series, whereas searching based solely on $Y_{n \times p}$ is useless, since these values are coordinates in different spaces. $\Lambda_{p \times 1}$ contains information about the shape of the time series and it may be used in conjunction with $A_{p \times p}$ for further distinguishing power.

The PCA representation of a data set $X_{n \times p}$ consists of the component weight matrix $A_{p \times p}$ and the variances matrix $\Lambda_{p \times 1}$. The data reduction may be substantial as long as the number of time instances n is much greater than the number of variables p. Moreover, a further data reduction can be achieved, if only m components are retained, where $m < p$. There are several criteria for determining the number of components to retain, such as the scree graph or the cumulative percentage of total variation [13]. According to the latter criterion, one could select that value for m, for which the first m components retain more than 90% of the total variation present in the original data.

Although PCA-based similarity search is complicated and usually requires expensive computations, it may improve the quality of similarity search providing at the same time useful information for post hoc analysis.

11.2.3 Related Work

Although there is a vast literature in univariate time series similarity search, the case of multivariate time series has not been extensively explored. Most of the papers concentrate on indexing multidimensional time series and provide an appropriate representation scheme and/or a similarity measure. In addition to that, most of the research interest lays on trajectories, which usually consist of two- or three-dimensional time series.

The authors of [34] and [5] suggest similarity measures based on the longest common subsequence (LCSS) model, whereas a modified version of the edit distance for real-valued series is provided in [8]. Bakalov et al. [2] extend the symbolic aggregate approximation (SAX) [23] and the corresponding distance measure for multivariate time series. Vlachos et al. [33] propose an indexing framework that supports multiple similarity/distance functions, without the need to rebuild the index. Several researchers approach similarity search by applying a measure and/or an indexing method on transformed data. Kahveci et al. [15] propose to convert a p-dimensional time series of length n to a univariate time series of length np by concatenation and then apply a representation scheme for the purpose of dimensionality reduction. Lee et al. [21] propose a scheme for searching a database, which in the pre-processing phase includes the representation (e.g., DFT) of each one of the p time series separately. Cai and Ng [6] approximate and index multidimensional time series with Chebyshev polynomials. In the latter three papers, the Euclidean distance is applied as a distance measure.

On the other hand, there are several PCA-based measures that have been proposed in order to compare two objects, which are in the form of multivariate time series. The main idea is to derive the principal components for each one and then to compare the produced matrices.

Suppose that we have two multivariate time series denoted $X_{n \times p}$ and $Q_{n \times p}$. Applying PCA on each one results in the matrices of component weights A_X and A_Q and variances Λ_X and Λ_Q, respectively. All the following measures assume that the number of variables p is the same for all series. This is a rational assumption, since usually the same process within a specific application generates these series.

One of the earliest measures has been proposed by Krzanowski [20]. This measure (Equation 11.2) is applicable to time series, although originally it was not applied on such type of data. The proposed approach is to retain m principal components and compare the angles between all the combinations of the first m components of the two objects

$$\text{SimPCA}(X,Q) = \text{trace}(A_X^T A_Q A_Q^T A_X) = \sum_{i=1}^{m} \sum_{j=1}^{m} \cos^2 \theta_{ij}, \quad 0 \leq \text{SimPCA} \leq m \quad (11.2)$$

where θ_{ij} is the angle between the ith principal component of X and the jth principal component of Q.

Johannesmeyer [12] modified the previous measure by weighting the angles with the corresponding variances as in Equation 11.3:

$$\text{S}\lambda\text{PCA}(X,Q) = \sum_{i=1}^{m} \sum_{j=1}^{m} (\lambda_{X_i} \cdot \lambda_{Q_j} \cdot \cos^2 \theta_{ij}) / \sum_{i=1}^{m} \lambda_{X_i} \cdot \lambda_{Q_j}, \quad 0 \leq \text{S}\lambda\text{PCA} \leq 1 \quad (11.3)$$

Yang and Shahabi [35] propose a similarity measure, Eros, which is based on the acute angles between the corresponding components from two objects X and Q (Equation 11.4). Contrary to the previous measures, all components are retained from each object and their variances form a weight vector w. More specifically, the variances obtained from all the objects in a database are aggregated into one weight vector, which is updated when objects are inserted or removed from the database. Finally, the authors provide lower and upper bounds for this measure:

$$\text{Eros}(X,Q,w) = \sum_{i=1}^{p} w(i) \cdot |\cos \theta_i|, \quad 0 \leq \text{Eros} \leq 1 \quad (11.4)$$

Li and Prabhakaran [22] propose a similarity measure for recognizing distinct motion patterns in motion streams in real time. This measure, which is called k Weighted Angular Similarity (kWAS), can be obtained by applying singular value decomposition on the transformed data sets, $X^T X$ and $Q^T Q$, and retaining the first m components. kWAS is based on the acute angles between the corresponding components weighted by the corresponding eigenvalues (Equation 11.5):

$$\Psi(X,Q) = \frac{1}{2} \sum_{i=1}^{m} ((\sigma_i / \sum_{i=1}^{n} \sigma_i + \lambda_i / \sum_{i=1}^{n} \lambda_i) |u_i \cdot v_i|), \quad 0 \leq \Psi(X,Q) \leq 1 \quad (11.5)$$

where σ_i and λ_i are the eigenvalues corresponding to the ith eigenvectors u_i and v_i of matrices $X^T X$ and $Q^T Q$. When the original data sets are mean centered, the above procedure is equivalent to applying PCA on the original data. The eigenvectors u_i and v_i are the corresponding principal components, while the eigenvalue-based weight in Equation 11.5 is equal to the one obtained if σ_i and λ_i are replaced by the variances of the corresponding components. The absolute value implies that the cosine of the acute angles is computed.

Singhal and Seborg [29] extend Johannesmeyer's [12] measure by incorporating an extra term, which expresses the distance between the original values of the two objects. This term is based on Mahalanobis distance and on the properties of the Gaussian distribution.

Another measure that incorporates the distance between the original values of two objects has been proposed by Otey and Parthasarathy [25]. The authors define a distance measure in terms of three dissimilarity functions that take into account the differences among the original values, the angles between the corresponding components, and the difference in variances. For the first term, the authors propose to use either the Euclidean or the Mahalanobis distance, whereas the second term is defined as the summation of the acute angles between the corresponding components, given that all components are retained. The third term accounts for the differences in the distributions of the variance over the derived components and is based on the symmetric relative entropy [9].

In the context of statistical process control, Kano et al. [16] propose a distance measure for the purpose of monitoring processes and identifying deviations from normal operating conditions. This measure is based on the Karhunen–Loeve expansion, which is mathematically equivalent to PCA. However, it involves applying eigenvalue decomposition twice during its calculation, which is the most computationally expensive part.

11.3 Proposed Approach

In this chapter, we propose a novel approach in multivariate time series similarity search that is based on principal component analysis. The main difference to other proposed methods is that it does not require applying PCA on the query object. Remember that an object is a multivariate time series that is expressed in the form of a matrix.

More specifically, PCA is applied on each object $X_{n \times p}$ of a database and the derived matrix of component weights $A_{p \times m}$ is stored (where m is the number of the retained components). Although this task is computationally expensive, it is performed only once during the pre-processing phase.

When a query object arrives, the objective is to identify the most similar object in the database. We propose a distance measure that relates to the squared prediction error (SPE), a well-known statistic in Multivariate statistical process control [19].

In particular, each time instance q_i of a query object $Q_{v \times p}$ is projected on the plane derived by PCA and its new coordinates (q_i') are obtained (Equation 11.6):

$$q_i' = q_i \cdot A, \quad i = 1, 2, \ldots, v \tag{11.6}$$

In order to determine the error that this projection introduces to the new values, we need to calculate the predicted values $(\hat{q}_i)$ of q_i (Equation 11.7):

$$\hat{q}_i = q_i' \cdot A^T, \quad i = 1, 2, \ldots, v \tag{11.7}$$

SPE is the sum of the squared differences between the original and the predicted values and represents the squared perpendicular distance of a time instance from the plane (Equation 11.8).

$$SPE_i = \sum_{j=1}^{p} (q_{ij} - \hat{q}_{ij})^2, \quad i = 1, 2, \ldots, v \tag{11.8}$$

This measure can be extended in order to incorporate all time instances of the query object $Q_{v \times p}$ (Equation 11.9). We call this new distance measure SPEdist (squared prediction error distance):

$$SPEdist\,(X, Q) = \sum_{i=1}^{v} \sum_{j=1}^{p} (q_{ij} - \hat{q}_{ij})^2 \tag{11.9}$$

SPE is particularly useful within statistical process control because it is very sensitive to outliers, and thus, it can efficiently identify possible deviations from the normal operating conditions of a process. However, this sensitivity may be problematic in other applications that require more robust measures. Therefore, we propose a variation of SPEdist that utilizes the absolute differences between the original and the predicted values (Equation 11.10). We call this measure APEdist (absolute prediction error distance):

$$APEdist\,(X, Q) = \sum_{i=1}^{v} \sum_{j=1}^{p} |q_{ij} - \hat{q}_{ij}| \tag{11.10}$$

The main concept is that the most similar object in a database is defined to be the one whose principal components describe more adequately the query object with respect to the reconstruction error. A similar approach can be found in the work of Barbic et al. [3], who propose a technique for the purpose of segmenting motion capture data into distinct motions. However, the authors utilize the squared error of the projected values and not the predicted values, as we propose in our work. Moreover, they focus on an application that involves one multivariate time series, which should be segmented.

As was mentioned earlier, the proposed approach does not apply the computationally expensive PCA technique on the query object. Moreover, we provide a method that further speeds up the calculation of APEdist, hopefully, without substantially affecting the quality of similarity search. This method involves applying a dimensionality reduction technique on each one of the time series that forms the query object, as a pre-processing step. The proposed technique is the piecewise aggregate approximation (PAA) that was introduced independently by Keogh et al. [18] and Yi and Faloutsos [36]. PAA is a well-known representation in the data mining community that can be extremely fast to compute. This technique segments a time series of length n into N consecutive sections of equal width and calculates the corresponding mean for each one. The series of these means is the new representation of the original series. According to this approach, a query object that consists of p time series of length n is transformed to an object of p time series of length N. Under this transformation, we only need a fraction (N/n) of the required calculations in order to compute APEdist. Equivalently, the required calculations will be executed n/N times faster than the original ones. The consequence of this method in the quality of similarity search depends mainly on the quality of PAA representation within a specific data set. Intuitively, APEdist is computed on a set of time instances, which may be considered as representatives of the original ones.

In general, our approach can be applied on data types other than time series. For example, suppose that we have customer data such as age, income, gender from several stores. A matrix whose rows correspond to customers and columns to their attributes represents each store. The objective is to identify similar stores with respect to their customer profiles. The PCA representation is based on the covariance matrix, which is independent of the order of the corresponding rows (time instances), and thus, the time dimension is ignored under the proposed representation.

In this chapter, we focus on time series because this type of data is generated at high rates and is of high dimensionality. Our approach has two advantages. First, PCA-based representation dramatically reduces the size of the database while retaining most of the important information present in the original data. Second, the proposed distance measure does not require applying the computationally expensive PCA technique on the query.

11.4 Experimental Methodology

The experiments are conducted on three real-world data sets and one synthetically created data set used extensively in the literature and described in Section 11.4.1. Section 11.4.2 presents the evaluation methods and Section 11.4.3 discusses the rival measures along with their corresponding settings.

11.4.1 Data Sets

The first data set relates to Australian Sign Language (AUSLAN), which contains sensor data gathered from 22 sensors placed on the hands (gloves) of a native AUS-LAN speaker. The objective is the identification of a distinct sign. There are 95 distinct signs, each one performed 27 times. In total, there are 2565 signs in the data set. More technical information can be found in [14].

The second data set, HUMAN GAIT, involves the task of identifying a person at a distance. Data is captured using a Vicon 3D motion capture system, which generates 66 values at each time instance. Fifteen persons participated in this experiment and were required to walk in three sessions, at four different speeds, three times for each speed. In total, there are 540 walk sequences. More technical information can be found in [30].

The third data set relates to EEG (electroencephalography) data that arises from a large study to examine EEG correlates of genetic predisposition to alcoholism [4]. It contains measurements from 64 electrodes placed on the scalp and sampled at 256 Hz (3.9-ms epoch) for 1 s. The experiments were conducted on 10 alcoholic and 10 control subjects. Each subject was exposed to 3 different stimuli, 10 times for each one. This data set is provided in the form of a train and a test set, both consisting of 600 EEG's. The test data was gathered from the same subjects as with the training data, but with 10 out-of-sample runs per subject per paradigm.

Finally, the transient classification benchmark (TRACE) is a synthetic data set designed to simulate instrumentation failures in a nuclear power plant [28]. There are 4 process variables, which generate 16 different operating states, according to their co-evolvement through time. There is an additional variable, which initially takes on the value of 0, until the start of the transient occurs and its value changes to 1. We retain only that part of data, where the transient is present. For each state, there are 100 examples. The data set is separated into train and test sets each consisting of 50 examples per state.

Table 11.1 summarizes the profile of the data sets.

Table 11.1 Description of data sets

DATA SET	Number of variables	Mean length	Number of classes	Size of class	Size of data set
AUSLAN	22	57	95	27	2565
HUMAN GAIT	66	133	15	36	540
EEG	64	256	2	600	1200
TRACE	4	250	16	100	1600

11.4.2 Evaluation Methods

In order to evaluate the performance of the proposed approach, we conduct several experiments in three phases.

First, we perform one-nearest neighbor classification (1-NN) and evaluate it by means of classification error rate. We use ninefold cross-validation for the AUSLAN and HUMAN GAIT data sets taking into account all the characteristics of the experiments, while creating the subsets. The observed differences in the error rates among the various methods are statistically tested. Due to the small number of subsets and to the violation of normality assumption in some cases, Wilcoxon signed-rank tests are performed at 5% significance level. For the EEG and TRACE data sets, we use the existing train and test sets.

Second, we perform leave-one-out k-NN similarity search and evaluate it by plotting the recall–precision graph [11]. In particular, every object in the data set is considered as a query. Then the r most similar objects are retrieved, where r is the smallest number of objects that should be retrieved in order to obtain k objects of the same class with the query ($1 \leq k \leq$ size_of_class-1). The precision and recall pairs corresponding to the values of k are calculated. Finally, the average values of precision and recall are computed for the whole data set. Precision is defined as the proportion of retrieved objects that are relevant to the query, whereas recall is defined as the proportion of relevant objects that are retrieved relative to the total number of relevant objects in the data set. In these experiments, the training and testing data sets of EEG and TRACE are merged.

Third, we evaluate the trade-off between classification accuracy and speed of calculating the proposed measure APEdist by applying 1-NN classification on objects that have been pre-processed as described in Section 11.3.

All the necessary codes and experiments are developed in MATLAB, whereas the statistical analysis is performed in SPSS.

11.4.3 Rival Measures

The similarity measures that are tested on our experiments are SimPCA, SλPCA, Eros, kWAS, SPEdist, and APEdist. We choose to omit the results for SPEdist, because it performs similarly or slightly worse than APEdist in most cases. For comparison reasons, we also include in the experiments the Euclidean distance. Since this measure requires data sets of equal number of time instances, we apply linear interpolation on the original data sets and set the length of the time series equal to the corresponding mean length (Table 11.1). The transformed data sets are utilized only when Euclidean distance is applied. The rest of the measures we review in Section 11.2.3 are not included in these experiments because they take into consideration the differences among the original values, whereas in our experiments, the measures are calculated on the mean centered values.

Regarding Eros, the weight vector w is computed by averaging the variances of each component across the objects of the training data set and normalizing them so that $\sum w_i = 1$, for $i = 1, 2, \ldots, p$. In [35] one can find alternative ways for computing the weight vector.

All other measures require determining the number of components m to be retained. For AUSLAN, HUMAN GAIT, and EEG, we are conducting classification for consecutive values of m between 1 and 20. For $m = 20$, at least 99% of the total variation is retained for all objects in AUSLAN and HUMAN GAIT, whereas at least 90% of the total variation is retained for all objects in EEG. For TRACE, we are conducting classification for all possible values of m ($m = 1-4$). Precision–recall graphs are plotted for the "best" value that it is observed in the classification experiments. In general, this value is different for each measure.

Principal component analysis is performed on the covariance matrices. For comparison reasons, the similarity measures kWAS and APEdist are computed on the mean centered values.

11.5 Results

We provide and discuss the results of 1-NN classification for each data set separately in Section 11.5.1. In particular, we present the classification error rates that the tested measures achieve across various values of m (the number of components retained), and we also report the m that corresponds to the lowest error rate for each measure. In Section 11.5.2, the results of performing leave-one-out k-NN similarity search are presented in precision–recall graphs for each data set. Finally, in Section 11.5.3, we provide and discuss the effect APEdist with PAA has on the classification accuracy for various degrees of speedup.

11.5.1 1-NN Classification

In Figs. 11.3, 11.4, 11.5, and 11.6, the classification error rates are presented graphically for various values of m (the number of components retained) for each data set. For the first three data sets, we show the error rates up to that value of m beyond which the behavior of similarity measures does not change significantly. For the TRACE data set, which has only four variables, we show error rates for values of m up to 3. For Euclidean distance (ED) and Eros the rates are constant across m.

Regarding the first three data sets (Fig. 11.3, 11.4, and Fig. 11.5), we observe that all measures seem to achieve the lowest error rate, when only a few components are retained. Moreover, as the number of components is further increased, the improvement in error rates seems to be negligible. In AUSLAN (Fig. 11.3), the performance of APEdist and SimPCA deteriorates with the increase of m. Note that these two measures do not take into account the variance that each component explains, contrary to the other three PCA-based measures. A second observation is that the performance of APEdist is comparable, if not better, to the "best" measure in each one of the three data sets. Regarding TRACE, which consist of only four variables, ED achieves considerably lower error rates than any other measure (Fig. 11.6).

In Table 11.2, the lowest classification error rates are presented along with the corresponding number of the retained components. First, we compare similarity/distance measures with respect to each data set separately.

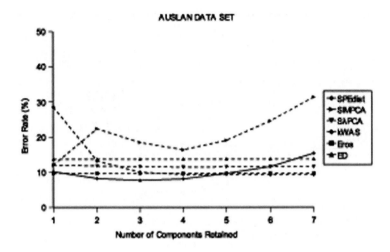

Fig. 11.3 1-NN classification error rates (AUSLAN data set)

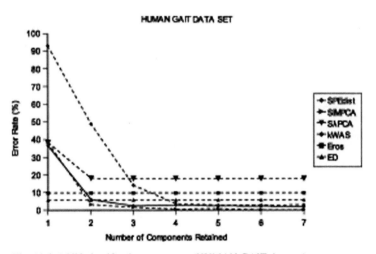

Fig. 11.4 1-NN classification error rates (HUMAN GAIT data set)

In AUSLAN, APEdist produces the lowest classification error rate. Statistically testing the differences across the specific subsets, APEdist produces better results than all measures ($p < 0.05$).

Regarding HUMAN GAIT, SimPCA, Eros, kWAS, and APEdist seem to provide the best results. Statistically testing their differences across the specific subsets, SimPCA produces better results than all ($p < 0.05$), whereas the performances of Eros, kWAS, and APEdist are statistically similar ($p > 0.05$).

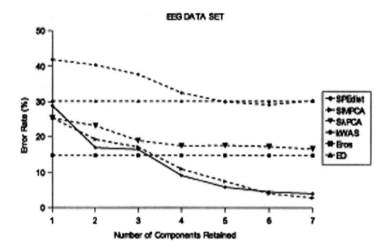

Fig. 11.5 1-NN classification error rates (EEG data set)

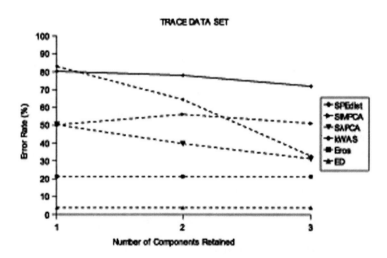

Fig. 11.6 1-NN classification error rates (TRACE data set)

For EEG, SimPCA and APEdist seem to provide considerably better results than other measures, with classification error rates of 0.00 and 1.83%, respectively, when the next best performing measure, Eros, has a classification error rate of 14.83%.

Finally, for TRACE that consists of only four variables, Euclidean distance, a non-PCA-based measure, performs essentially better than all measures with 3.9% classification error rate. The next best performing measures are Eros and kWAS with classification error rates of 21.38 and 21.88%, respectively.

Table 11.2 Classification error rates (%) [numbers in parentheses indicate the number of principal components retained. Lack of number indicates measures that exploit all components]

Measure	ASL	HG	EEG	TRC
ED	13.76	5.74	30.17	**3.88**
SimPCA	(1) 12.05	(8) **0.00**	(14) **0.00**	(1) 50.25
SλPCA	(4) 11.46	(3) 17.78	(10) 16.50	(3) 31.38
Eros	9.71	2.96	14.83	21.38
kWAS	(6) 9.24	(12) 2.59	(17) 25.33	(4) 21.88
APEdist	(3) **7.68**	(7) 1.85	(14) 1.83	(3) 72.00

11.5.2 k-NN Similarity Search

In the following figures, the precision–recall graphs are presented for each data set separately. The number of retained components is set equal to the one for which the corresponding measure provided the lowest classification error rates (Table 11.2). Regarding AUSLAN (Fig. 11.7), all measures seem to perform similarly to each other and better than the Euclidean distance. APEdist provides better results than all; however, the differences cannot be considered significant.

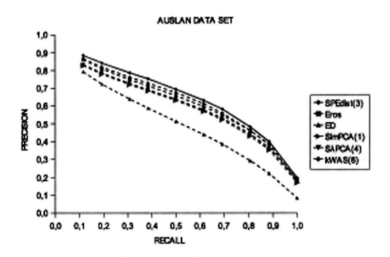

Fig. 11.7 Precision–recall graph for various measures (AUSLAN data set)

In HUMAN GAIT (Fig. 11.8) and EEG (Fig.11.9), however, SimPCA and APEdist perform better than all. As mentioned in the previous section, these two measures do not take into consideration the explained variance of the retained components. This fact may imply that for these specific data sets, the variance information may not be significant. On the other hand, in AUSLAN, where this information may be important, APEdist provides comparable results to other measures.

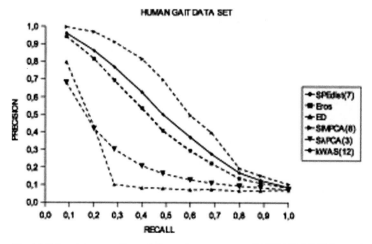

Fig. 11.8 Precision–recall graph for various measures (HUMAN GAIT data set)

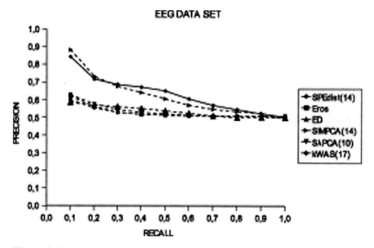

Fig. 11.9 Precision–recall graph for various measures (EEG data set)

In the final data set, TRACE, Euclidean distance performs better for recall values up to 0.3, whereas kWAS performs better for greater recall values (Fig. 11.10). Compared to other measures, APEdist seems to improve its performance for recall values greater than 0.6.

11.5.3 Speeding Up the Calculation of APEdist

The idea is to apply PAA on each one of the time series that comprises the query object (see Section 11.3), in order to speed up the calculations of APEdist. We

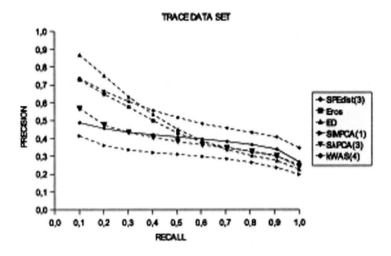

Fig. 11.10 Precision–recall graph for various measures (TRACE data set)

experiment with various degrees of dimensionality reduction by using PAA to retain 10, 20, and 30% of the original dimensions of the query object, thus expecting a $10\times$, $5\times$, and $3.33\times$ speed up of the calculations, respectively.

Table 11.3 presents the effect the speedup has on the classification error rate. The number of the retained components is different among data sets and is set equal to the optimal value obtained in Section 11.5.1 (Table 11.2).

As expected, the classification error rate increases as the speedup increases. Nevertheless, in all data sets, we are able to achieve similar classification error rates by doing at most 20% of the required calculations (a $5\times$ speedup). More specifically, for AUSLAN, even a $5\times$ speedup provides better results than rival measures (Table 11.2). Regarding HUMAN GAIT, a $10\times$ speedup results into exactly the same classification error rate as the one observed when full calculations were applied. In EEG, although the error rates differ significantly for the various degrees of speedup, the $10\times$ speedup provides lower error rate than rival measures (except from Sim-PCA). Regarding TRACE, a $5\times$ speedup results into almost the same classification error rate as the one observed when full calculations were applied.

Table 11.3 1-NN classification error rates for various degrees of dimensionality reduction on the query object

Percentage of retained dimensions	10%	20%	30%	100%
Speedup	$10\times$	$5\times$	$3.33\times$	$1\times$
AUSLAN	10.80	8.50	8.27	7.68
HUMAN GAIT	1.85	1.85	1.85	1.85
EEG	8.00	3.67	2.33	1.83
TRACE	72.12	74.50	71.38	72.00

11.6 Conclusion

The main contribution of this chapter is the introduction of a novel approach in multivariate time series similarity search for the purpose of improving the efficiency of data mining techniques without affecting the quality of the corresponding results. We investigate the usefulness of our approach, mainly in the context of query by content and 1-NN classification.

Experiments are conducted on four widely utilized data sets and various measures are tested with respect to 1-NN classification and precision/recall. There are three key observations with respect to the results of these experiments. First, there is no measure that can be clearly considered as the most appropriate one for any data set. Second, in three data sets, our approach provides significantly better results than the Euclidean distance, whereas its performance is at least comparable to the four other PCA-based measures that are tested. Third, there is strong evidence that the application of the proposed approach can be accelerated with little cost in the quality of similarity search. In all data sets, one-tenth up to one-third of the required calculations is adequate in order to achieve similar results to the full computation case.

A secondary contribution of this chapter is the review of several PCA-based similarity/distance measures that have been recently proposed from diverse fields, not necessarily within the data mining context. A more general conclusion is that principal component analysis has not been extensively explored in the context of similarity search in multivariate time series, and hence, it has the potential to offer more in the data mining field.

Future work will focus on improving the speedup of the proposed approach during the pre-processing stage by exploiting the features of other dimensionality reduction techniques. We also intend to conduct experiments on more data sets in order to further validate our approach.

References

1. Agrawal, R., Faloutsos, C., Swami, A.: Efficient similarity search in sequence databases. In: Proc. 4th Int. Conf. FODO, Evanston, IL, pp. 69–84, (1993).
2. Bakalov, P., Hadjieleftheriou, M., Keogh, E., Tsotras, V.J.: Efficient trajectory joins using symbolic representations. In: Proc. 6th Int. Conf. on Mobile data management, Ayia Napa, Cyprus, pp. 86–93, (2005).
3. Barbic, J., Safonova, A., Pan, J.Y., Faloutsos, C., Hodgins, J.K., Pollard, N.S.: Segmenting motion capture data into distinct behaviors. In: Proc. Graphics Interface Conf, London, Ontario, Canada, pp. 185–194, (2004).
4. Begleiter, H.: The UCI KDD Archive [http://kdd.ics.uci.edu]. Irvine, CA: University of California, Department of Information and Computer Science, (1999).
5. Buzan, D., Sclaroff, S., Kollios, G.: Extraction and clustering of motion trajectories in video. In: Proc. 17th ICPR, Boston, MA, vol. 2, pp. 521–524, (2004).
6. Cai, Y., Ng, R.: Indexing spatio-temporal trajectories with Chebyshev polynomials. In: Proc. ACM SIGMOD, Paris, France, pp. 599–610, (2004).

7. Chapman, L., Thornes, J.E.: The use of geographical information systems in climatology and meteorology. Progress in Physical Geography, 27(*3*), pp. 313–330, (2003).
8. Chen, L., Ozsu, M.T., Oria, V.: Robust and fast similarity search for moving object trajectories. In: Proc. ACM SIGMOD Int. Conf. on Management of Data, Baltimore, MD, pp. 491–502, (2005).
9. Cover, T.M., Thomas, J.A.: Elements of Information Theory. John Wiley & Sons Inc., New York, (1991).
10. Gower, J.C.: Multivariate Analysis and Multidimensional Geometry. The Statistician, 17(*1*), pp. 13–28, (1967).
11. Hand, D., Mannila, H., Smyth, P.: Principles of Data Mining. MIT Press, Cambridge, MA (2001).
12. Johannesmeyer, M.C.: Abnormal situation analysis using pattern recognition techniques and historical data. M.S. thesis, UCSB, Santa Barbara, CA, (1999).
13. Jolliffe, I.T.: Principal Component Analysis. Springer, New York, Chapter 1, (2004).
14. Kadous, M.W.: Temporal Classification: extending the classification paradigm to multivariate time series. Ph.D. Thesis, School of Computer Science and Engineering, University of New South Wales, (2002).
15. Kahveci, T., Singh, A., Gurel, A.: Similarity searching for multi-attribute sequences. In: Proc. 14th SSDBM, Edinburg, Scotland, pp. 175–184, (2002).
16. Kano, M., Nagao, K., Ohno, H., Hasebe, S., Hashimoto, I.: Dissimilarity of process data for statistical process monitoring. In: Proc. IFAC Symp. ADCHEM, Pisa, Italy, vol. I, pp. 231–236, (2000).
17. Kano, M., Nagao, K., Hasebe, S., Hashimoto, I., Ohno, H., Strauss, R., Bakshi, B.R.: Comparison of multivariate statistical process monitoring methods with applications to the Eastman challenge problem. Computers & Chemical Engineering, 26(2), pp. 161–174, (2002).
18. Keogh, E., Chakrabarti, K., Pazzani, M., Mehrotra, S.: Dimensionality Reduction for Fast Similarity Search in Large Time Series Databases. Knowledge and Information Systems, 3(*3*), pp. 263–286, (2001).
19. Kresta, J., MacGregor, J.F., Marlin, T.E.: Multivariate statistical monitoring of process operating performance. The Canadian Journal of Chemical Engineering, 69, pp. 35–47, (1991).
20. Krzanowski, W.: Between-groups comparison of principal components. JASA, 74(*367*), pp. 703–707, (1979).
21. Lee, S.L., Chun, S.J., Kim, D.H., Lee, J.H., Chung, C.W.: Similarity search for multidimensional data sequences. In: Proc. ICDE, San Diego, CA, pp. 599–608, (2000).
22. Li, C., Prabhakaran, B.: A similarity measure for motion stream segmentation and recognition. In: Proc.6th Int. Workshop MDM/KDD, Chicago, IL, pp. 89–94, (2005).
23. Lin, J., Keogh, E., Lonardi, S., Chiu, B.: A symbolic representation of time series, with implications for streaming algorithms. In: Proc. 8th ACM SIGMOD Workshop on Research Issues in Data Mining and Knowledge Discovery, San Diego, CA, pp. 2–11, (2003).
24. Moeslund, T.B., Granum, E.: A survey of computer vision-based human motion capture. Computer Vision and Image Understanding, 81(*3*), pp. 231–268, (2001).
25. Otey, M.E., Parthasarathy, S.: A dissimilarity measure for comparing subsets of data: application to multivariate time series. In: Proc. ICDM Workshop on Temporal Data Mining, Houston, TX, (2005).
26. Quinlan, J.R.: C4.5 – Programs for machine learning. Morgan Kaufmann Publishers, San Mateo, (1993).
27. Ratanamahatana, C.A., Lin, J., Gunopulos, D., Keogh, E., Vlachos, M., Das, G.: Data Mining and Knowledge Discovery Handbook, chapter 51, Mining Time Series Data. Springer US, pp. 1069–1103, (2005).
28. Roverso, D.: Plant diagnostics by transient classification: the Aladdin approach. International Journal of Intelligent Systems, 17(*8*), pp. 767–790, (2002).
29. Singhal, A., Seborg, D.E.: Clustering multivariate time-series data. Journal of Chemometrics, 19(*8*), pp. 427–438, (2005).

30. Tanawongsuwan, R., Bobick, A.: Performance analysis of time-distance gait parameters under different speeds. In: Proc. 4th Int. Conf. AVBPA, Guilford, UK, pp. 715–724, (2003).
31. Valera, M., Velastin, S.A.: Intelligent distributed surveillance systems: a review. In: IEE Proc. Vision Image and Signal Processing, 152(2), pp. 192–204, (2005).
32. Vapnik, V.: The Nature of Statistical Learning Theory. Springer, New York, (1995)
33. Vlachos, M., Hadjieleftheriou, M., Gunopoulos, D., Keogh, E.: Indexing multidimensional time-series with support for multiple disatance measures. In: Proc. 9th ACM SIGKDD, Washington, D.C., pp. 216–225, (2003).
34. Vlachos, M., Hadjieleftheriou, M., Gunopoulos, D., Keogh, E.: Indexing multidimensional time-series. VLDB Journal, 15(1), pp. 1–20, (2006).
35. Yang, K., Shahabi, C.: A PCA-based similarity measure for multivariate time series. In: Proc. 2nd ACM MMDB, Washington, D.C., pp. 65–74, (2004).
36. Yi, B.K., Faloutsos, C.: Fast time sequence indexing for arbitrary Lp Norms. In: Proc. VLDB-2000: Twenty-Sixth International Conference on Very Large Databases, Cairo, Egypt, (2000).

Chapter 12
Evolutionary Optimization of Least-Squares Support Vector Machines

Arjan Gijsberts, Giorgio Metta, and Léon Rothkrantz

Abstract The performance of kernel machines depends to a large extent on its kernel function and hyperparameters. Selecting these is traditionally done using intuition or a costly "trial-and-error" approach, which typically prevents these methods from being used to their fullest extent. Therefore, two automated approaches are presented for the selection of a suitable kernel function and optimal hyperparameters for the least-squares support vector machine. The first approach uses evolution strategies, genetic algorithms, and genetic algorithms with floating point representation to find optimal hyperparameters in a timely manner. On benchmark data sets the standard genetic algorithms approach outperforms the two other evolutionary algorithms and is shown to be more efficient than grid search. The second approach aims to improve the generalization capacity of the machine by evolving combined kernel functions using genetic programming. Empirical studies show that this model indeed increases the generalization performance of the machine, although this improvement comes at a high computational cost. This suggests that the approach may be justified primarily in applications where prediction errors can have severe consequences, such as in medical settings.

Arjan Gijsberts
Italian Institute of Technology, Via Morego, 30 – Genoa 16163, Italy; Delft University
of Technology, Mekelweg 4, 2628 CD Delft, The Netherlands,
e-mail: arjan.gijsberts@iit.it

Giorgio Metta
Italian Institute of Technology, Via Morego, 30 – Genoa 16163, Italy; University of Genoa, Viale
F. Causa, 13 – Genoa 16145, Italy, e-mail: giorgio.metta@iit.it

Léon Rothkrantz
Delft University of Technology, Mekelweg 4, 2628 CD Delft, The Netherlands; Netherlands Defence Academy, Nieuwe Diep 8, 1781 AT Den Helder, The Netherlands,
e-mail: l.j.m.rothkrantz@ewi.tudelft.nl

R. Stahlbock et al. (eds.), *Data Mining*, Annals of Information Systems 8,
DOI 10.1007/978-1-4419-1280-0_12, © Springer Science+Business Media, LLC 2010

12.1 Introduction

Kernel machines allow the construction of powerful, non-linear classifiers using relatively simple mathematical and computational techniques [35]. As such, they have successfully been applied in fields as diverse as data mining, economics, biology, medicine, and robotics. Much of the success of the kernel machines is due to the *kernel trick*, which can best be described as an implicit mapping of the input data into a high-dimensional feature space. In this manner, the algorithms can be applied in a high-dimensional space, without the need to explicitly map the data points. This implicit mapping is done by means of a *kernel function*, which represents the inner product for the specific hypothetical feature space.

The performance of kernel machines is highly dependent on the chosen kernel function and parameter settings. Unfortunately, there are no analytical methods or strong heuristics that can guide the user in selecting an appropriate kernel function and good parameter values. The common way of finding optimal hyperparameters is to use a costly grid search, which scales exponentially with the number of parameters. Additionally, it is usually necessary to manually determine the region and resolution of the search to ensure computational feasibility. Selection of the kernel function is done similarly, i.e., either trial and error or only considering the default Gaussian kernel function. Consequently, tuning the techniques may be arduous, such that less than optimal performance is achieved. For a successful integration in real-life information systems, kernel machines should be combined with an automated, efficient optimization strategy for both hyperparameters and kernel function.

Two distinct approaches are proposed for the automated selection of the parameters and the kernel function itself. These models are based on techniques that fall in the class of evolutionary computation, which are techniques inspired by neo-Darwinian evolution. The first approach uses evolutionary algorithms to optimize the hyperparameters of a kernel machine in a time-efficient manner. The second aims to increase the generalization performance by constructing combined, problem-specific kernel functions using genetic programming. Implementations of both approaches have been evaluated on seven benchmark data sets, for which traditional grid search was used as a reference.

Kernel machines and the kernel trick are presented in Section 12.2. We emphasize on one particular type of kernel machine, namely the least-squares support vector machine. In Section 12.3, an introduction is given into the evolutionary algorithms that are used in the models. A review of related work on hyperparameter optimization and kernel construction is given in Section 12.4. The two approaches are presented in Section 12.5, after which the experimental results are presented in Section 12.6. This chapter is finalized in Section 12.7 with the conclusions and suggestions for future work.

12.2 Kernel Machines

All kernel machines rely on a kernel function to transform a non-linear problem into a linear one by mapping the input data into a hypothetical, high-dimensional feature space. This mapping – the *kernel trick* – is not done explicitly, as the kernel function calculates the inner product in the corresponding feature space. The kernel trick is explained together with the *least-squares support vector machine* (LS-SVM), which is a particular type of kernel machine.

12.2.1 Least-Squares Support Vector Machines

Assume a set of ℓ labeled training samples, i.e., $S = \{(\mathbf{x}_i, y_i)\}_{i=1}^{\ell}$, where $\mathbf{x} \in X \subseteq \mathbb{R}^n$ is an input vector of n features and $y \in \mathcal{Y}$ is the corresponding label. In the case $\mathcal{Y}$ denotes a set of discrete classes, e.g., $\mathcal{Y} \subseteq \{-1, 1\}$, then the problem is considered a *classification* problem. Conversely, if $\mathcal{Y} \subseteq \mathbb{R}$, then we are dealing with a *regression* problem. The LS-SVM aims to construct a linear function [37]

$$f(\mathbf{x}) = \langle \mathbf{x}, \mathbf{w} \rangle + b \tag{12.1}$$

which is able to predict an output value y given an input sample $\mathbf{x}$. Note that for binary classification purposes it is necessary to apply the sign function on the predicted output value. The error in the prediction for each sample i is defined as

$$y_i - (\langle \mathbf{x_i}, \mathbf{w} \rangle + b) = \varepsilon_i \qquad \text{for } 1 \leq i \leq \ell \tag{12.2}$$

The optimization problem in LS-SVM is analogous to that of traditional support vector machines (SVM) [38]. The goal is to minimize both the norm of the weight vector $\mathbf{w}$ (i.e., maximize the margin) and the sum of the squared errors. In contrast to SVM, LS-SVM uses *equality constraints* for the errors instead of *inequality constraints*. Combining the optimization problem with the equality constraints for the errors (12.2), one obtains

$$\text{minimize} \quad \frac{1}{2}\|\mathbf{w}\|^2 + \frac{1}{2}C\sum_{i=1}^{\ell} \varepsilon_i^2 \tag{12.3}$$

$$\text{subject to} \quad y_i = \langle \mathbf{x}_i, \mathbf{w} \rangle + b + \varepsilon_i \qquad \text{for } 1 \leq i \leq \ell$$

where C is the regularization parameter. Reformulating this optimization problem as a Lagrangian gives the unconstrained minimization problem

$$\frac{1}{2}\|\mathbf{w}\|^2 + \frac{1}{2}C\sum_{i=1}^{\ell} \varepsilon_i^2 - \sum_{i=1}^{\ell} \alpha_i \left(\langle \mathbf{x}_i, \mathbf{w} \rangle + b + \varepsilon_i - y_i\right) \tag{12.4}$$

where $\alpha_i \in \mathbb{R}$ for $1 \leq i \leq \ell$. Note that the Lagrange multipliers α_i can be either positive or negative, due to the equality constraints in the LS-SVM algorithm. The optimality conditions for this problem can be obtained by setting all derivatives equal to zero. This yields a set of linear equations

$$\sum_{j=1}^{\ell} \alpha_j \langle \mathbf{x}_j, \mathbf{x}_i \rangle + b + C^{-1}\alpha_i = y_i \qquad \text{for } 1 \leq i \leq \ell \tag{12.5}$$

12.2.2 Kernel Functions

We observe that the training samples are only present within the inner products in
(12.5). The kernel function used to compute an inner product is defined as

$$k(\mathbf{x}, \mathbf{z}) = \langle \phi(\mathbf{x}), \phi(\mathbf{z}) \rangle \tag{12.6}$$

where $\phi(\mathbf{x})$ is the mapping of the input samples into a feature space. If we substitute the standard inner product with a kernel function in (12.5), we obtain the
"kernelized" variant

$$\sum_{j=1}^{\ell} \alpha_j k(\mathbf{x}_j, \mathbf{x}_i) + b + C^{-1} \alpha_i = y_i \qquad\qquad \text{for } 1 \le i \le \ell \tag{12.7}$$

Usually it is convenient to define a symmetric kernel matrix as $\mathbf{K} = (k(\mathbf{x}_i, \mathbf{x}_j))_{i,j=1}^{\ell}$,
so that the system of linear equations can be rewritten as

$$\begin{bmatrix} \mathbf{K} + C^{-1}\mathbf{I} & \mathbf{1} \\ \mathbf{1}^T & 0 \end{bmatrix} \begin{bmatrix} \alpha \\ b \end{bmatrix} = \begin{bmatrix} \mathbf{y} \\ 0 \end{bmatrix} \tag{12.8}$$

Note that the bottom row and rightmost column have been added to integrate the
bias b in the system of linear equations. Other than the sign function, the algorithm
is identical for both regression and classification. After the optimal Lagrange multipliers and bias have been obtained using (12.8), unseen samples can be predicted
using

$$f(\mathbf{x}) = \sum_{i=1}^{\ell} \alpha_i k(\mathbf{x}_i, \mathbf{x}) + b \tag{12.9}$$

12.2.2.1 Conditions for Kernels

It is important to obtain functions that correspond to an inner product in some feature space. *Mercer's theorem* states that valid kernel functions must be symmetric,
continuous, and positive semi-definite [38], formalized as the following condition:

$$\int_{\mathcal{X} \times \mathcal{X}} k(\mathbf{x}, \mathbf{z}) f(\mathbf{x}) f(\mathbf{z}) \, d\mathbf{x} d\mathbf{z} \ge 0 \qquad\qquad \text{for all } f \in L_2(\mathcal{X}) \tag{12.10}$$

Kernel functions that satisfy these conditions are referred to as *admissible kernel functions*. If this condition is satisfied, then the kernel matrix is accordingly
positive semi-definite [4]. Unfortunately, it is not trivial to verify that a kernel function satisfies Mercer's condition, nor whether the kernel matrix is positive semi-
definite. There are, however, certain functions that have analytically been proven
to be admissible. Common kernel functions – for classification and regression purposes – include the *polynomial* (12.11), the *RBF* (12.12), and the *sigmoid* function

(12.13). Note that the sigmoid kernel function is only admissible for certain parameter values.

$$k(\mathbf{x}, \mathbf{z}) = (\langle \mathbf{x}, \mathbf{z} \rangle + c)^d \qquad \text{for } d \in \mathbb{N},\ c \geq 0 \qquad (12.11)$$

$$k(\mathbf{x}, \mathbf{z}) = \exp\left(-\gamma \|\mathbf{x} - \mathbf{z}\|^2\right) \qquad \text{for } \gamma > 0 \qquad (12.12)$$

$$k(\mathbf{x}, \mathbf{z}) = \tanh\left(\gamma \langle \mathbf{x}, \mathbf{z} \rangle + c\right) \qquad \text{for some } \gamma > 0, c \geq 0 \qquad (12.13)$$

All these functions are parameterized, allowing for adjustments with respect to the training data. The kernel parameter(s) and the regularization parameter C are the *hyperparameters*. The performance of an LS-SVM (or an SVM, for that matter) is *critically dependent* on the selection of hyperparameters.

Mercer's condition can be used to infer simple operations for creating combined kernel functions, which are also admissible. For instance, assume that k_1 and k_2 are admissible kernel functions, then the following combined kernels are admissible [35]:

$$k(\mathbf{x}, \mathbf{z}) = c_1 k_1(\mathbf{x}, \mathbf{z}) + c_2 k_2(\mathbf{x}, \mathbf{z}) \qquad \text{for } c_1, c_2 \geq 0 \qquad (12.14)$$

$$k(\mathbf{x}, \mathbf{z}) = k_1(\mathbf{x}, \mathbf{z}) k_2(\mathbf{x}, \mathbf{z}) \qquad (12.15)$$

$$k(\mathbf{x}, \mathbf{z}) = a k_2(\mathbf{x}, \mathbf{z}) \qquad \text{for } a \geq 0 \qquad (12.16)$$

Moreover, these operations allow modular construction of kernel functions. Increasingly complex kernel functions can be constructed by recursively applying these operations.

12.3 Evolutionary Computation

Several biologically inspired techniques have been developed over the years for search, optimization, and machine learning under the collective term *Evolutionary Computation* (EC) [40]. The key principle in EC is that potential solutions are generated, evaluated, and reproduced iteratively. Between iterations, individuals are subject to certain forms of mutation and can reproduce with a probability that is proportional to their *fitness*. A selection procedure removes individuals with low fitness from the population, so that the more fit ones are more likely to "survive." Three of the main branches within EC are *genetic algorithms*, *evolution strategies*, and *genetic programming*.

12.3.1 Genetic Algorithms

Probably the most recognized form of EC is the class of *genetic algorithms* (GA), popularized by Holland [15]. Genetic algorithms mainly operate in the realm of the genotype, which is commonly represented as a bit string. This means that all parameters need to be converted to a binary representation and are then concatenated to form the chromosome. Various types of bit encoding may be used, such as *gray codes* or even floating point representations.

Reproduction of individuals is usually emphasized in preference to mutation in GA. Two or more parents exchange part of their chromosome, resulting in offspring that contains genetic information from each of the parents. The common implementation is *crossover recombination*, in which two parents exchange a fragment of their chromosome. The size of the fragment is determined by a randomly selected crossover point. Mutation, on the other hand, is implemented by flipping the bits in the chromosome with a certain probability. Note that the implementation of both reproduction and mutation operators may depend on the specific representation that is used. For instance, reproduction of floating point chromosomes is done by blending the parents [10].

In addition to the mutation and recombination operators, the other key element in GA is the selection mechanism. The selection procedure selects the individuals that will be subject to mutation and reproduction with a probability proportional to their fitness. Further, offspring can be created on a *generational* interval or, alternatively, individuals can be replaced one by one (i.e., *steady-state* GA).

12.3.2 Evolution Strategies

Evolution strategies (ES) operate in the realm of the phenotype and use real-valued representations for the individuals [2]. An optimization problem with three parameters is represented as a vector $\mathbf{c} = (x_1, x_2, x_3)$, where the parameters $x_i \in \mathbb{R}$ are the *object parameters*. There are two main types of ES, namely $(\mu + \lambda)$-ES and (μ, λ)-ES. In these notations, μ is the size of the parent population and λ is the size of the offspring population. In $(\mu + \lambda)$-ES, the new parent population is chosen from *both* the current parent population and the offspring. In contrast, in (μ, λ)-ES the new parent population is chosen only from the offspring population, which requires that $\lambda \geq \mu$.

The canonical ES rely solely on the mutation operation for diversifying the genetic material. The mutation operation is typically implemented as a random perturbation of the parameters according to a probability distribution. More formally

$$x_i' = x_i + \mathcal{N}_i(0, \sigma_i) \tag{12.17}$$

where $\mathcal{N}$ denotes a logarithmic normal distribution. Note that this mutation mechanism requires the user to specify a standard deviation σ_i (i.e., the *strategy parameters*) for each object parameter in the chromosome. The common approach is to not define these standard deviations explicitly, but to integrate them in the chromosome. This is known as *self-adaptation*, as certain parameters of the algorithm are subject to the algorithm itself. An example of a chromosome with three object parameters and the additional *endogenous strategy parameters* is $\mathbf{c} = (x_1, x_2, x_3, \sigma_1, \sigma_2, \sigma_3)$ [3].

12.3.3 Genetic Programming

A vastly different paradigm within EC is that of *genetic programming* (GP) [22]. GP should rather be considered a form of automated programming than a parameter optimization technique. It aims to solve a problem by breeding a population of computer programs, which – when executed – are direct solutions to the problem. Obviously, this gives much more freedom in the structure of the solutions and it can therefore be applied to a wide variety of problems. The common way to represent programs in GP is by means of *syntax trees*, as shown in Fig. 12.1. Other types of genotype representations, e.g., graphs or linear structures, may be preferred for certain problem domains.

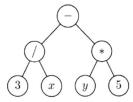

Fig. 12.1 An example tree representation for the mathematical function $(3/x) - (y * 5)$

GP includes recombination and mutation operators that are similar to their GA counterparts. In crossover recombination, two parents swap a subtree rooted at a random crossover point. Traditional mutation in GP involves randomly selecting a mutation point in the tree and replacing the subtree rooted at this point with a new, randomly generated tree.

For some problems it may be desirable to impose restrictions on the structure of the syntax tree, as to ensure that non-terminals operate only on appropriate data types. Consider, for instance, a binary equality function, which takes two integers as its children and returns a boolean. *Strongly typed genetic programming* has been proposed as an enhanced version of GP that enforces this type of constraint [28]. This influences both the representation of the individuals and the chromosome altering operators. First, while defining the terminals and non-terminals, the user also has to specify the types of the terminals, and the parameter and return types of non-terminals. Second, the recombination and mutation operators must be altered in such a way that they respect the type constraints.

12.4 Related Work

Hyperparameters and the kernel function are usually selected using a *trial and error* approach. Trial runs are performed using various configurations, the best of which is selected. This approach is generally considered time consuming and does not scale well with the number of parameters. Furthermore, the process often yields less than optimal performance in situations where time is limited. More elaborate approaches have been suggested for both selection problems, which will be summarized below.

12.4.1 Hyperparameter Optimization

An analytical technique that has been proposed for hyperparameter optimization is that of *gradient descent* [5, 20], which finds a local minimum by taking steps in the negative-gradient direction. This approach has been used for hyperparameter selection with a non-spherical RBF function, which means that each feature has a distinct scaling factor. Accordingly, there are more hyperparameters than there are features, demonstrating the scalability of the approach. The gradient descent method is shown to be able to find reasonable hyperparameters more efficiently than grid search. However, the method requires a continuous differentiable kernel and objective function, which may not be satisfiable for specific types of problems (e.g., non-vectorial kernel functions). Approaches based on *pattern search* have been proposed to overcome this problem [27]. In this method the neighborhood of a parameter vector is investigated in order to *approximate* the gradient empirically. However, the whole class of gradient descent methods has the inherent disadvantage that they may find local minima.

One of the first mentions of the use of EC for hyperparameter optimization can be found in the work of Fröhlich et al. [12], in which GA is primarily used for feature selection. However, the optimization of the regularization parameter C is done in parallel. Other GA-based approaches focus mainly on the optimization of the hyperparameters. The objective function in these type of approaches is either the error on a validation set [18, 26, 29], the radius-margin bound [7], or k-fold cross-validation [6, 33]. Some studies make use of a real-valued variant of GA [17, 42], although it is not clear whether the real-valued representation performs significantly better than a binary representation. All these studies suggest that GA can successfully be applied for hyperparameter optimization. However, there are some caveats, such as heterogeneity of the solutions and the selection of a reliable *and* efficient objective function.

ES have only scarcely been used for hyperparameter optimization [11]. In this approach, ES optimize not only the scaling but also the orientation of the RBF kernel. An improvement on the generalization performance is achieved over the kernel parameters that were found using grid search. This result should be interpreted with care, as the optimal grid search parameters are used as the initial solutions for the evolutionary algorithm. The classification error on separate test sets is used as the empirical objective function.

The main advantage of evolutionary algorithms in comparison to grid search is that they usually find good parameter settings efficiently and that the technique scales well with the number of hyperparameters. An advantage compared to gradient descent methods is that they cope better with local minima. Furthermore, they do not impose requirements on the kernel and objective functions, such as differentiability.

12.4.2 Combined Kernel Functions

It is intuitive that combined kernel functions are capable of improving the generalization performance, as the implicit feature mapping can be tuned for a specific problem. Several methods have been proposed for the composition of kernel functions. One of the first manifestations of combined kernel optimization was investigated by Lanckriet et al. [23]. This work considers linear combinations of kernels, i.e., $\mathbf{K} = \sum_{i=0}^{m} a_i \mathbf{K}_i$ for $a > 0$ and $\mathbf{K}_i$ chosen from a predefined set of kernel functions. The optimization of weight factors $\mathbf{a}$ is done using *semi-definite programming*, which is an optimization method that deals with convex functions over the convex cone of positive semi-definite matrices. This method can be applied to kernel matrices, since these need to be semi-definite to satisfy Mercer's condition. However, other methods may be used for the optimization of the weights, such as the so called *hyperkernels* [30], the *Lagrange multiplier* method [19], or a *generalized eigenvalue* approach [36].

Lee et al. argue that during the combination of kernels some potentially useful information is lost [24]. They propose a method for combining kernels that aims to prevent this loss of information. Instead of combining various kernel matrices into one, their method creates a large kernel matrix that contains all original kernel matrices and all possible mixtures of kernel functions, e.g., $k_{i,j}(\mathbf{x},\mathbf{z}) = \langle \phi_i(\mathbf{x}), \phi_j(\mathbf{z}) \rangle$, where ϕ_i is the mapping that belongs to kernel function k_i and ϕ_j the mapping that belongs to kernel k_j. This eliminates the requirement to optimize the weight factor for each kernel, as this is done implicitly by the SVM algorithm. However, special mixture functions need to be provided for the combination of two kernel functions. Furthermore, the spatial and temporal requirements of the algorithm increase drastically, as the kernel matrix is enlarged in both dimensions in proportion to the number of kernels in the combination.

Other EC-inspired approaches have been proposed to combine kernel functions. Most of these optimize a linear combination of weighted kernels using either GA or ES. The distinguishing elements are the set of kernel functions that are considered and the type of combination operators. Some only consider linear combinations (i.e., the addition operator) [31, 9], while others may allow both addition and multiplication [25]. These studies suggest that combining kernel functions can improve the generalization performance of the machine. However, the combinations are restricted to a predefined size and structure.

Howley and Madden propose a method to construct complete kernel functions using GP [16]. In this method, a kernel function is evolved for use with an SVM classifier. They use a tree-structured genotype, with the operators $+$, $-$, and $\times$ in both scalar and vector variants as the non-terminals. The terminals in their approach are the two vectors $\mathbf{x}_1$ and $\mathbf{x}_2$. Since the kernels are constructed using simple arithmetic, they are not guaranteed to satisfy Mercer's condition. Nonetheless, the technique still keeps up with (or outperforms) traditional kernels for most data sets. It is emphasized that techniques such as GP require a sufficiently large data set. Dioşan et al. have proposed some enhancements; their method differs from the original approach by an enriched operator set (e.g., various norms are included) and small changes to certain operators [8]. Similar modifications are presented for kernel nearest-neighbor classification by Gagné et al., who also use co-evolution to keep the approach computationally tractable [14]. Besides a species that evolves kernel functions, there are two other species for the training and validation sets. The training set species cooperates with the kernel function on minimizing the error and thus maximizing the fitness, whereas the species for the validation set is competitive and tries to maximize the error of the kernel functions.

12.5 Evolutionary Optimization of Kernel Machines

Two methods for the evolutionary optimization of hyperparameters and the kernel function are proposed. The first approach uses ES, GA, and GA with floating point representation to optimize the hyperparameters for a given kernel function (EvoKM^{ES}, EvoKM^{GA}, and EvoKM^{GAflt}, respectively). The aim is to find optimal hyperparameters more efficiently than using traditional grid search. Our second model uses GP to evolve combined kernel functions (EvoKM^{GP}), with the aim to increase the generalization performance.

12.5.1 Hyperparameter Optimization

In the hyperparameter optimization models, ES and GA are used to optimize the hyperparameters θ. Two variants of the GA model have been implemented: one that uses the traditional bit string representation with gray coding and another that uses a floating point representation. Evolutionary algorithms are highly generalized and their application on this specific problem is straightforward.

In EvoKM^{ES}, the chromosomes contain the real-valued hyperparameters and the corresponding endogenous strategy parameters σ, which yields for the RBF kernel the chromosome $\mathbf{c} = \left[\gamma, C, \sigma_\gamma, \sigma_C \right]$. Note that all the models use the hyperparameters on a logarithmic scale with base 2. Each hyperparameter is initialized to the center of its range and mutated according to the initial standard deviation $\sigma_i = 1.0$. An interesting issue is whether to use $(\mu + \lambda)$-ES or (μ, λ)-ES in the model. Both types have their own specific advantages and disadvantages. Typical application areas of $(\mu + \lambda)$-ES are discrete finite size search spaces, such as combinatorial optimization problems [3]. When the problem has an unbounded, typically real-valued search space, then (μ, λ)-ES is preferred [34]. Furthermore, Whitley presents empirical evidence that indicates that (μ, λ)-ES generally performs better than $(\mu + \lambda)$-ES [41]. We prefer to follow both the heuristic and the empirical indications and adopted (μ, λ)-ES for our model. Unfortunately, there is no guarantee that the search process will converge, as would have been the case with $(\mu + \lambda)$-ES. For our model, we have empirically selected $\mu = 3$ and $\lambda = 12$ based on preliminary experimentations.

EvoKM^{GA} and EvoKM^{GAflt} differ from EvoKM^{ES} in terms of the operators and the genotype representation. EvoKM^{GA} uses a gray code of 18 bits for each parameter and *one-point crossover recombination* and *bit-flip mutation* operators. One disadvantage of GA, as compared to ES, is that there are many more parameters that need to be set. The population size of 10 is relatively low for GA standards. However, one must take into account that the maximum number of evaluations is limited to several hundreds up to a few thousand and, moreover, the goal is to see convergence to good solutions within the first hundred evaluations. Large population sizes, e.g., larger than 50, would have a disadvantage in this context, as the algorithm can only perform one or two generations within this range. Further, preliminary experiments have shown that a population size of 10 shows similar convergence to larger population sizes. Other parameters of EvoKM^{GA} have been tuned using a coarse grid search as well. One-point crossover recombination occurs with a probability of $p_c = 0.2$. During mutation, each bit in the chromosome is inverted with a probability of $p_m = 0.1$. The number of participants in tournament selection is 5. Further, the *steady state* variant of GA has been used.

EvoKM^{GAflt}, on the other hand, uses a floating point representation. Crossover recombination in this model is performed by blending two individuals using the BLX-α method [10]. This recombination operator is applied with a probability of $p_c = 0.3$ and with $\alpha = 0.5$. Additionally, each parameter has a probability of $p_m = 0.4$ of being mutated using a random perturbation according to the normal distribution $\mathcal{N}(\mu, \sigma)$, where $\mu = 0$ and $\sigma = 0.5$. All other settings are equal to those for EvoKM^{GA}.

12.5.2 Kernel Construction

The second optimization method constructs complete kernel functions using GP. In this model, the functions are represented using syntax trees. The syntactic structure of the trees is based on the combination operations that guarantee admissible kernel functions, cf. (12.14), (12.15), and (12.16). These operations form the set of non-terminals, whereas the polynomial and RBF kernels form the set of terminals. This is formalized in the context-free grammar shown in Fig. 12.2. The model makes use of strongly typed GP, as it needs to ensure that the syntactic structure is enforced for all individuals. An example chromosome of a kernel function using the tree representation is shown in Fig. 12.3. Note that the regularization parameter C is omitted in this figure; it is included in a separate real-valued chromosome.

A common heuristic with regard to population size in GP is that difficult problems require a large population. As time efficiency is not of primary concern for

$\langle kernel \rangle$	$\rightarrow$	$\langle add\_kernels \rangle \mid \langle multiply\_kernels \rangle \mid$ $\langle weighted\_kernel \rangle \mid \langle polynomial \rangle \mid \langle rbf \rangle$	
$\langle add\_kernels \rangle$	$\rightarrow$	$\langle kernel \rangle$ '+' $\langle kernel \rangle$	
$\langle multiply\_kernels \rangle$	$\rightarrow$	$\langle kernel \rangle$ '×' $\langle kernel \rangle$	
$\langle weighted\_kernel \rangle$	$\rightarrow$	a '×' $\langle kernel \rangle$	for $a \in \mathbb{R}^+$
$\langle polynomial \rangle$	$\rightarrow$	'$(\langle \mathbf{x}, \mathbf{z} \rangle + c)^d$'	for $d \in \mathbb{N}, c \in \mathbb{R}^+$
$\langle rbf \rangle$	$\rightarrow$	'$\exp(-\gamma \|\mathbf{x} - \mathbf{z}\|^2)$'	for $\gamma \in \mathbb{R}^+$

Fig. 12.2 Context-free grammar – in Backus–Naur form – that constrains the generated expressions for the GP model

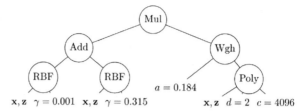

Fig. 12.3 An example of a tree generated by the GP model

this model, the population size is set to 2000 and each run spans 13 generations.[1]
Further, the following operators and settings are used within the EvoKM^{GP} model:

1. *Reproduction* occurs with probability $p_r = 0.05$, i.e., an individual is directly
   copied into the offspring population, without any kind of mutation.
2. *Crossover recombination* occurs with probability $p_c = 0.2$, i.e., two parents exchange a subtree at a random crossover point; the two new individuals are *both*
   inserted in the offspring population.
3. *Random mutation* occurs with probability $p_m = 0.15$, i.e., substituting a subtree
   of the individual with a new random subtree.
4. *Shrink mutation* occurs with probability $p_s = 0.05$, i.e., replacing a subtree with
   one of the branches of this subtree in order to reduce the size of the tree.
5. *Swap mutation* occurs with probability $p_w = 0.05$, i.e., replacing a subtree in
   the individual with another subtree, effectively swapping two branches of the
   same tree.
6. *PDF parameter mutation* occurs with probability $p_p = 0.5$, i.e., mutating a hyperparameter according to a probability density function.

The PDF mutation operator is specially crafted for our model. This operator
ensures that the optimization includes the hyperparameters, as well as evolving
the structure of the kernel functions. The common GP operators would only be
able to mutate these parameters by substituting them for another randomly selected
parameter.

12.5.3 Objective Function

When applying EC techniques it is important to decide which objective function
to use, as this is the actual measure that is being optimized. It should, therefore,
measure the "quality" of a solution for the given domain. In the context of this
study quality is best described as the generalization performance of the machine.
A very important aspect is that the fitness function must prevent overfitting of the
machine to the training data. This is especially true for EvoKM^{GP}, as this model

---

[1] The total number of evaluations will thus be less than 26,000, as unmodified individuals are not
reevaluated.

tunes both the hyperparameters and the kernel function for the specific data set. There are several methods to estimate this generalization performance, of which cross-validation can be applied to practically any learning method. Both k-fold and leave-one-out cross-validation have been shown to be approximately unbiased in terms of estimating the true expected error [21]. However, k-fold cross-validation usually exhibits a lower variance on the error than the leave-one-out measure. For this reason, k-fold cross-validation is used as fitness function for both approaches.

12.6 Results

All models have been validated experimentally on a standard set of benchmark problems. An LS-SVM has been implemented in C++ using the efficient Atlas library for linear algebra [39]. This implementation uses an approximate variant of the LS-SVM kernel machine [32], so as to reduce the computational demands of the experiments. The size of the subset that is used to describe the model is set to 10% of the total data set. Although LS-SVM is only one specific type of kernel machine, all relevant aspects of the models have been kept generalized, so that extension to other types of kernel machines (e.g., SVM) is straightforward. Two kernel functions have been considered in these experiments. The first is the RBF kernel function, cf. (12.12), which is commonly regarded the "default" choice for kernel machines. The second is the polynomial function, cf. (12.11).

The evolutionary algorithms in the models have been implemented using the *OpenBeagle* framework for EC [13]. The objective function is, as explained, k-fold cross-validation with $k = 5$. This value gives an adequate trade-off between accuracy and computational expenses. For classification problems, the error measure is the normalized classification error; in case of regression problems the *mean squared error* (MSE) is used.

12.6.1 Data Sets

Seven different benchmark data sets have been selected for the empirical validation. Five of these data sets are regression problems, whereas the remaining two are binary classification problems. The data sets *Concrete, Diabetes, Housing*, and *Wisconsin* are well-known benchmark data sets obtained from the UCI machine learning repository [1]. The data sets *Reaching* 1, 2, and 3 are obtained internally from the LiraLab of the University of Genoa.[2] These data sets concern orienting the head of a humanoid robot in the direction of its reaching arm. The features are the traces of four arm encoders, whereas the outputs are the corresponding actuator values for three head joints. Table 12.1 shows standard characteristics of the data

---

[2] These data sets can be obtained from `http://eris.liralab.it/wiki/Reaching_Data_Sets`.

Table 12.1 Basic characteristics of the data sets used in the experiments

Name	Type	#Samples	#Features	Positive (%)
Concrete	Regression	1005	8	n/a
Diabetes	Classification	768	8	65.1
Housing	Regression	506	13	n/a
Reaching 1	Regression	1126	4	n/a
Reaching 2	Regression	2534	4	n/a
Reaching 3	Regression	2737	4	n/a
Wisconsin	Classification	449	9	52.6

sets after preprocessing. The exact preprocessing steps that have been performed on the data sets are as follows:

1. All features have been (independently) standardized, i.e., rescaling to zero mean and unit standard deviation.
2. For regression problems, output values have been standardized in the same manner as the features. For classification problems, labels have been set to $+1$ for positive labels and -1 for negative labels.
3. Duplicate entries have been removed from the data sets.
4. The order of the samples in the data set has been randomized.

12.6.2 Results for Hyperparameter Optimization

The models for hyperparameter optimization have been verified using the following scenario: a very coarse grid search has been performed to identify an interesting region for the parameter ranges for each data set and kernel function. Subsequently, a very dense grid search is performed on this region to establish a reference for our models. For the polynomial kernel function, which has two parameters, the degree has been kept fixed at $d = 3$ in order to keep the search computationally tractable. This reference contains the number of evaluations used for grid search[3] and the corresponding minimum error, which serves as the *target* for our models.

The evolutionary models have been used on the same parameter ranges as the grid search. The only exception is that for the polynomial kernel we have *not* kept the degree fixed at $d = 3$; instead it is set within a range of $d = \{1,\ldots,8\}$. This exception is made to investigate the scaling properties of the EC-based approaches, i.e., to see whether evolutionary optimization can yield better solutions by optimizing more parameters. The evolutionary search is terminated after the same number of evaluations as used for the grid search.

---

[3] Note that the number of evaluations directly translates into time, as solving the LS-SVM problem is independent of the chosen parameters.

Table 12.2 Comparison of the minimum errors of grid search and the evolutionary optimization methods. Note that the results of the latter are averages over 25 runs

Name	Kernel	Grid search ε_{min}	Eval.	EvoKM^{ES} $\bar{\varepsilon}_{min}$	EvoKM^{GA} $\bar{\varepsilon}_{min}$	EvoKM^{GAflt} $\bar{\varepsilon}_{min}$
Concrete	RBF	0.1607	1221	0.1591 ± 0.0000	0.1590 ± 0.0000	0.1590 ± 0.0000
	Poly.	0.1700	899	0.1741 ± 0.0000	0.1698 ± 0.0001	0.1778 ± 0.0235
Diabetes	RBF	0.2200	621	0.2201 ± 0.0004	0.2202 ± 0.0006	0.2207 ± 0.0015
	Poly.	0.2213	777	0.2226 ± 0.0003	0.2215 ± 0.0012	0.2231 ± 0.0016
Housing	RBF	0.1676	2793	0.1674 ± 0.0000	0.1674 ± 0.0000	0.1674 ± 0.0000
	Poly.	0.1675	1739	0.1641 ± 0.0016	0.1661 ± 0.0015	0.1646 ± 0.0022
Reaching 1	RBF	0.0683	3185	0.0683 ± 0.0000	0.0683 ± 0.0000	0.0683 ± 0.0000
	Poly.	0.0720	1517	0.0670 ± 0.0002	0.0670 ± 0.0001	0.0677 ± 0.0029
Reaching 2	RBF	0.0042	561	0.0042 ± 0.0000	0.0042 ± 0.0000	0.0042 ± 0.0000
	Poly.	0.0063	399	0.0045 ± 0.0004	0.0043 ± 0.0001	0.0045 ± 0.0003
Reaching 3	RBF	0.0019	561	0.0019 ± 0.0000	0.0019 ± 0.0000	0.0019 ± 0.0000
	Poly.	0.0032	399	0.0022 ± 0.0001	0.0021 ± 0.0001	0.0023 ± 0.0003
Wisconsin	RBF	0.0423	3185	0.0467 ± 0.0025	0.0423 ± 0.0000	0.0432 ± 0.0017
	Poly.	0.0401	2337	0.0433 ± 0.0031	0.0400 ± 0.0017	0.0424 ± 0.0017

A comparison of the generalization performance of the grid search and the evolutionary models is shown in Table 12.2. The overall impression is that all the evolutionary algorithms are able to find competitive solutions. In particular, EvoKM^{GA} shows stable performance, as it finds equal or better solutions for all of the data sets. The only minor exception is the Diabetes data set, for which it finds solutions that are only marginally worse than those found using grid search. Another observation is that for the majority of the data sets the inclusion of the degree of the polynomial kernel indeed decreases the generalization error. This suggests that the methods scale well with the number of parameters and, moreover, that the extra degree of freedom is used to decrease the error. Furthermore, EvoKM^{ES} and EvoKM^{GAflt} perform worse than that of the GA-based model on this real-valued optimization problem, suggesting that real-valued chromosomes are not necessarily beneficial for hyperparameter optimization.

More interesting than the optimal solutions is the rate of convergence of the various methods. This has been analyzed by considering the number of evaluations that were needed to reach an error that is close to the target, cf. Table 12.3. These results confirm the previous observation that EvoKM^{GA} outperforms the two other models in most situations. The GA method converges to the target error in only a fraction of the number of evaluations used for grid search, with the exception of the Diabetes data set. Furthermore, in almost all situations, it is able to find solutions within a range of 5% of the target within the first 100 evaluations.

The ES and GAflt methods converge slower than EvoKM^{GA}, although EvoKM^{ES} outperforms the others on a number of regression data sets. Conversely, it performs much worse on the Wisconsin classification data set. One of the reasons for this

Table 12.3 Comparison of the convergence of the evolutionary models. The column *ETT* (*Evaluations To Target*) denotes the number of evaluations that the average run needs to reach the target error. Analogously, the column $ETT_{5\%}$ denotes the number of evaluations needed to reach an error that is at most 5% higher than the target error

Name	Kernel	Grid search ε_{min}	Eval.	EvoKM^{ES} ETT	$ETT_{5\%}$	EvoKM^{GA} ETT	$ETT_{5\%}$	EvoKM^{GAflt} ETT	$ETT_{5\%}$
Concrete	RBF	0.1607	1221	243	135	98	79	274	134
	Poly.	0.1700	899	>899	39	513	102	> 899	285
Diabetes	RBF	0.2200	621	>621	3	>621	10	>621	10
	Poly.	0.2213	777	>777	3	>777	10	>777	10
Housing	RBF	0.1676	2793	267	123	306	64	333	135
	Poly.	0.1675	1739	291	27	550	19	558	39
Reaching 1	RBF	0.0683	3185	435	207	165	35	446	84
	Poly.	0.0720	1517	39	15	28	19	18	10
Reaching 2	RBF	0.0042	561	63	51	128	71	237	136
	Poly.	0.0063	399	39	39	44	44	82	44
Reaching 3	RBF	0.0019	561	75	51	183	88	278	130
	Poly.	0.0032	399	39	39	28	28	65	65
Wisconsin	RBF	0.0423	3185	>3185	>3185	802	28	>3185	80
	Poly.	0.0401	2337	>2337	>2337	2158	270	>2337	>2337

behavior is that ES use the mutated offspring to sample the proximity of the parent individuals. This information is then used to find a direction in which the error is decreasing, in a manner similar to gradient descent or pattern search. The difficulty with classification problems is that the error surface incorporates plateaus. Offspring individuals in the proximity of a parent are thus likely to have an identical fitness score and the algorithm will perform a random search on the plateau. Smoothness of the fitness landscape may be regarded as a prerequisite to efficient optimization using ES [3]. The situation is somewhat similar for EvoKM^{GAflt}, as this model also incorporates a random perturbation operator for mutation. However, this model has a larger population size and a recombination operator, which can "diversify" the population when progress is ceased on a plateau.

The problematic behavior of EvoKM^{ES} can be verified in the error convergences depicted in Fig. 12.4. Albeit the ES method shows a steep initial convergence, the search in these situations stagnates, indicating a random search. Further, in Fig. 12.4(c) and (d) it can be seen that EvoKM^{ES} has a considerably higher initial position. This can be attributed to the smaller initial population size, as these individuals are used as the starting points for the search. Additionally, the individuals in EvoKM^{ES} are initialized near the center of the range, in contrast to the two other methods. We have verified that, for this data set only, the results of EvoKM^{ES} can be improved by initializing the individuals uniformly over the search space, as is done in the other two models.

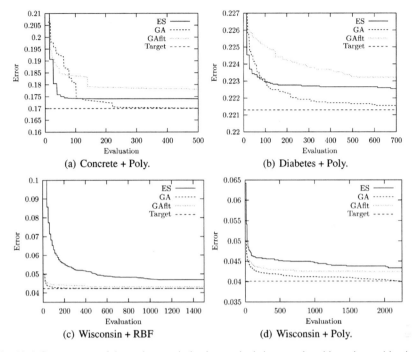

Fig. 12.4 Convergence of the various optimization methods in several problematic combinations of data sets and kernels

Inspection of the solutions confirms the observation that all the evolutionary models produce heterogeneous "optimal" solutions. This is not necessarily problematic, given that variance in the quality of the solutions is limited. Further, although the presented results give some insight regarding the performance of various evolutionary algorithms, it must be taken into account that there is a variety of parameters and operators – in particular, for EvoKM^{GA} and EvoKM^{GAflt} – that influence the speed of convergence. It is likely that additional fine-tuning of these parameters can improve the performance of these models.

12.6.3 Results for EvoKM^{GP}

The results from grid search have also been used as a performance benchmark for EvoKM^{GP}. However, for this model we consider only the quality of the solution and ignore the temporal aspects (i.e., number of evaluations). The minimum errors of both grid search and EvoKM^{GP} are shown in Table 12.4. It can be observed that EvoKM^{GP} increases the generalization performance for all data sets. However, the minimum errors are only marginally lower than those obtained by grid search. This indicates that the combined kernel functions perform only slightly better than singular kernel functions.

Table 12.4 The minimum errors as obtained with EvoKM^{GP}. Note that $\bar{\varepsilon}_{min}$ indicates the average minimum error over 10 runs, whereas ε_{min} indicates the absolute minimum error

Name	Grid search ε_{min}	EvoKM^{GP} $\bar{\varepsilon}_{min}$	ε_{min}
Concrete	0.1607	0.1513 ± 0.0010	0.1490
Diabetes	0.2200	0.2176 ± 0.0032	0.2096
Housing	0.1675	0.1633 ± 0.0006	0.1620
Reaching 1	0.0683	0.0592 ± 0.0004	0.0587
Reaching 2	0.0042	0.0038 ± 0.0000	0.0037
Reaching 3	0.0019	0.0018 ± 0.0000	0.0018
Wisconsin	0.0401	0.0358 ± 0.0012	0.0333

It is difficult to provide strict interpretations of this result, since not finding any combined kernel functions that drastically improves the generalization performance does not necessarily mean that they will not exist at all. This relates directly to the difficulty of finding good configurations for the GP method, as seen with the GA models as well. There are many parameters that need to be set and one has to find a suitable evolver model (i.e., the set of individual altering operators and their order). Unfortunately, there is no structured approach for optimizing the configuration. Therefore, it remains mostly a task that has to be solved using loose heuristics or even intuition. This problem is particularly evident in this GP context, as the computational demand does not allow for an empirical verification of multiple possible configurations, as was done for the ES and GA models.[4]

12.7 Conclusions and Future Work

Two approaches for the evolutionary optimization of LS-SVM have been presented. The distinction is that the first aims to find optimal hyperparameters more efficiently than traditional methods (i.e., grid search) and the second aims to increase the generalization performance by means of combined kernel functions. The models for the first approach are based on ES, GA, and GA with a floating point representation. In particular, the standard GA model has shown to be an efficient and generalized method for performing hyperparameter optimization for LS-SVM. It was able to find solutions comparable to optimal grid search solutions in only a fraction of the computational demands. Furthermore, the method scales well with the number of parameters. The ES and floating point GA models performed worse than GA, although they are still preferable to grid search for regression problems. Classification problems, on the other hand, are more challenging particularly for the

---

[4] The experiments that we presented for EvoKM^{GP} need more than half a year of CPU time on a Pentium 4 class computer running at 3 GHz.

ES model, as the error surface is discontinuous. ES use the offspring individuals to sample the neighborhood in order to find a direction that minimizes the error. The plateaus found in the error surface of classification problem interfere with this strategy, as offspring are likely to have a fitness that is identical to that of the parent. This problem may be avoided by using the squared error loss function also for classification problems, such that the error surface becomes continuous. Further, the performance of all models may be improved upon by fine-tuning the variety of parameters. In future work, it would be interesting to compare the evolutionary algorithms with various gradient descent methods in terms of solution quality and convergence rate.

The genetic programming approach for the generation and selection of kernel functions increases the generalization performance of the kernel machine only marginally. This suggests that combined kernel functions may not improve the performance as much as one may expect. In most circumstances, this slight improvement will not justify the high computational demands of this model. The fact that we have not found kernel functions that considerably improve on the generalization performance does not necessarily mean that such kernel functions will not exist at all. The configuration of GP, in terms of the evolver model and parameters, influences to a great extent the results. However, the numerous options and the high computational demand make it very difficult to find an optimal configuration for our model. It is worth investigating whether more advanced variants of GP and further tuning of the configuration can improve the presented results.

Acknowledgments This study has partially been funded by EU projects *RobotCub* (IST-004370) and *CONTACT* (NEST-5010). The authors gratefully acknowledge Francesco Orabona for his constructive comments and Francesco Nori and Lorenzo Natale for supplying the Reaching data sets.

References

1. Arthur Asuncion and David J. Newman. UCI machine learning repository, 2007.
2. Hans-Georg Beyer. *The Theory of Evolution Strategies*. Springer-Verlag New York, Inc., New York, NY, USA, 2001.
3. Hans-Georg Beyer and Hans-Paul Schwefel. Evolution strategies – a comprehensive introduction. *Natural Computing: An International Journal*, 1(1):3–52, 2002.
4. Christopher J. C. Burges. A tutorial on support vector machines for pattern recognition. *Data Mining and Knowledge Discovery*, 2(2):121–167, 1998.
5. Olivier Chapelle, Vladimir N. Vapnik, Olivier Bousquet, and Sayan Mukherjee. Choosing multiple parameters for support vector machines. *Machine Learning*, 46(1–3):131–159, 2002.
6. Peng-Wei Chen, Jung-Ying Wang, and Hahn-Ming Lee. Model selection of svms using ga approach. *Proceedings of the 2004 IEEE International Joint Conference on Neural Networks*, 3(2):2035–2040, July 2004.
7. Zheng Chunhong and Jiao Licheng. Automatic parameters selection for svm based on ga. In *WCICA 2004: Fifth World Congress on Intelligent Control and Automation*, volume 2, pages 1869–1872, June 2004.
8. Laura Dioşan and Mihai Oltean. Evolving kernel function for support vector machines. In Cagnoni C., editor, *The 17th European Conference on Artificial Intelligence, Evolutionary Computation Workshop*, pages 11–16, 2006.

9. Laura Dioşan, Mihai Oltean, Alexandrina Rogozan, and Jean Pierre Pecuchet. Improving svm performance using a linear combination of kernels. In *ICANNGA '07: International Conference on Adaptive and Natural Computing Algorithms*, number 4432 in LNCS, pages 218–227. Springer, 2007.
10. Larry J. Eshelman and J. David Schaffer. Real–coded genetic algorithms and interval-schemata. In L. Darrell Whitley, editor, *Proceedings of the Second Workshop on Foundations of Genetic Algorithms*, pages 187–202, San Mateo, 1993. Morgan Kaufmann.
11. Frauke Friedrichs and Christian Igel. Evolutionary tuning of multiple svm parameters. In *ESANN 2004: Proceedings of the 12th European Symposium on Artificial Neural Networks*, pages 519–524, April 2004.
12. Holger Fröhlich, Olivier Chapelle, and Bernhard Schölkopf. Feature selection for support vector machines by means of genetic algorithms. In *ICTAI '03: Proceedings of the 15th IEEE International Conference on Tools with Artificial Intelligence*, page 142, Washington, DC, USA, 2003. IEEE Computer Society.
13. Christian Gagné and Marc Parizeau. Genericity in evolutionary computation software tools: Principles and case study. *International Journal on Artificial Intelligence Tools*, 15(2): 173–194, April 2006. 22 pages.
14. Christian Gagné, Marc Schoenauer, Michele Sebag, and Marco Tomassini. Genetic program-ming for kernel-based learning with co-evolving subsets selection. In *Parallel Problem Solv-ing from Nature – PPSN IX*, volume 4193 of *LNCS*, pages 1008–1017, Reykjavik, Iceland, September 2006. Springer-Verlag.
15. John H. Holland. *Adaptation in Natural and Artificial Systems*. University of Michigan Press, Ann Arbor, MI, 1975.
16. Tom Howley and Michael G. Madden. The genetic kernel support vector machine: Descrip-tion and evaluation. *Artificial Intelligence Review*, 24(3–4):379–395, 2005.
17. Chin-Chia Hsu, Chih-Hung Wu, Shih-Chien Chen, and Kang-Lin Peng. Dynamically opti-mizing parameters in support vector regression: An application of electricity load forecasting. In *HICSS '06: Proceedings of the 39th Annual Hawaii International Conference on System Sciences*, page 30.3, Washington, DC, USA, 2006. IEEE Computer Society.
18. Cheng-Lung Huang and Chieh-Jen Wang. A ga-based feature selection and parameters op-timization for support vector machines. *Expert Systems with Applications*, 31(2):231–240, 2006.
19. Jaz Kandola, John Shawe-Taylor, and Nello Cristianini. Optimizing kernel alignment over combinations of kernels. Technical Report 121, Department of Computer Science, Royal Holloway, University of London, UK, 2002.
20. S. Sathiya Keerthi, Vikas Sindhwani, and Olivier Chapelle. An efficient method for gradient-based adaptation of hyperparameters in svm models. In B. Schölkopf, J. Platt, and T. Hoff-man, editors, *Advances in Neural Information Processing Systems 19*, pages 673–680. MIT Press, Cambridge, MA, USA, 2007.
21. Ron Kohavi. A study of cross-validation and bootstrap for accuracy estimation and model se-lection. In *International Joint Conference on Artificial Intelligence*, pages 1137–1145, 1995.
22. John R. Koza. *Genetic Programming: On the Programming of Computers by Means of Nat-ural Selection*. MIT Press, Cambridge, MA, USA, 1992.
23. Gert R. G. Lanckriet, Nello Cristianini, Peter L. Bartlett, Laurent El Ghaoui, and Michael I. Jordan. Learning the kernel matrix with semidefinite programming. *Journal of Machine Learning Research*, 5:27–72, 2004.
24. Wan-Jui Lee, Sergey Verzakov, and Robert P. W. Duin. Kernel combination versus classifier combination. In *MCS 2007: Proceedings of the 7th International Workshop on Multiple Classifier Systems*, pages 22–31, May 2007.
25. Stefan Lessmann, Robert Stahlbock, and Sven F. Crone. Genetic algorithms for support vector machine model selection. In *IJCNN '06: International Joint Conference on Neural Networks*, pages 3063–3069. IEEE Press, July 2006.
26. Sung-Hwan Min, Jumin Lee, and Ingoo Han. Hybrid genetic algorithms and support vector machines for bankruptcy prediction. *Expert Systems with Applications*, 31(3):652–660, 2006.

27. Michinari Momma and Kristin P. Bennett. A pattern search method for model selection of support vector regression. In *Proceedings of the Second SIAM International Conference on Data Mining*. SIAM, April 2002.
28. David J. Montana. Strongly typed genetic programming. *Evolutionary Computation*, 3(2):199–230, 1995.
29. Syng-Yup Ohn, Ha-Nam Nguyen, Dong Seong Kim, and Jong Sou Park. Determining optimal decision model for support vector machine by genetic algorithm. In *CIS 2004: First International Symposium on Computational and Information Science*, pages 895–902, Shanghai, China, December 2004.
30. Cheng S. Ong, Alexander J. Smola, and Robert C. Williamson. Learning the kernel with hyperkernels. *Journal of Machine Learning Research*, 6:1043–1071, 2005.
31. Tanasanee Phienthrakul and Boonserm Kijsirikul. Evolutionary strategies for multi-scale radial basis function kernels in support vector machines. In *GECCO '05: Proceedings of the 2005 Conference on Genetic and Evolutionary Computation*, pages 905–911, New York, NY, USA, 2005. ACM Press.
32. Ryan Rifkin, Gene Yeo, and Tomaso Poggio. Regularized least squares classification. In *Advances in Learning Theory: Methods, Model and Applications*, volume 190, pages 131–154, Amsterdam, 2003. VIOS Press.
33. Sergio A. Rojas and Delmiro Fernandez-Reyes. Adapting multiple kernel parameters for support vector machines using genetic algorithms. In *Proceedings of the 2005 IEEE Congress on Evolutionary Computation*, volume 1, pages 626–631, Edinburgh, Scotland, UK, September 2005. IEEE Press.
34. Hans-Paul Schwefel. Collective phenomena in evolutionary systems. In *Problems of Constancy and Change – The Complementarity of Systems Approaches to Complexity*, volume 2, pages 1025–1033. International Society for General System Research, 1987.
35. John Shawe-Taylor and Nello Cristianini. *Kernel Methods for Pattern Analysis*. Cambridge University Press, June 2004.
36. Jian-Tao Sun, Ben-Yu Zhang, Zheng Chen, Yu-Chang Lu, Chun-Yi Shi, and Wei-Ying Ma. Ge-cko: A method to optimize composite kernels for web page classification. In *WI '04: Proceedings of the IEEE/WIC/ACM International Conference on Web Intelligence*, pages 299–305, Washington, DC, USA, 2004. IEEE Computer Society.
37. Johan A. K. Suykens, Tony Van Gestel, Jos De Brabanter, Bart De Moor, and Joost Vandewalle. *Least Squares Support Vector Machines*. World Scientific Publishing Co., Pte, Ltd., Singapore, 2002.
38. Vladimir N. Vapnik. *The nature of statistical learning theory*. Springer-Verlag New York, Inc., New York, NY, USA, 1995.
39. R. Clint Whaley and Antoine Petitet. Minimizing development and maintenance costs in supporting persistently optimized BLAS. *Software: Practice and Experience*, 35(2):101–121, February 2005.
40. Darrell Whitley. An overview of evolutionary algorithms: practical issues and common pitfalls. *Information and Software Technology*, 43(14):817–831, 2001.
41. Darrell Whitley, Marc Richards, Ross Beveridge, and Andre' da Motta Salles Barreto. Alternative evolutionary algorithms for evolving programs: evolution strategies and steady state gp. In *GECCO '06: Proceedings of the 8th annual conference on Genetic and evolutionary computation*, pages 919–926, New York, NY, USA, 2006. ACM Press.
42. Chih-Hung Wu, Gwo-Hshiung Tzeng, Yeong-Jia Goo, and Wen-Chang Fang. A real-valued genetic algorithm to optimize the parameters of support vector machine for predicting bankruptcy. *Expert Systems with Applications*, 32(2):397–408, 2007.

Chapter 13
Genetically Evolved kNN Ensembles

Ulf Johansson, Rikard König, and Lars Niklasson

Abstract Both theory and a wealth of empirical studies have established that ensembles are more accurate than single predictive models. For the ensemble approach to work, base classifiers must not only be accurate but also diverse, i.e., they should commit their errors on different instances. Instance-based learners are, however, very robust with respect to variations of a data set, so standard resampling methods will normally produce only limited diversity. Because of this, instance-based learners are rarely used as base classifiers in ensembles. In this chapter, we introduce a method where genetic programming is used to generate kNN base classifiers with optimized *k*-values and feature weights. Due to the inherent inconsistency in genetic programming (i.e., different runs using identical data and parameters will still produce different solutions) a group of independently evolved base classifiers tend to be not only accurate but also diverse. In the experimentation, using 30 data sets from the UCI repository, two slightly different versions of kNN ensembles are shown to significantly outperform both the corresponding base classifiers and standard kNN with optimized *k*-values, with respect to accuracy and AUC.

13.1 Introduction

Most data mining techniques consist of a two-step process: first an *inductive step*, where a model is constructed from data, and then a second, *deductive*, step where the model is applied to test instances. An alternative approach is, however, to omit

---

Ulf Johansson and Rikard König are equal contributors.

Ulf Johansson · Rikard König
School of Business and Informatics, University of Borås, Borås, Sweden,
e-mail: ulf.johansson@hb.se;rikard.konig@hb.se

Lars Niklasson
Informatics Research Centre, University of Skövde, Skövde, Sweden,
e-mail: lars.niklasson@his.se

R. Stahlbock et al. (eds.), *Data Mining,* Annals of Information Systems 8,
DOI 10.1007//978-1-4419-1280-0_13, © Springer Science+Business Media, LLC 2010

the model building and directly classify novel instances based on available training instances. Such approaches are called *lazy learners* or *instance-based learners*. The most common lazy approach is *nearest neighbor classification*. The nearest neighbor algorithm classifies a test instance based on the majority class among the k closest (according to some distance measure) training instances. The value k is a parameter to the algorithm, and the entire technique is known as *k-Nearest Neighbor* (kNN).

kNN, consequently, does not, in contrast to techniques like neural networks and decision trees, use a global model covering the entire input space. Instead, classification is based on local information. This use of neighboring instances for the actual classification makes it, in theory, possible for kNN to produce arbitrarily shaped decision boundaries, while decision trees and rule-based learners are constrained to rectilinear decision boundaries.

Standard kNN is a straightforward classification technique that normally performs quite well, in spite of its simplicity. Often, standard kNN is used as a first choice just to obtain a lower bound estimation of the accuracy that should be achieved by more powerful methods, like neural networks or ensemble techniques, see, e.g., [2].

Unfortunately, the performance of standard kNN is extremely dependent on the parameter value k. If k is too small, the algorithm becomes very susceptible to noise. If k is too large, the locality aspect becomes less important, typically leading to test instances being misclassified based on training instances quite different from the test instance.

Needless to say, different problems will require different k-values, so in practice, data miners often have to determine k by means of cross-validation on the training data. The use of cross-validation to optimize k, introduces, however, another problem. Even for a single data set, the optimal value for k probably varies over the particular regions of the input space, so the cross-validation will most likely sacrifice performance in some regions to obtain better overall performance.

A basic procedure, trying to reduce the importance of the parameter k, is to use *weighted voting*, i.e., each vote is weighted based on the distance to the test instance, see, e.g., [21]. Nevertheless, techniques using weighted voting are still restricted to using a global k, i.e., a single k-value for the entire data set. Another similar option is to use no k at all, i.e., all instances affect the decision, typically proportionally to their proximity to the test instance. Yet another possibility is to employ *axes scaling* (or *feature weighting*), where more important features will have greater impact on the neighbor selection process and the voting. Clearly, these methods are slightly less sensitive to the actual value of k, but they still use a single, global k-value, instead of allowing local optimization, i.e., the use of different k-values in different regions of the input space.

A totally different approach, often used for other machine learning schemes in order to boost accuracy and reduce the variance, would be to somehow combine several kNN models into an ensemble. In this chapter, we will look into kNN ensembles, focusing, in particular, on how the necessary diversity can be achieved.

13.2 Background and Related Work

An *ensemble* is a composite model, aggregating multiple *base models* into one predictive model. An ensemble prediction, consequently, is a function of all included base models. The main motivation for using ensembles is the fact that combining several models using averaging will eliminate uncorrelated base classifier errors, see, e.g., [8]. This reasoning requires the base classifiers to commit their errors on different instances – clearly there is no point in combining identical models. Informally, the key term *diversity* is therefore used to denote the extent to which the base classifiers commit their errors on different instances.

The vital finding that ensemble error depends not only on the average accuracy of the base models, but also on their diversity was formally derived in [15]. From this, the overall goal when designing ensembles seems to be fairly simple, i.e., somehow combine models that are highly accurate but diverse. Base classifier accuracy and diversity are, however, highly correlated, so maximizing diversity will most likely reduce the average base classifier accuracy. Moreover, diversity is not uniquely defined for classification, further complicating the matter. As a matter of fact, numerous different diversity measures have been suggested, often in combination with quite technical and specialized ensemble creation algorithms. So, although there is strong consensus that ensemble models will outperform even the most accurate single models, there is no widely accepted solution to the problem of how to maximize ensemble accuracy.

The most renowned ensemble techniques are probably *bagging* [4], *boosting* [17], and *stacking* [20], all of which can be applied to different types of models and be used for both regression and classification. Most importantly, bagging, boosting, and stacking will, almost always, increase predictive performance compared to a single model. Unfortunately, machine learning researchers have struggled to understand why these techniques work, see, e.g., [19]. In addition, it should be noted that these general techniques must be regarded as schemes rather than actual algorithms, since they all require several design choices and parameter settings.

Brown et al. [7] introduced a taxonomy of methods for creating diversity. The first obvious distinction made is between *explicit* methods, where some metric of diversity is directly optimized, and *implicit* methods, where the method is likely to produce diversity without actually targeting it. The most common implicit methods strive for diversity by splitting the training data in order to train each base classifier using a slightly different training set. Such methods, called *resampling techniques*, can divide the available data either by *features* or by *instances*.

Historically, kNN models have only rarely been combined into ensembles. The main reason for this is that kNN turns out to be very robust with respect to data set variations, i.e., resampling methods dividing the data by instances will normally produce only limited diversity. kNN is, in contrast, sensitive to the features and the distance function used, see, e.g., [9]. This, consequently, is an argument for using either *feature weighting* or *feature sampling* (reduction). It must be noted, however, that feature reduction most often leads to lower base classifier accuracy.

Furthermore, techniques using locally optimized kNN are not very common either. One reason is probably the fact that when Wettschereck and Dietterich [18] investigated locally adaptive nearest neighbor algorithms, they found that local kNN methods, on real-world data sets, performed no better than standard kNN.

We have recently introduced a novel instance-based learner, specifically designed to avoid the drawbacks related to the choice of the parameter value k; see [13]. The suggested technique, named G-kNN, optimizes the number of neighbors to consider for each specific test instance, based on its position in input space; i.e., the algorithm uses several, locally optimized k's, instead of just one global. More specifically, G-kNN uses genetic programming (GP) to build decision trees, partitioning the input space in regions, where each leaf node (region) contains a kNN classifier with a locally optimized k. In the experimentation, using 27 data sets from the UCI repository, the basic version of G-kNN was shown to significantly outperform standard kNN, with respect to accuracy.

In this chapter, we will build on the previous study and use GP to evolve two different kinds of classification models, both based on kNN. The first group of models has a single, global k-value, together with a global set of feature weights. The second group of models is similar to G-kNN trees in the previous study, but here leaf nodes contain not only a locally optimized k-value but also a vector of feature weights. Naturally, neither of these models require the number of neighbors to consider as a parameter, but instead evolution is used to find optimal (global or local) k-values and feature weights. The main contribution of this chapter is, however, that we will combine our kNN models into ensembles, showing that the inherent inconsistency in GP will produce enough diversity to make the ensembles significantly more accurate than the base classifiers. In the experimentation, the ensembles produced will be compared to standard kNN, where k is found by cross-validation.

13.3 Method

The overall purpose of this study is to evaluate kNN ensembles, where each individual kNN base classifier is genetically evolved. In this study, all base classifiers have optimized k-values and optimized feature weights. Feature weighting will "stretch" the significant input axes, making them more important in the neighbor selection process. Here, all feature weights are between 0 and 1. Setting a weight to 0 will eliminate that dimension altogether.

For the actual evaluation, we used 4-fold cross-validation, measuring *accuracy* and *AUC*. Accuracy is, of course, the proportion of instances classified correctly when the model is applied to novel data, i.e., it is based only on the final classification. AUC, which is defined as the area under the ROC curve, can, on the other hand, measure the model's ability to rank instances based on how likely they are to belong to a certain class. Often, AUC is interpreted as the probability of an instance that do belong to the class being ranked ahead of an example that do not belong to the class; see, e.g., [10].

Obviously, the key concept in this study is the use of GP, which not only facilitate straightforward evaluation of different representation languages and optimization criteria, but also makes it easy to introduce diversity among the base classifiers. As a matter of fact, in this specific context, the inherent inconsistency of GP (meaning that different runs on identical data can produce quite different models) must be regarded as an asset, since it makes it uncomplicated to obtain a number of accurate, but still diverse, base classifiers. In the experimentation, all ensembles consisted of ten base classifiers, where each base classifier was optimized using all available training data. It must be noted that each base classifier was individually evolved, and that all parameters, including fitness function, were identical between the runs.

In the experimentation, we evaluated two different kNN models as base classifiers. In the first version (called *kNN ensemble*) each base classifier had a single, global *k*-value, together with a global set of feature weights. As expected, these two properties were, for each fold, simultaneously optimized using GP. Figure 13.1 shows a sample, evolved model, where the GP settled for $k = 3$. Naturally, the nine numbers represent the evolved relative weights of the features.

```
[ 0.7 0.4 0.3 0.2 0.5 0.8 0.8 0.7 0.4 ] KNN3
```

Fig. 13.1 Sample kNN ensemble base classifier. WBC data set

The second version (called *G-kNN ensemble*) used G-kNN trees as base classifiers. Here, the GP was used to evolve classification trees, where interior nodes represent splits, similar to decision trees like CART [5] and C5.0 [16]. Leaf nodes, however, did not directly classify an instance, but instead used a locally optimized kNN for all test instances reaching that leaf node. More specifically, each kNN leaf node contained a *k*-value and a vector of feature weights. Having said that, it must be noted that a G-kNN tree in itself is globally optimized, i.e., the GP used only the performance of the entire tree during evolution. Figure 13.2 shows a sample G-kNN base classifier.

```
if age > 54.7
|T: if pedi < 0.27
|  |T: if mass > 27.94
|  |  |T: [ 0.7 0.7 0.9 0.6 0.4 0.7 0.1 0.0 ] KNN1(5)
|  |  |F: [ 0.2 0.6 0.2 0.9 0.8 0.9 0.8 0.3 ] KNN5(8)
|  |F: [ 0.5 0.9 0.4 0.8 0 0.9 0.9 0.0 ] KNN17(27)
|F: if preg > 6.664
|  |T: [ 0.5 0.6 0.7 0.1 0.9 0 0.2 0.8 ] KNN3(116)
|  |F: [ 0.5 0.9 0.4 0.8 0 0.9 0.9 0.0 ] KNN7(420)
```

Fig. 13.2 Sample G-kNN ensemble base classifier. Diabetes data set

As seen in Fig. 13.2, the GP evolved a tree with three interior nodes and five different kNN leaf nodes. In this tree, the number of neighbors used varies from 1 to 17, and an inspection of the feature weight vectors shows that they are quite

```
F = { if, ==, <, > }
T = { i₁, i₂, ..., iₙ, ℜ, 1, 3, ..., 21 }

DTree              :-      (if RExp Dtree Dtree) | kNN node
RExp               :-      (ROp ConI ConC) | (== CatI CatC)
ROp                :-      < | >
CatI               :-      Categorical input variable
ConI               :-      Continuous input variable
CatC               :-      Categorical attribute value
ConC               :-      ℜ
kNN node           :-      Feature weights kNN k-value
k-value            :-      1, 3, ..., 21
Feature weights    :-      [FW₁, FW₂, ..., FWₙ]
FWᵢ                :-      ℜ in [0, 1]
```

Fig. 13.3 G-kNN representation language

different. From this, it seems reasonable to assume that a G-kNN tree using this representational language is more than able to capture local properties of a data set.

Figure 13.3 describes the representation language used by G-kNN. The sets *F* and *T* include the available functions and terminals, respectively. The functions are an *if-statement* and three relational operators. The terminals, in addition to a kNN node containing a *k*-value and a feature weight vector, are attributes from the data set and random real numbers. The exact grammar used is also presented, using Backus–Naur form.

The previous study showed some risk of overfitting. With this in mind, together with the fact that all evolved kNN models in this study were to be used in ensembles, we decided to strive for relatively weak and small models. More specifically, the number of generations and the number of individuals in the GP population were kept quite small. In addition, we included a small length penalty in all fitness functions used, in order to encourage smaller and potentially more general models. The length penalty used is much smaller than the cost of misclassifying an instance. Nevertheless, it will put some parsimony pressure on the evolution, resulting in less complex models, on average. For the exact GP parameters used in the experimentation, see Table 13.1.

Table 13.1 GP parameters

Parameter	kNN base classifier	G-kNN base classifier
Crossover rate	0.8	0.8
Mutation rate	0.01	0.01
Population size	500	300
Generations	10	30
Creation depth	N/A	5
Creation method	N/A	Ramped half-and-half
Fitness function	Training performance – length penalty	Training performance – length penalty
Selection	Roulette wheel	Roulette wheel
Elitism	Yes	Yes

Regarding fitness functions, three different fitness functions were evaluated: *Accuracy*, *AUC*, and *Brier score*. A Brier score measures the accuracy of a set of probability assessments. Proposed by Brier [6], Brier score is the average deviation between predicted probabilities for a set of events and their outcomes, i.e., a lower score represents higher accuracy. Using Brier score for classification requires a probability estimation for each class. In this study, the probability estimations used, when calculating both Brier score and AUC, are based directly on the number of votes for each class, i.e., no correctional function like a Laplace estimate was used. The reason for this is a rather recent study, investigating Random Forests of probability estimation trees, where it is shown that using non-corrected probability estimates, like the relative frequencies, is actually better than using Laplace estimates or m-estimates; see [3]. Naturally, using a specific fitness function corresponds to optimizing that performance measure on the training data.

13.3.1 Data sets

The 30 data sets used are all publicly available from the UCI repository [1]. When preprocessing, all attributes were linearly normalized to the interval [0, 1]. For numerical attributes, missing values were handled by replacing the missing value with the mean value of that attribute. For nominal attributes, missing values were replaced with the mode, i.e., the most common value.

It is, of course, very important how distance is measured when using instance-based learners. In this study, standard Euclidean distance between feature vectors was used. For nominal attributes, the distance is 0 for identical values and 1 otherwise. Nominal and ordered categorical attributes could potentially be handled differently, since ordered attributes often use the same distance function as continuous attributes. In this study we, for simplicity, settled for treating all attributes marked as categorical as nominal. For a summary of data set characteristics, see Table 13.2. *Inst.* is the total number of instances in the data set. *Class* is the number of classes, *Con.* is the number of continuous input variables, and *Cat.* is the number of categorical input variables.

All experimentation was carried out using G-REX [12]. G-REX was initially used for extracting rules from opaque models, like neural networks and ensembles, and has been thoroughly extended and evaluated in several papers; for a summary see [11]. Lately, G-REX has been substantially modified, with the aim of becoming a general data mining framework based on GP; see [14].

In Experiment 1, standard kNN, with k-values optimized using cross-validation was evaluated.[1] In this experiment, the number of neighbors to consider was optimized, for each fold, based on performance on the training data. During experimentation, all odd k-values between 1 and 21 were tried, and the k-value obtaining the highest training score was applied to the test data. All three criteria (accuracy,

---

[1] This technique is from now on referred to as *kNN-cv*.

Table 13.2 Data sets

Dataset	Inst.	Class	Con.	Cat.
Breast cancer	286	2	0	9
Colic	368	2	7	15
CMC	1473	3	2	7
Credit-A	690	2	6	9
Credit-G	1000	2	7	13
Cylinder	512	2	20	20
Dermatology	366	6	1	32
Diabetes	768	2	8	0
Ecoli	336	8	7	0
Glass	214	7	9	0
Haberman	306	2	3	0
Heart-C	303	2	6	7
Heart-S	270	2	6	7
Hepatitis	155	2	6	13
Iono	351	2	34	0
Iris	150	3	4	0
Labor	57	2	8	8
Liver	345	2	6	0
Lymph	148	4	3	15
Postoperative patient (Postop)	90	3	0	9
Primary tumor	339	22	0	17
Sick	2800	2	7	22
Sonar	208	2	60	0
TAE	151	3	1	4
Tic–Tac–Toe (TTT)	958	2	0	9
Vehicle	846	4	18	0
Vote	435	2	0	16
Wisconsin breast cancer (WBC)	699	2	9	0
Wine	178	3	13	0
Zoo	100	7	0	16

Brier score, and AUC) were evaluated as score (fitness) functions. In addition, the experiment also compared standard majority voting to the use of weighted voting. It should be noted that one explicit purpose of this experiment was to find the best "simple" procedure to compare our suggested techniques against Experiment 2.

The overall purpose of the second experiment was to evaluate the more advanced procedures. More specifically, kNN ensembles and G-kNN ensembles were compared to each other and to the best simple kNN procedure found in Experiment 1. Here, a kNN ensemble consisted of ten kNN models. Each base model was independently optimized on the specific training data, using GP and one of the three performance measures (accuracy, Brier score, or AUC) as fitness function. During evolution, the k-value and a global vector containing feature weights were simultaneously optimized. A G-kNN ensemble also consisted of ten base models, each independently optimized using GP. When using G-kNN, however, the base models were G-kNN trees, where each leaf contains a k-value and a vector of feature weights.

13.4 Results

Table 13.3 shows the accuracy results for Experiment 1. Unsurprisingly, it is best to use accuracy as score function when optimizing accuracy. Regarding the use of weighted voting, the results show that it is in fact better to use straightforward majority voting. Finally, it is interesting to note that using Brier score for the optimization is more successful than using AUC. All in all, however, the setup producing the most accurate models was clearly using majority voting while optimizing accuracy.

Table 13.3 Accuracy results Experiment 1

| Model type | kNN-cv majority voting | | | | | | kNN-cv weighted voting | | | | | |
| Score function | ACC | | Brier | | AUC | | ACC | | Brier | | AUC | |
Data	Train	Test	Train	Test	Train	Test	Train	Test	Train	Test	Train	Test
Breast cancer	75.06	71.68	72.73	72.73	72.49	72.73	74.48	**74.48**	72.73	72.73	72.61	72.73
Colic	83.97	**83.42**	82.16	83.15	82.07	81.52	83.97	83.15	82.16	83.15	82.16	81.79
CMC	47.39	46.64	46.69	**47.73**	46.98	47.25	47.34	47.46	46.21	47.25	46.37	46.64
Credit-A	86.96	85.96	86.52	85.81	86.23	85.38	86.96	86.10	86.76	**86.24**	86.47	85.81
Credit-G	73.87	74.20	73.40	73.90	73.30	**74.30**	73.87	74.20	73.40	73.90	73.30	**74.30**
Cylinder	77.84	**77.41**	70.99	69.63	73.27	71.48	72.10	70.56	70.99	69.63	72.10	70.56
Dermatology	96.81	95.36	96.72	95.36	95.90	**96.18**	96.90	95.63	96.90	95.36	96.08	**96.18**
Diabetes	75.00	**74.61**	74.44	74.09	74.61	73.96	75.00	74.09	74.44	74.09	74.61	73.96
Ecoli	85.91	**87.20**	85.52	86.31	83.53	83.93	86.21	86.01	85.91	86.01	84.52	84.52
Glass	67.76	67.26	63.54	64.07	61.53	59.84	66.66	**69.64**	64.16	65.93	62.92	61.71
Haberman	73.97	73.86	73.42	**74.19**	72.11	70.59	74.07	71.89	73.42	**74.19**	71.57	70.59
Heart-C	83.61	82.51	83.17	82.18	82.84	82.51	83.61	82.51	83.17	82.18	82.95	**82.84**
Heart-S	82.71	82.58	82.47	82.58	82.47	82.58	82.71	**82.95**	82.47	82.58	82.47	82.58
Hepatitis	84.95	**84.51**	84.09	80.65	83.01	82.57	84.95	**84.51**	84.09	80.65	82.79	81.93
Iono	86.80	84.33	85.75	**85.76**	81.57	83.20	85.18	84.62	84.80	**85.76**	81.10	82.35
Iris	97.11	94.68	97.11	94.68	96.67	**95.34**	97.11	94.03	97.11	94.68	96.67	**95.34**
Labor	90.63	87.74	89.47	85.95	85.35	**89.40**	88.88	87.74	88.30	87.74	83.61	**89.40**
Liver	64.35	61.74	62.71	62.32	63.09	61.45	64.45	60.58	62.71	62.32	62.80	**63.48**
Lymph	82.88	80.41	81.98	81.08	81.98	79.05	83.11	79.05	82.43	**82.43**	81.53	81.08
Postop	71.12	**71.15**	71.12	**71.15**	67.41	63.39	71.12	**71.15**	71.12	**71.15**	71.12	**71.15**
Primary tumor	41.50	42.17	40.12	41.88	39.82	41.00	41.00	**42.18**	39.82	41.00	39.73	41.30
Sick	96.25	**96.42**	96.12	96.26	95.48	95.47	96.25	**96.42**	96.12	96.26	95.56	95.49
Sonar	85.26	**85.10**	82.53	78.85	82.53	78.85	83.33	80.29	82.53	78.85	80.61	79.33
TAE	56.51	**62.23**	50.99	49.68	51.20	48.35	56.06	55.64	52.32	51.65	55.62	54.98
TTT	98.40	**98.54**	96.97	97.70	98.40	**98.54**	98.40	**98.54**	96.97	97.70	98.05	**98.54**
Vehicle	70.72	67.49	69.66	67.38	69.82	**68.80**	71.20	67.73	70.76	68.08	69.98	68.09
Vote	94.18	92.65	93.64	92.88	92.80	92.65	94.02	92.42	93.87	**93.11**	92.80	92.65
WBC	96.90	96.57	96.85	**96.71**	96.57	96.42	96.90	96.57	96.33	96.57	96.28	96.57
Wine	97.75	**97.20**	96.44	96.07	96.63	96.07	97.75	**97.20**	96.44	96.07	96.63	96.07
Zoo	95.37	**95.08**	95.04	**95.08**	90.10	90.15	93.40	93.08	93.07	91.08	91.08	92.12
MEAN	**80.72**	**80.02**	79.41	78.86	**78.66**	78.10	80.23	79.35	79.38	78.95	78.80	78.80
Mean rank	2.47		3.07		3.90		2.60		3.03		3.07	

Table 13.4 shows the AUC results for Experiment 1. Again, it turns out to be beneficial to use the criterion that we actually would like to optimize as score function. For AUC, weighted voting turned out to be clearly better than majority voting.

Using Brier score was here better than optimizing on accuracy, but clearly worse than using AUC as score function. So, the best procedure for finding models with high AUC was to use weighted voting while optimizing AUC.

Table 13.4 AUC results Experiment 1

Model type	kNN-cv majority voting						kNN-cv weighted voting					
Score function	ACC		Brier		AUC		ACC		Brier		AUC	
Data	Train	Test	Train	Test	Train	Test	Train	Test	Train	Test	Train	Test
Breast cancer	62.76	57.20	66.96	63.73	67.08	**64.74**	63.14	64.50	67.08	63.64	67.31	64.37
Colic	86.11	86.01	86.99	85.25	87.06	84.39	86.51	**86.28**	87.18	85.37	87.27	84.60
CMC	62.66	64.11	63.35	64.56	63.25	64.22	62.66	63.94	63.36	**64.58**	63.33	64.20
Credit-A	90.85	90.94	91.14	91.00	91.30	91.04	90.93	90.96	91.19	90.95	91.29	**91.10**
Credit-G	73.03	74.86	73.96	75.41	74.00	75.44	73.28	75.15	74.17	**75.70**	74.15	75.63
Cylinder	76.49	75.84	77.85	76.33	78.81	76.21	79.30	**78.48**	78.75	77.28	79.30	**78.48**
Dermatology	99.75	99.81	99.73	99.62	99.88	**99.90**	99.73	99.67	99.75	99.64	99.89	**99.90**
Diabetes	80.23	80.11	80.53	80.06	80.52	80.30	79.86	79.15	80.63	80.15	80.65	**80.40**
Ecoli	95.20	95.21	95.39	95.68	96.09	**96.22**	95.38	95.55	95.40	95.69	96.05	96.12
Glass	79.63	79.76	86.25	85.47	86.50	86.06	85.65	86.28	86.57	85.72	86.89	**86.40**
Haberman	57.84	59.45	58.94	59.55	62.75	60.69	55.88	56.58	59.15	59.82	62.37	**61.84**
Heart-C	90.33	89.98	90.63	90.27	90.71	90.29	90.30	89.98	90.60	90.20	90.68	**90.51**
Heart-S	88.39	88.17	88.55	88.11	88.55	88.11	88.13	**88.43**	88.42	88.10	88.42	88.10
Hepatitis	80.96	80.56	83.77	82.71	84.23	83.40	80.76	80.51	83.59	82.70	83.96	**84.53**
Iono	86.65	87.27	88.05	88.68	91.90	**92.92**	88.94	89.89	89.13	89.42	92.01	92.80
Iris	99.60	98.51	99.61	98.53	99.71	**99.62**	99.59	98.51	99.61	98.52	99.73	99.52
Labor	89.82	91.28	92.09	91.69	96.44	91.08	92.30	**94.39**	93.16	93.31	96.52	92.43
Liver	65.09	**66.12**	65.46	64.33	65.72	62.94	65.05	64.87	65.68	64.63	65.76	65.89
Lymph	90.61	88.55	90.78	90.27	90.86	89.66	90.12	90.27	90.91	**90.49**	90.89	90.37
Postop	37.40	45.81	41.05	43.37	47.25	39.51	31.88	35.36	39.47	42.57	50.00	**50.00**
Primary tumor	74.32	75.50	74.91	76.49	75.26	74.57	74.69	75.90	75.21	**76.60**	75.46	74.71
Sick	89.64	89.98	92.18	91.56	94.55	**95.33**	89.61	89.92	92.20	91.59	94.58	95.17
Sonar	84.93	84.80	90.13	89.53	90.13	89.53	90.01	89.64	90.68	**90.70**	90.62	89.57
TAE	67.41	71.68	67.09	66.54	69.07	68.29	71.63	71.29	69.48	67.96	71.81	**71.91**
TTT	99.85	99.92	99.63	99.82	99.85	99.92	99.88	**99.95**	99.60	99.80	99.85	99.91
Vehicle	88.56	87.42	89.41	88.41	89.61	89.17	89.40	88.12	89.59	88.57	89.85	**89.28**
Vote	97.23	97.43	97.42	97.68	97.45	97.71	97.33	97.49	97.39	97.72	97.49	**97.75**
WBC	99.10	98.61	99.08	98.53	99.19	**98.62**	85.65	84.73	99.11	96.76	99.14	97.66
Wine	99.72	99.77	99.87	99.45	99.87	99.77	99.78	**99.80**	99.89	99.45	99.89	99.79
Zoo	97.41	98.25	97.57	98.20	99.28	98.56	98.18	98.77	98.01	**98.80**	99.41	98.56
MEAN	**83.05**	83.43	84.28	84.03	85.23	84.27	83.18	83.48	84.50	84.21	85.49	85.05
Mean rank	**4.47**		**4.07**		**3.00**		**3.70**		**3.33**		**2.27**	

In summary, the results for Experiment 1 show that it is most favorable to determine the number of neighbors to consider based on training performance measured using the criterion we would like to optimize. Having said that, it must be noted that models optimized on accuracy will normally have relatively poor AUC and vice versa. This is, of course, not necessarily a problem in itself, it is just a consequence of the fact that the two criteria actually measure quite different properties.

Table 13.5 shows the accuracy results for Experiment 2. The most important result is, of course, that the two ensemble approaches, when using accuracy as fitness function, clearly outperform kNN-cv. On the other hand, using either Brier score or AUC as score (fitness) function is again not very successful when targeting accuracy. As a matter of fact, kNN-cv is actually more accurate than three of the four ensemble techniques optimized using either Brier score or AUC.

Table 13.5 Accuracy results Experiment 2

Model type	kNN-cv majority vote	kNN ensemble			G-kNN ensemble		
Fitness function	ACC	ACC	Brier	AUC	ACC	Brier	AUC
Breast cancer	71.68	**74.11**	**74.11**	71.33	73.08	73.08	73.42
Colic	83.42	82.88	83.42	81.52	**83.70**	83.15	82.88
CMC	46.64	50.04	49.08	**51.39**	50.17	47.25	50.65
Credit-A	85.96	85.09	85.52	85.66	85.52	**86.24**	86.10
Credit-G	74.20	73.30	74.60	74.60	73.50	73.90	**74.80**
Cylinder	**77.41**	75.37	71.30	72.04	74.81	69.63	74.26
Dermatology	95.36	95.63	95.90	95.90	**96.17**	95.36	95.63
Diabetes	**74.61**	73.57	73.70	73.70	74.22	74.09	**74.61**
Ecoli	**87.20**	86.01	85.12	85.42	85.12	86.01	84.82
Glass	67.26	68.71	67.76	66.39	**69.65**	66.86	67.31
Haberman	73.86	73.20	72.88	69.30	**74.51**	**74.51**	70.27
Heart-C	82.51	82.84	82.83	**84.49**	83.50	81.86	84.16
Heart-S	82.58	**82.59**	79.98	80.73	82.22	82.58	81.83
Hepatitis	84.51	**85.81**	83.87	80.03	**85.81**	81.95	82.61
Iono	84.33	**87.74**	86.33	82.63	**87.74**	85.76	82.06
Iris	94.68	96.00	95.34	95.34	**96.66**	94.68	95.34
Labor	87.74	92.98	94.64	91.07	**96.43**	87.74	89.29
Liver	61.74	**67.25**	**67.25**	66.09	**67.25**	62.32	66.68
Lymph	80.41	82.43	83.11	83.78	83.78	**84.46**	81.08
Postop	71.15	64.43	**71.49**	55.43	67.79	71.15	63.34
Primary tumor	42.17	42.48	41.89	44.54	43.06	41.30	**44.84**
Sick	**96.42**	95.65	95.60	95.55	95.71	96.26	95.55
Sonar	**85.10**	82.21	77.88	78.37	82.21	79.81	79.33
TAE	62.23	64.85	50.25	59.58	**66.20**	54.94	61.58
TTT	98.54	**99.17**	**99.17**	99.06	**99.17**	97.70	98.22
Vehicle	67.49	69.03	68.44	67.14	**69.26**	68.20	66.44
Vote	92.65	94.26	95.18	**94.72**	**94.72**	93.11	94.26
WBC	96.57	96.85	97.00	**97.14**	96.85	96.71	**97.14**
Wine	**97.20**	96.09	95.52	95.52	96.64	96.64	95.52
Zoo	**95.08**	93.04	93.04	93.04	93.04	91.08	93.04
MEAN	80.02	80.45	79.74	79.05	80.95	79.28	79.57
Mean rank	3.97	3.33	3.80	4.37	2.60	4.63	4.10

In order to determine if there are any statistically significant differences between the two ensemble techniques and kNN-cv, standard sign tests were used. When using 30 data sets, 20 wins are required for a statistically significant difference, at $\alpha = 0.05$. Table 13.6 shows number of wins, ties, and losses for the row technique against the column technique. Statistically significant differences are given in bold. Consequently, the tests show that G-kNN ensembles were significantly more accurate than both kNN ensembles and the best kNN-cv found in Experiment 1. In addition, the difference between kNN ensembles and kNN-cv is almost significant.

Table 13.6 Accuracy: wins–ties–losses

	kNN-cv	kNN ensemble
kNN ensemble	18-0-12	–
G-kNN ensemble	**20-0-10**	**20-4-6**

The AUC results for Experiment 2 are shown in Table 13.7. Here, both ensemble techniques outperform the best kNN-cv from Experiment 1, but only when maximizing AUC. So, the picture is again that the best optimization criterion when aiming for high AUC is to use AUC as the fitness function.

Table 13.7 AUC results Experiment 2

Model type Fitness function	kNN-cv weighted vote AUC	kNN ensemble			G-kNN ensemble		
		ACC	Brier	AUC	ACC	Brier	AUC
Breast cancer	**64.37**	61.92	60.73	58.77	62.54	63.43	59.82
Colic	84.60	86.15	85.94	85.35	85.74	85.28	**86.40**
CMC	64.20	67.03	66.99	67.93	66.91	64.58	**68.15**
Credit-A	91.10	91.27	91.19	91.16	91.29	91.01	**91.40**
Credit-G	75.63	73.09	75.60	75.53	74.73	75.70	**76.00**
Cylinder	78.48	77.12	78.76	78.63	77.11	77.31	**80.99**
Dermatology	**99.90**	99.84	99.82	**99.90**	99.85	99.64	99.87
Diabetes	**80.40**	79.26	78.84	79.03	79.60	80.15	79.89
Ecoli	**96.12**	96.02	95.79	96.00	96.04	96.11	95.95
Glass	86.40	85.69	86.13	86.68	86.38	**87.16**	86.87
Haberman	61.84	**68.10**	65.62	67.05	65.43	61.16	65.57
Heart-C	**90.51**	89.93	90.20	90.34	90.17	90.16	90.47
Heart-S	88.10	88.13	87.57	87.75	**88.47**	88.14	88.18
Hepatitis	84.53	84.45	86.32	86.53	84.38	83.02	**87.63**
Iono	92.80	90.84	92.35	93.30	91.50	91.30	**93.36**
Iris	**99.52**	99.37	98.09	98.94	99.26	98.52	99.05
Labor	92.43	97.57	99.12	96.28	**99.32**	92.97	97.30
Liver	65.89	67.56	68.80	**68.95**	67.48	64.52	67.96
Lymph	90.37	91.75	**93.49**	92.89	92.78	90.43	92.23
Postop	50.00	35.59	50.12	46.32	37.01	42.37	**51.05**
Primary tumor	74.71	76.74	76.58	76.38	**77.13**	76.81	76.74
Sick	95.17	94.86	94.34	94.22	94.20	91.59	**95.22**
Sonar	89.57	**91.22**	88.33	89.21	87.97	91.07	87.90
TAE	71.91	75.71	68.65	73.77	**77.16**	71.85	74.60
TTT	99.91	99.95	99.96	**99.97**	99.95	99.80	99.91
Vehicle	**89.28**	88.01	89.10	88.75	88.16	88.59	89.09
Vote	97.75	98.06	97.89	97.73	97.89	97.68	**98.07**
WBC	97.66	98.64	**99.03**	98.90	98.91	96.76	99.02
Wine	**99.79**	99.68	99.34	99.36	99.71	99.74	99.67
Zoo	98.56	98.39	98.39	**99.65**	98.42	98.79	98.75
MEAN	85.05	85.06	85.44	85.51	85.18	84.52	85.90
Mean rank	3.90	4.23	4.17	3.83	4.00	4.87	2.80

Table 13.8 shows that G-kNN ensembles obtained significantly higher AUC than both kNN-cv and the kNN ensembles.

Table 13.8 AUC: wins–ties–losses

	kNN-cv	kNN ensemble
kNN ensemble	16-1-13	–
G-kNN ensemble	**20-1-9**	**22-0-8**

Table 13.9, finally, compares ensemble results to average base classifier results.

Table 13.9 Comparing ensembles and base classifiers

Measure	Accuracy				AUC			
Technique	KNN		G-kNN		KNN		G-kNN	
Model	Base	Ens.	Base	Ens.	Base	Ens.	Base	Ens.
Breast cancer	73.55	**74.11**	71.74	73.08	57.85	**58.77**	58.60	**59.82**
Colic	**83.10**	82.88	82.74	**83.70**	**85.64**	85.35	85.89	**86.40**
CMC	49.08	**50.04**	48.98	50.17	67.08	**67.93**	66.87	**68.15**
Credit-A	**85.09**	85.09	85.01	85.52	91.28	91.16	**91.52**	91.40
Credit-G	**73.30**	73.30	73.34	73.50	75.47	**75.53**	75.28	**76.00**
Cylinder	75.06	**75.37**	74.59	74.81	77.71	**78.63**	76.77	**80.99**
Dermatology	94.89	**95.63**	95.66	**96.17**	99.88	**99.90**	99.84	**99.87**
Diabetes	73.22	**73.57**	73.72	74.22	78.64	**79.03**	79.18	**79.89**
Ecoli	84.52	**86.01**	84.61	85.12	**96.06**	96.00	**96.00**	95.95
Glass	67.68	**68.71**	68.32	69.65	86.04	**86.68**	85.60	**86.87**
Haberman	**73.69**	73.20	73.62	74.51	66.80	**67.05**	64.41	**65.57**
Heart-C	82.28	**82.84**	82.87	83.50	90.14	**90.34**	90.13	**90.47**
Heart-S	82.29	**82.59**	81.51	82.22	87.66	**87.75**	87.45	**88.18**
Hepatitis	84.79	**85.81**	84.46	85.81	85.23	**86.53**	86.33	**87.63**
Iono	87.03	**87.74**	86.61	87.74	91.96	**93.30**	91.78	**93.36**
Iris	**96.14**	96.00	95.73	**96.66**	98.80	**98.94**	**99.22**	99.05
Labor	92.79	**92.98**	93.50	96.43	95.24	**96.28**	95.73	**97.30**
Liver	67.22	**67.25**	66.79	67.25	67.51	**68.95**	66.03	**67.96**
Lymph	80.68	**82.43**	80.81	83.78	90.59	**92.89**	89.39	**92.23**
Postop	**64.86**	64.43	64.78	67.79	46.09	**46.32**	47.91	**51.05**
Primary tumor	42.00	**42.48**	41.74	43.06	75.11	**76.38**	75.42	**76.74**
Sick	95.58	**95.65**	95.60	95.71	93.87	**94.22**	94.13	**95.22**
Sonar	81.92	**82.21**	80.48	82.21	84.83	**89.21**	84.08	**87.90**
TAE	63.30	**64.85**	61.75	66.20	69.27	**73.77**	71.09	**74.60**
TTT	98.71	**99.17**	98.69	99.17	99.79	**99.97**	99.77	**99.91**
Vehicle	67.13	**69.03**	67.57	69.26	88.04	**88.75**	88.22	**89.09**
Vote	94.24	**94.26**	94.10	94.72	97.15	**97.73**	97.28	**98.07**
WBC	96.41	**96.85**	96.51	96.85	98.79	**98.90**	98.92	**99.02**
Wine	**96.20**	96.09	96.53	96.64	**99.43**	99.36	99.47	**99.67**
Zoo	**93.74**	93.04	93.54	93.04	96.94	**99.65**	97.43	**98.75**
MEAN	80.02	80.45	79.86	80.95	84.63	85.51	84.66	85.90
Wins–Ties–Losses		**22-2-6**		**29-0-1**		**26-0-4**		**27-0-3**

From Table 13.9, it is obvious that the base models are clearly weaker than the ensembles. Or, put in another way, the ensemble approach really works. Standard sign tests also show, in all four comparisons, that the ensembles are significantly more accurate than their base classifiers. It is noteworthy that, for both accuracy and AUC, the differences between the base classifiers and the ensembles are greater for G-kNN than for kNN. Most likely, the explanation for this is that G-kNN, with its more complex models, obtains greater diversity among base models. Another interesting observation is that genetically evolved base classifiers, on several data sets, have worse performance than the best kNN-cv from Experiment 1.

13.5 Conclusions

We have in this chapter suggested a novel technique producing ensembles of instance-based learners. The technique is based on GP, making it straightforward to modify both the representation language and the score function. Most importantly, GP has an inherent ability to produce several, quite different models, all having similar individual performance. Naturally, this is exactly what is sought when building ensembles, i.e., accurate yet diverse base classifiers. One specific advantage for the GP approach is the fact that each model is still built using all features and all instances. This is in contrast to resampling methods, where implicit diversity is often introduced mainly by making all base classifiers less accurate, i.e., they do not have access to all relevant data. The different solutions produced by the GP, on the other hand, are all models of the original problem.

In this study, two different kinds of genetically evolved base classifiers were evaluated. The first was a global kNN model, similar to standard kNN, but with optimized feature weights and an optimized k-value. The second was a decision tree where each leaf node contains optimized feature weights and an optimized k-value. In the experimentation, the ensemble models significantly outperformed both the corresponding base classifiers and standard kNN with optimized k-values, with respect to both accuracy and AUC. Comparing the two different versions, ensembles using the more complex decision tree models, potentially utilizing several locally optimized k-values, obtained significantly higher accuracy and AUC. Interestingly, this superior performance was achieved despite the fact that base classifiers had similar performance, indicating that the ensembles built from the more complex models had greater diversity.

On a lesser note, the results also show that it is generally best to use the criterion that we would like to optimize as score function. Although this may sound obvious, it should be noted that it becomes more important for more powerful models. In this study, the genetically evolved models had, of course, many more degrees of freedom than standard kNN, resulting in more specialized models. In particular, models optimized on accuracy turned out to have relatively poor AUC and vice versa.

Acknowledgments This work was supported by the Information Fusion Research Program (University of Skövde, Sweden) in partnership with the Swedish Knowledge Foundation under grant 2003/0104 (URL: http://www.infofusion.se).

References

1. Asuncion, A., Newman, D.J.: UCI machine learning repository (2007)
2. Bishop, C.M.: Neural Networks for Pattern Recognition. Oxford University Press, Oxford (1995)
3. Boström, H.: Estimating class probabilities in random forests. In: ICMLA '07: Proceedings of the Sixth International Conference on Machine Learning and Applications, pp. 211–216. IEEE Computer Society, Washington, DC, USA (2007)
4. Breiman, L.: Bagging predictors. Machine Learning 24(2), 123–140 (1996)
5. Breiman, L., Friedman, J., Stone, C.J., Olshen, R.A.: Classification and Regression Trees. Chapman & Hall/CRC, Boca Raton, FL (1984)
6. Brier, G.: Verification of forecasts expressed in terms of probability. Monthly Weather Review 78, 1–3 (1950)
7. Brown, G., Wyatt, J., Harris, R., Yao, X.: Diversity creation methods: a survey and categorisation. Journal of Information Fusion 6(1), 5–20 (2005)
8. Dietterich, T.G.: Machine-learning research: Four current directions. The AI Magazine 18(4), 97–136 (1998)
9. Domeniconi, C., Yan, B.: Nearest neighbor ensemble. In: 17th International Conference on Pattern Recognition, vol. 1, pp. 228–231. IEEE Computer Society, Los Alamitos, CA, USA (2004)
10. Fawcett, T.: Using rule sets to maximize roc performance. In: Proceedings of the 2001 IEEE International Conference on Data Mining, ICDM'01, pp. 131–138. IEEE Computer Society, Washington, DC, USA (2001)
11. Johansson, U.: Obtaining Accurate and Comprehensible Data Mining Models: An Evolutionary Approach. PhD-thesis. Institute of Technology, Linköping University (2007)
12. Johansson, U., König, R., Niklasson, L.: Rule extraction from trained neural networks using genetic programming. In: 13th International Conference on Artificial Neural Networks, supplementary proceedings, pp. 13–16 (2003)
13. Johansson, U., König, R., Niklasson, L.: Evolving a locally optimized instance based learner. In: 4th International Conference on Data Mining – DMIN'08, pp. 124–129. CSREA Press (2008)
14. König, R., Johansson, U., Niklasson, L.: G-REX: A versatile framework for evolutionary data mining, ieee international conference on data mining (icdm'08), demo paper. in press (2008)
15. Krogh, A., Vedelsby, J.: Neural network ensembles, cross validation, and active learning. Advances in Neural Information Processing Systems 2, 231–238 (1995)
16. Quinlan, J.R.: C4.5: Programs for Machine Learning. Morgan Kaufmann Publishers Inc., San Fransisco, CA (1993)
17. Schapire, R.E.: The strength of weak learnability. Machine Learning 5(2), 197–227 (1990)
18. Wettschereck, D., Dietterich, T.G.: Locally adaptive nearest neighbor algorithms. Advances in Neural Information Processing Systems 6, 184–191 (1994)
19. Witten, I.H., Frank, E.: Data Mining: Practical Machine Learning Tools and Techniques, Second Edition (Morgan Kaufmann Series in Data Management Systems). Morgan Kaufmann, San Fransisco, CA (2005)
20. Wolpert, D.H.: Stacked generalization. Neural Networks 5, 241–259 (1992)
21. Zavrel, J.: An empirical re-examination of weighted voting for k-nn. In: Proceedings of the 7th Belgian-Dutch Conference on Machine Learning, pp. 139–148 (1997)

Part V
Web-Mining

Chapter 14
Behaviorally Founded Recommendation Algorithm for Browsing Assistance Systems

Peter Géczy, Noriaki Izumi, Shotaro Akaho, and Kôiti Hasida

Abstract We present a novel recommendation algorithm for browsing assistance systems. The algorithm efficiently utilizes a priori knowledge of human interactions in electronic environments. The human interactions are segmented according to the temporal dynamics. Larger behavioral segments – sessions – are divided into smaller segments – subsequences. The observations indicate that users' attention is primarily focused on the starting and the ending points of subsequences. The presented algorithm offers recommendations at these essential navigation points. The recommendation set comprises of suitably selected desirable targets of the observed subsequences and the consecutive initial navigation points. The algorithm has been evaluated on a real-world data of a large-scale organizational intranet portal. The portal has extensive number of resources, significant traffic, and large knowledge worker user base. The experimental results indicate satisfactory performance.

14.1 Introduction

The effective browsing assistance services should aim at satisfying the objective navigational needs of the users. Rather than focusing on predicting the next page in a user's navigation stream, it is of higher benefit to the users to be offered direct access to the desired resource. Thus, the users can skip all the essentially unwanted transitional pages and reach the desired resource immediately. This potentially saves users' time, servers' computational resources, and networks' bandwidth.

Peter Géczy · Noriaki Izumi · Shotaro Akaho · Kôiti Hasida
National Institute of Advanced Industrial Science and Technology (AIST), Tokyo and Tsukuba, Japan

R. Stahlbock et al. (eds.), *Data Mining,* Annals of Information Systems 8,
DOI 10.1007/978-1-4419-1280-0_14, © Springer Science+Business Media, LLC 2010

The pertinent questions arising in this context are "How to identify the desired resources?" and "How to appropriately present them to users?" Constructive answers to these questions lie in the detailed analysis of human–web interactions. Behavioral analytics provide a suitable base for deeper understanding of human behavior in digital environments and translate to actionable knowledge vital for designing effective browsing assistance systems.

The browsing assistance systems are in high demand in web-based portals. The web portals utilize the browsing assistance services for usability improvements. Improved usability of web-based information systems brings economic benefits to organizations and time benefits to users. The progress in advancing organizational information systems has been slow. Knowledge-intensive organizations increasingly rely on advanced information technology and infrastructure [1]. The information systems should facilitate higher operating efficiency of organizations and their members [2]. This necessitates well-deployed organizational knowledge portals [3, 4].

It has been observed that organizational knowledge portals are underutilized despite the vast amount of resources and services they provide [5]. The underutilization is mainly due to the misalignment between the design and implementation of business processes and services, and the usability characteristics of users. The assistance services should incorporate recommender systems that help users navigate in intranet environments more efficiently. This requires a new generation of recommender systems [6–8], those that effectively utilize behavioral analytics of users.

14.1.1 Related Works

The human web behavior analytics have been attracting a spectrum of research activity. The body of reported work includes mining clickstream data of page transitions [9, 10], improving web search ranking utilizing information on user behavior [11, 12], web page traversing employing eye-tracking studies and devices [13, 14], and commercial aspects of user behavior analysis [15–18].

A major portion of the research has been focused on deriving models of user navigation with predictive capabilities – targeting the next visited page. Such automated predictors have applicability in pre-caching systems of web servers, collaborative filtering engines, and also recommender systems. Applied statistical approaches in this area have been favoring Markov models [19]. However, higher order Markov models become exceedingly complex and computationally expensive. The computationally less-intensive cluster analysis methods [20], adaptive machine learning strategies [21, 22], and fuzzy inference [23] suffer from scalability drawbacks.

The local domain heuristics have also been explored for improving collaborative filtering techniques for browsing assistance systems at corporate knowledge portals [24]. The frequent pattern mining reduces the computational complexity and improves the speed, however, at the expense of substantial data loss [25, 26]. The latent semantic indexing approach in collaborative recommender systems has been reported to reduce the execution times [27]. New advancements call for more efficient approaches built on deeper quantitative and qualitative understanding of human behavior in electronic environments.

14.1.2 Our Contribution and Approach

The presented work advances the state of the art in browsing assistance systems by employing detailed behavioral analytics of the target user population. It has been observed that human behavior in digital environments displays particular dynamics. Shorter periods of rapid activity are followed by longer periods of passivity [28, 29]. This is attributed to perceptually prioritized task execution.

Human temporal dynamics are utilized for identifying the activity periods and elucidating the web interactions and usability features [5]. Extracted actionable knowledge of human behavior permits formulation of the essential strategic elements for designing a novel recommendation algorithm. The algorithm is behaviorally founded, computationally efficient, and scalable. It complies with the core requirements for effective deployment in browsing assistance systems of large-scale organizational portals.

The novel algorithm provides recommendations on potentially desired target resources, rather than just the following page in the navigational sequence. The recommendations are offered only at the specific navigation points. This enhances efficiency and saves computing and bandwidth resources. The points are identified based on the analysis of user's browsing interactions. As the user's browsing behavior evolves and the interaction characteristics change, the algorithm naturally reflects the changes and adjusts its recommendations.

14.2 Concept Formalization

We introduce the essential terminology and formal description of the approach. The presented terms are accompanied by intuitive and illustrative explanations. This helps us to understand and comprehend the concept at both practical and higher order abstract levels.

The analytic framework utilizes a segmentation of human behavior in digital environments. The basic framework has been introduced in [5]. We recall only applicable constructs and expand the framework for concepts relevant to this study.

Human behavior in electronic environments is analyzed from the recorded interactions. The interactions are represented as sequences of page transitions – sometimes referred to as *click streams*. The long interaction sequences are divided into smaller segments: sessions and subsequences. These two essential segments of human interactions underline the tasks of different complexities undertaken in the web environments. More complex tasks constitute sessions. The sessions are split into the subsequences. The subsequences exemplify the elemental browsing tasks.

Human interactions display the bursts of activity followed by the longer periods of inactivity. The segmentation of sequences accounts for the observed temporal dynamics. The suitable separation points are determined based on the delays between transitions.

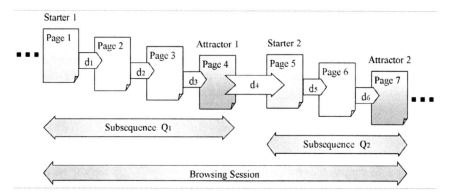

Fig. 14.1 Depiction of segmentation of page transition sequences. The sequences are divided into the browsing sessions and subsequences based on inactivities d_i between transitions. The first and the last elements of subsequences are the important navigation points. The first point is a starter, and the last point is an attractor

The page transition sequences are recorded as indexed sequences $\{(p_i, t_i)\}_i$ of pairs (p_i, t_i), where p_i denotes the visited page URL_i at the time t_i. The sequences are converted into the form $\{(p_i, d_i)\}_i$, where $d_i = t_{i+1} - t_i$ represents a delay between the consecutive page transitions $p_i \rightarrow p_{i+1}$. This facilitates direct observation of the transitional delays.

Depending on the observed transitional delays we segment the long clickstream sequences into smaller constituents: sessions and subsequences. An intuitive illustration of the segmentation concept is shown in Fig. 14.1. The sessions and the subsequences are the following:

Browsing session *is a sequence* $B = \{(p_i, d_i)\}_i$, *where each delay between page views,* d_i, *is shorter than the predetermined interval* T_B. *Browsing session is often referred to simply as a* **session**.

Subsequence *of an individual browsing session B is a sequence* $Q = \{(p_i, d_i)\}_i$, *where each transitional delay,* d_i, *is less than or equal to the dynamically calculated value* T_Q.

The sessions generally represent longer human–web interactions and contain several subsequences. They correspond to larger and more complex browsing tasks. Users accomplish these tasks via several smaller subtasks indicated by the subsequences. Consider the following example of user interactions on the organizational intranet portal. Upon arrival to the office an employee logins into the intranet system

(Subsequence 1). A successful login procedure will present the user with the opening portal page. Navigating from the initial page, the user accesses the attendance monitoring service and records the starting time of the current working day (Subsequence 2). Completing the attendance process successfully, the user proceeds to the organizational announcements page where he/she finds an interesting information concerning the latest overtime remuneration policy (Subsequence 3). Then the user leaves computer for another work-related activity.

The example of employee interactions with the organizational intranet system describes a potentially typical "morning session" – consisting of three subsequences indicating distinct tasks: entering the system, recording the attendance, and accessing the latest announcements. Each subsequence displays agile transitions at the end of which user's attention is required. The attention takes time. Thus there are longer delays recorded prior to initiating the next task.

The pertinent issue in segmenting the human browsing interactions into sessions and subsequences is the appropriate determination of separating delays – the values of T_B and T_Q. Elucidation of students web behavior revealed that their browsing sessions last on average 25.5 min [30]. Analysis of knowledge workers' browsing interactions on the organizational intranet portal exposed longer session duration: 48.5 min on average [5]. The study used empirically determined minimum intersession delay $T_B = 1$ h. The subsequence delay separator T_Q was calculated dynamically as an average delay in the session bounded from below by 30 s. This determination proved to be appropriate in the studied case.

A deeper understanding of human interactions in web environments involves observing where the users initiate their actions and which resources they target. This translates to the elucidation of the starting and the ending points of subsequences. The starting navigation points of subsequences are refereed to as **starters**, whereas the ending navigation points are refereed to as **attractors**. A set of starters is denoted as S and a set of attractors as A.

The concept of starters and attractors is illustrated in Fig. 14.1. The starters correspond to the initial navigation points of subsequences, that is, the pages 1 and 5 (points p_1 and p_5). Considering the formerly staged example of employee interactions, the starter p_1 would be the login page of the portal. The attractors are the terminal navigation points of subsequences, that is, the pages 4 and 7 (points p_4 and p_7). In our example, the first subsequence – login – terminates with displaying the opening portal page to the user. Thus, the opening portal page is the attractor of the first subsequence.

Assume that the user bookmarks the opening portal page for easier access later. After some browsing, e.g., reading news, he/she would like to return to the opening portal page and navigate from there to another resource. Simply by going to bookmarks and clicking on the record, the user can access the opening portal page and start navigating from there. Note that the opening portal page has been the attractor in the example from the previous paragraph, and in this one it is the starter. A single navigation point can be both starter and attractor.

Navigation using hotlists such as bookmarks and/or history [31] often leads to the single point subsequences. The points detected in such subsequences are called

singletons. The singletons relate to the single actions followed by longer inactivity. This is frequently the case when using hotlists. The user accesses the desired resource directly, rather than navigating through the link structure.

An individual point can be used for navigating to numerous different targets. Analogously, from a single target, users can transition to the several different initial points of the following subtasks. It is desirable to identify the attractors to which users navigate from the given starters, as well as the starters to which they transition from the given attractors. This is expressed by the mappings *starter → set of attractors* and *attractor → set of starters*.

Starter–attractor mapping $\omega : S \rightarrow A$ is a mapping where for each starter $s \in S$, $\omega(s)$ is a set of attractors of subsequences having the identical starter s.

Attractor–starter mapping $\psi : A \rightarrow S$ is a mapping where for each attractor $a \in A$ of the subsequences, $\psi(a)$ is a set of starters of the detected consecutive subsequences.

The starter–attractor mapping underlines the range of different attractors the users accessed when initiating their browsing interactions from the given starter. It does not quantify the number of available links on the starter page. Instead, it exposes the range of detected abstract browsing patterns: starter → set of attractors. Between the starter and the attractor there may be several intermediate pages in the observed subsequences. The starter–attractor mapping outlines an important "long-range" browsing pattern indicator. On the other hand, the attractor–starter mapping delineates an important "close-range" interaction pattern indicator: attractor → set of starters. The transition from the attractor to the next starter is direct. The attractor–starter mapping relates more closely to the spectrum of links exposed on the given attractor page (static or dynamic) and/or the utilization of hotlists.

A single starter can map to a large number of attractors in starter–attractor mapping. Analogously, a single attractor can map to a large number of starters in attractor–starter mapping. Among the points in the mapped sets, some may be more important than the others. The importance is determined by a suitably defined ordering function with respect to which the points can be ranked. Various ordering functions can be defined. A relative frequency of occurrences can be a simple yet suitable ordering function, for instance.

It is useful to select a limited number of the most viable candidates among the navigation points in mapped sets. Selection of the best candidates is done with respect to the ordering. The selected subsets containing a limited number of candidates are called *top sets*. They are denoted by a superscript and expressed as follows:

Top-n sets $\omega^{(n)}$ and $\psi^{(n)}$ are the ordered sets of the first n points selected with respect to an ordering function f defined on the mapped sets.

The top sets describe the sampling from the image sets of the starter and attractor mappings with respect to the ordering function. They are extracts from the image set and contain a number of the highest ranking elements. Consider, for example, a starter s with $\omega(s) = \{a_1, \ldots, a_x\}$, $x \in N$. Top-n set $\omega^{(n)}(s) = \{a_1, \ldots, a_n\}$, $n \leq x$, can be a selection of the first n attractor points according to a ranking function f defining ordering on the set $S \cup A$.

14.3 System Design

The conceptual design of the presented system efficiently employs valuable a priori information obtained from the analysis of knowledge worker browsing interactions on a large corporate intranet portal. The observations have pertinent implications to the architecture of the assistance system.

14.3.1 A Priori Knowledge of Human–System Interactions

The exploratory analysis of the knowledge worker browsing behavior and the usability of the organizational information system highlighted numerous relevant issues [5]. Several of them are directly or indirectly applicable to the browsing assistance system design. Following is a concise list of the important observations.

- Knowledge workers form repetitive browsing and behavioral patterns.
- Complex interaction tasks are divided into three subtasks on average.
- General browsing strategy can be expressed as knowledge of the starting point and familiarity with the navigational pathway to the target.
- Extended use of the information system leads to the habitual interaction behavior.
- Knowledge workers navigate rapidly in the subsequences – within seconds.
- Users have relatively short attention span for elemental tasks – approximately 7 min on average.
- Knowledge workers utilize a small set of starting navigation points and target a small number of resources.

The knowledge workers have generally focused browsing interests. Their browsing tasks are mainly related to their work description. Thus, they effectively utilize only a relatively small subset of resources from a large pool of available ones. Knowledge workers' browsing habitually focuses on the initial navigation points and the traversal path to the desired resource. As they get used to the system, their navigation from a starter to an attractor is progressively rapid.

14.3.2 Strategic Design Factors

The essential requirements on the recommendation algorithm for intranet browsing assistance system fall into three main categories:

1. **Recommendation quality**: The algorithm should provide reasonably accurate and suitable recommendations.
2. **Diverse user population accountability**: While focusing on the local knowledge workers, the algorithm should encompass diversity in the user population.

3. **Computational efficiency and scalability**: The algorithm should be computationally efficient and scalable in the dimensions of user population and resource number.

The adequate coverage of these three domains demands formulating effective strategies for algorithm design. In devising the strategic elements, we utilize the findings of the human–web interactions on a large-scale organizational intranet portal. They provide actionable a priori knowledge. Building upon these observations enables us to determine the core strategic design factors.

Exploit starters and attractors for assistance services.

The starters and attractors should be the primary navigation points for appropriate assistance services. The observed knowledge worker browsing strategy relies on knowing the right starters for reaching their goals. The attractors are the desired targets and transition initiators to the subsequent starters. These are the navigation points where users pay the most attention to the content (and spend their time at). The intermediate points between starters and attractors in the subsequences are transitional. They are passed through relatively rapidly – within seconds. Thus, the users do not pay sufficient attention to the content of these pages and proceed straight to the known link in the navigational pathway to the target. If the assistance service is provided on these pages, it is unlikely the users would notice it, not to mention use it within such a short time. It would simply be an inefficient use of computing resources.

Provide recommendations on relevant attractors and consecutive starters.

The former strategic point proposes to provide assistance services only at the starter and attractor pages. When a user reaches the starter, his/her desired target is the corresponding attractor. Analogously, when a user arrives at the attractor, he/she would like to transit to the appropriate starter. Hence, the effective browsing assistance service should be recommending suitable attractors and starters.

Limit the prediction depth to less than three levels.

There is practically no need to go beyond three levels of depth in predicting the appropriate attractors and starters. This implies from the empirical evidence obtained when analyzing knowledge workers' browsing interactions. The knowledge workers divided their browsing tasks into three subtasks – on average. Their browsing sessions thus contained three subsequences. Each subsequence has its starter s_i and attractor a_i. Consider the following generic session:

$$s_1 \xrightarrow{\ 1\ } a_1, s_2 \xrightarrow{\ 2\ } a_2, s_3 \xrightarrow{\ 3\ } a_3$$

where the numbers above the right arrows denote the depth. Assume the user is at the beginning of a session, that is, the point s_1. The desired elements in the first depth level are a_1, s_2; in the second level a_2, s_3; and in the third only a_3. The recommendation set $r = \{a_1, s_2, a_2, s_3, a_3\}$ would be sufficient for the whole session, in principle. Hence, to cover the generic session, it is sufficient to limit the prediction depth to less than or equal to three. It may be practical to focus just on the next level, since when the user reaches the desired attractor or starter, the recommendations on the next level attractors or starters will be provided again. This strategic design consideration may lead to computationally more efficient and scalable algorithms.

The fast responsiveness of the assistance system should also be among the high-priority issues. It has been observed that the knowledge workers have relatively short attention span in the electronic environments. The extended waiting times may result in negative browsing experiences. The secondary effect of unfavorable experiences leads to relatively low usability perceptions. The responsiveness factor directly relates to the computational efficiency. The recommendation algorithm of the assistance system should be computationally inexpensive.

14.3.3 Recommendation Algorithm Derivation

The design of the recommendation algorithm for browsing assistance system utilizes the presented strategic concepts and accounts for the essential system requirements. The recommendations are provided on the starter and attractor pages. The system aims at supplying a list of viable resources comprising of both starter and attractor pages. The recommendations are based on the first-level predictions.

The recommendation algorithm has several phases. First, it identifies the reached navigation point. If the point is starter and/or attractor, it proceeds to the generation of the initial recommendation set. The initial recommendation set is generated in two stages (see Fig. 14.2). In the first stage, a set of top-n elements according to the appropriate starter–attractor or attractor–starter mappings is generated. The selected top-n points are used as seeds for the second-stage expansion. The two-stage process produces the initial set of $n(1 + m)$ elements. The initial set contains an appropriate mix of starters and attractors. The final recommendation set is selected from the initially generated set. The elements in the initial set are ranked with respect to the ordering function. Then, the set of the highest ranking points is chosen.

Recall that a navigation point can be starter, attractor, singleton, simple point, or any combination of these. The two-stage generation of the initial recommendation set varies depending on whether the detected navigation point is starter, attractor, or both. If the point is both starter and attractor, it is prioritized as a starter. The details of the algorithm for the relevant cases are described in the following paragraphs.

Assume the reached navigation point is a starter s. The algorithm maps the starter s to the set of attractors, $\omega(s)$, according to the starter–attractor mapping ω, $s \rightarrow \omega(s)$. The top-n attractors, $\omega^{(n)}(s)$, are selected from the set $\omega(s)$. The selection is done with respect to the suitable ranking/ordering function. The selected

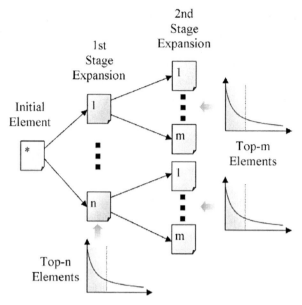

Fig. 14.2 Illustration of the two-stage generation of the initial recommendation set

top-n attractors in $\omega^{(n)}(s)$ are used for generating n additional sets by the attractor–starter mapping. The corresponding set of top-m starters, $\psi^{(m)}(a_i)$, is obtained for each attractor $a_i \in \omega^{(n)}(s)$, and the collection of the resulting points is the union: $u_s = \bigcup_{a_i \in \omega^{(n)}(s)} \psi^{(m)}(a_i)$. The initial recommendation set $r(s)$ is then obtained as the union of $\omega^{(n)}(s)$ and u_s, that is, $r(s) = \omega^{(n)}(s) \cup u_s$. It is intentionally larger than the required final recommendation set. Hence, the set $r(s)$ undergoes further selection. The subset $r^{(w)}(s)$, having the best w elements of $r(s)$, is chosen according to the proper ordering function.

Analogous process is repeated when the user reaches the attractor navigation point. Given the attractor a, the top-n set, $\psi^{(n)}(a)$, is generated according to the attractor–starter mapping ψ. Sampling of $\psi(a)$ is done with respect to the given ordering. This is the first-stage expansion: $a \rightarrow \psi^{(n)}(a)$. The obtained top-n set, $\psi^{(n)}(a)$, is used for the second-stage expansion. The corresponding sets of the top-m attractors, $\omega^{(m)}(s_i)$, are derived for each starter $s_i \in \psi^{(n)}(a)$ and the collection of the resulting points is the union: $u_a = \bigcup_{s_i \in \psi^{(n)}(a)} \omega^{(m)}(s_i)$. The initial recommendation set, $r(s)$, is then again obtained as the union: $r(a) = \psi^{(n)}(a) \cup u_a$. The acquired initial recommendation set, $r(a)$, is correspondingly sampled. The top w elements are selected according to the ordering function. The resulting final recommendation set, $r^{(w)}(a)$, is then obtained.

The important element of the algorithm is the right choice of the ordering function f. The function should provide qualitatively appropriate ranking of the navigation points. In addition, it should be computationally inexpensive, in order to enable on-the-fly recommendations and scalability of the algorithm.

The suitable ordering function is the relative frequency. The navigation points are evaluated according to their relative utilization frequency detected during the knowledge worker interactions. This facilitates the reuse of the analytic data and efficient implementation. It also permits easy extensions to various domains of definition. As knowledge workers utilize the intranet portal resources more frequently, the relative frequency becomes more accurate and convergent.

Multiple categories and multiplicity of navigation points present a slight difficulty. The sets of starters, attractors, and singletons are not necessarily disjunct. This raises an important question: How to compute the relative frequency of a point that has been detected as starter, attractor, and singleton (or any valid combination of these)? Simple and effective solution to this problem is to compute the average of the applicable relative frequencies:

$$f(p) = avrg(f_S(p) + f_A(p) + f_Z(p)) \; ; \; f_S(p) \neq 0, \; f_A(p) \neq 0, \; f_Z(p) \neq 0 \quad (14.1)$$

where f_S denotes the starter relative frequency, f_A stands for the attractor relative frequency, and f_Z indicates the singleton relative frequency. This evaluation accounts for the average combined relative frequency value of a point.

At this stage we are ready to present the complete algorithm. Simplified, but intuitively understandable, flowchart illustration of the derived recommendation algorithm for browsing assistance system is presented in Fig. 14.3. At the beginning, the reached navigation point p is examined. If the point p is neither starter nor attractor, the algorithm exits. In parallel with the point examination, the initial parameters are set:

n – the first level expansion range,
m – the second level expansion range,
w – the recommendation window size.

If the point p has been detected to be a starter, the algorithm calculates the appropriate recommendation set $r^{(w)}(p)$. This applies also to the case where point p has been identified as both, starter and attractor. It is then preferentially treated as a starter. If the reached navigation point p is identified as an attractor, the algorithm calculates the recommendation set $r^{(w)}(p)$ accordingly. The averaged relative frequency ordering function (14.1) is employed in all cases. The obtained recommendation set r of the size w is then suitably presented to the user on the fly at the given page p.

14.4 Practical Evaluation

The evaluation of the introduced recommendation algorithm for browsing assistance system has been performed on the real-world data of a large-scale organizational information system. The information system incorporates an intranet portal providing access to a large number of organizational resources and services. The primary users of the portal are skilled knowledge workers. The knowledge workers

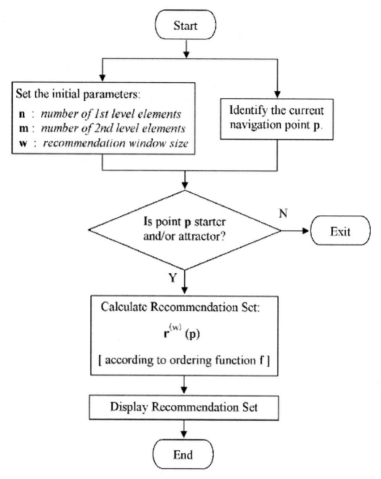

Fig. 14.3 Intuitive flowchart presentation of recommendation algorithm

are significantly diverse in terms of interaction characteristics, accessed resources, and utilization style. We start with a concise introduction of the intranet portal and then proceed to the evaluation of the algorithm.

14.4.1 Intranet Portal

The target intranet portal of this study is a large-scale system implemented at the National Institute of Advanced Industrial Science and Technology. The system is significantly complex and distributed. The intranet portal is a gateway to a large number of resources and services. Its web core comprises of six servers connected to the high-speed backbone in a load-balanced configuration. The user access points are located on the local subnets. The subnet infrastructures range from high-speed optical to wireless.

The institute has a number of branches throughout the country. Thus, the services and resources are decentralized and distributed. The intranet portal incorporates a rich set of resources such as documents (in various formats), multimedia, software. The extensive spectrum of the implemented services supports the institutional business processes; management of cooperation with industry, academia, and other institutes; localization of internal resources; etc. Blogging and networking services within organization are also implemented. The visible web space is in excess of 1 GB, and the deep web space is considerably larger; however, it is difficult to estimate its size due to the distributed architecture and alternating back-end data.

The intranet portal traffic is considerable – primarily during the working hours. Knowledge worker interactions on the portal are recorded by the web servers and stored in web logs. The web log data is voluminous and contains relatively rich information about the knowledge workers' browsing features and portal utilization. The web logs, however, contain both human- and machine-generated traffic (e.g., automated software and hardware monitoring systems, crawlers and indexers, download managers). The data needs preprocessing and cleaning before the human traffic can be extracted and analyzed. The preprocessing, elimination of the machine-generated traffic, and segmentation of the detected human interactions into sessions and subsequences have been presented in [5] and are not detailed here. The resulting working data together with the essential portal statistics are described in Table 14.1.

Table 14.1 Information and basic data statistics of the organizational intranet portal

Web log volume	~60 GB
Average daily volume	~54 MB
Number of servers	6
Number of log files	6814
Average file size	~9 MB
Time period	1 year
Log records	315,005,952
Resources	3,015,848
Sessions	3,454,243
Unique sessions	2,704,067
Subsequences	7 335 577
Unique subsequences	3,547,170
Valid subsequences	3,156,310
Unique valid subsequences	1,644,848
Users	~10,000

14.4.2 System Evaluation

The presented recommendation algorithm has been evaluated using the processed data of the large-scale target intranet portal. The main goal of the evaluation has been to examine the correctness of the algorithm's recommendations given the actual interactions of knowledge workers during their browsing experiences. The recommendation correctness of the algorithm has been tested for various sizes of the recommendation window.

The segmentation of the knowledge worker interactions on the portal has been performed. The essential navigation points (i.e., starters and attractors) have been extracted from the identified sessions and subsequences. The relative frequency values for the detected starters and/or attractors have been calculated using the obtained access data. All the processed data was stored in a database, in order to facilitate efficient storage, retrieval, and manipulation.

The individual users were associated with the distinct IP addresses. The set of detected unique IP addresses contained both statically and dynamically assigned addresses. Smaller number of the distinct IP addresses were static and larger number of the addresses were dynamic. This was due to widespread use of dynamic addressing in the organization. It should be noted that the exact identification of the individual users was generally not possible for dynamically assigned IP addresses. However, the detected IP address space proportionally corresponded to the number of portal users.

We identified IP addresses with more than 50 sessions originating from them. This represents approximately at least once per week interaction activity on the intranet portal. There were 8739 such IPs. A random sample of subsequences originating from these addresses was obtained. Ten subsequences were selected from each IP address. The test points were selected from the subsequence samples. If the test point was a starter, the desired target was the corresponding attractor of the original subsequence. In case of the attractor test point, the desired target was the starter of the consecutive subsequence. The testing set consisted of the pairs (p, y): point $p \rightarrow$ desired target y. The cardinality of the testing set was 87,390.

Given the navigation point p_i in the testing set $\{(p_i, y_i)\}_i$, the introduced algorithm generated the recommendation set $r^{(w)}(p_i)$. The generated recommendation set, $r^{(w)}(p_i)$, was scanned for the corresponding desired target element y_i. If the set $r^{(w)}(p_i)$ contained the actual desired point y_i, the recommendation was considered correct, otherwise it was considered incorrect. The correctness of the recommendation algorithm was measured by a simple indicator function of y_i on $r^{(w)}(p_i)$.

The recommendation correctness of the introduced algorithm was evaluated for different sizes of the recommendation window $w \in\ <1, 30>$. The range $<1, 10>$ was examined with the increment 1 and the range $<10, 30>$ with the increment 5. The first- and the second-stage expansion parameters were set as $m = n = 5$. Thus, the cardinality of the initial recommendation set r was $|r| = n(1 + m) = 30$. The top-w candidates, $r^{(w)}$, were selected from the initial recommendation set according to the averaged combined relative frequency (14.1). The obtained correctness results are graphically presented in Fig. 14.4.

Correctness [%]

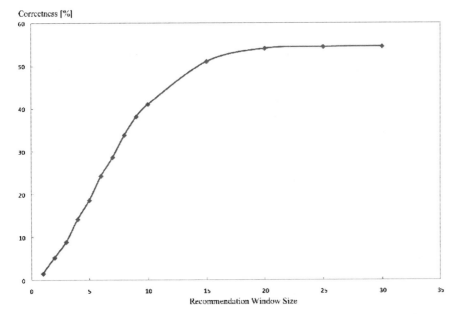

Fig. 14.4 Recommendation correctness evaluated for varying sizes of recommendation window

The recommendation performance of the derived algorithm was rising approximately linearly up to the window size ten. In this range the correctness, as a function of window size w, indicated the steepest gain. At the window size $w = 10$, the correctness was approximately 41% – which is significant. Then the recommendation correctness of the algorithm started saturating. The saturating range is noticeable between the window size values $w \in\, < 10, 20 >$. The recommendation correctness at $w = 20$ was over 54%. The algorithm's performance started stabilizing from window size values greater than 20. The performance gains in the window size interval $w \in\, < 20, 30 >$ were relatively minor.

14.4.3 Practical Implications and Limitations

The results indicate that the appropriate size of the recommendation window is between 10 and 20, $w \in\, < 10, 20 >$. The performance of the algorithm in this range is around 50% and the number of recommended items is not excessive. In practice, this may be the suitable range for window size. It represents a reasonable balance between the recommendation correctness and the variety of choices. Expanding the recommendation window size beyond 20 is not justifiable on the performance grounds. It does not offer a viable increase in recommendation correctness given the extra choices. Depending on the implementation and application, it may also offset the computational cost.

The range for the recommendation window (between 10 and 20) offers a sufficient space for adjustability according to other characteristics of user interactions. The users with a short attention span may prefer less recommendations, whereas the more exploratory users may appreciate more recommendations. The inpatient users may be well served by 10 recommendations, while the exploratory ones even 20.

The recommendation window size adjustments may be managed by users or by adaptive methods. The system can incorporate individual or user group profiles for adjusting the recommendation window size. Naturally, this optionality is in practice strongly influenced by the target implementation and by users themselves. If the automated and/or adaptive user profiling is applicable (given privacy, legal, and other concerns), it may be a suitable functionality option. In practice, it is often the best to provide users with the appropriate choices.

The presented browsing assistance system may be applicable more broadly than just within the organizational intranet portals. Numerous web sites and other portals have similar characteristics. It is reasonable to presume that behavior of their users displays similar features. Hence, the system design and implementation can be transferable directly or with minor adjustments.

The limiting aspect of the presented approach, in our opinion, is that it utilizes the most frequent navigation points for generating and sampling the recommendation set. Human behavior in electronic environments is characterized by the long-tail distributions [32, 33]. The long tails may extend to over 90% of the elements. This suggests that there is potentially a significant amount of information in the long tails that can be explored. Combining the information extracted from the long tails with the current approach may further improve the recommendation correctness of the system.

14.5 Conclusions and Future Work

A novel recommendation algorithm for browsing assistance systems has been presented. The algorithm provides recommendations on the desirable resources during the browsing interactions. It benefits users by shortening the navigation paths and the browsing time. The algorithm is computationally efficient and scalable.

The recommendation algorithm utilizes a priori knowledge of human interactions in digital environments. The human browsing behavior is divided into the activity segments reflecting the complexness and the temporal dynamics of the interactions. Sessions and subsequences are obtained. The sessions represent larger segments comprising smaller elemental segments – subsequences. The initial navigation points of the subsequences – starters – and the ending points – attractors – are the pages where users pay the greatest attention. The intermediate points are essentially transitional. Users pass through them rapidly. Hence, the starters and attractors are the most appropriate navigation points for providing browsing assistance. The recommendation algorithm offers the selected set of points constructed from the potentially desirable attractors and starters.

The algorithm has been evaluated on a real-world data of a large-scale organizational information system. The primary users were skilled knowledge workers. The performance of the algorithm was examined for varying sizes of the recommendation window – ranging from 1 to 30. The detected optimal range was between 10 and 20. The recommendation correctness of the algorithm in this range was $50\pm5\%$. The algorithm indicated satisfactory performance.

The future work will target further improvements in the recommendation correctness. Two initial dimensions shall be explored: personalization and mining the long tails of observed web interaction attributes of users. The personalization domain presents an opportunity for utilizing the observed behavior analytics to create behaviormetric user profiles. The profiles may be used for generating and sampling the recommendation sets. More challenging task is mining the long tails. Information extracted from the long tails, in combination with the presented approach and user profiling, may be beneficial for improving the assistance system in terms of both recommendation correctness and user friendliness.

Acknowledgments The authors would like to thank Tsukuba Advanced Computing Center (TACC) for providing raw web log data.

References

1. M. Alvesson. *Knowledge Work and Knowledge-Intensive Firms*. Oxford University Press, Oxford, 2004.
2. T.H. Davenport. *Thinking for a Living – How to Get Better Performance and Results from Knowledge Workers*. Harvard Business School Press, Boston, MA, 2005.
3. D. Sullivan. *Proven Portals: Best Practices for Planning, Designing, and Developing Enterprise Portal*. Addison-Wesley, Boston, MA, 2004.
4. H. Collins. *Enterprise Knowledge Portals*. Amacom, New York, NY, 2003.
5. P. Géczy, S. Akaho, N. Izumi, and K. Hasida. Knowledge worker intranet behaviour and usability. *International Journal of Business Intelligence and Data Mining*, 2:447–470, 2007.
6. G. Adomavicius and A. Tuzhilin. Toward the next generation of recommender systems: A survey of the state-of-the-art and possible extensions. *IEEE Transactions on Knowledge and Data Engineering*, 17:734–749, 2005.
7. S. Perugini, M.A. Gonçalves, and E.A. Fox. Recommender systems research: A connection-centric survey. *Journal of Intelligent Information Systems*, 23(2):107–143, 2004.
8. R. Burke. Hybrid recommender systems: Survey and experiments. *User Modeling and User-Adapted Interaction*, 12(4):331–370, 2002.
9. O. Nasraoui, C. Cardona, and C. Rojas. Using retrieval measures to assess similarity in mining dynamic web clickstreams. In *Proceedings of KDD*, pp. 439–448, Chicago, IL, USA, 2005.
10. K. Ali and S.P. Kechpel. Golden path analyzer: Using divide-and-conquer to cluster web clickstreams. In *Proceedings of KDD*, pp. 349–358, Washington, DC, USA, 2003.
11. N. Kammenhuber, J. Luxenburger, A. Feldmann, and G. Weikum. Web search clickstreams. In *Proceedings of The 6th ACM SIGCOMM on Internet Measurement*, pp. 245–250, Rio de Janeriro, Brazil, 2006.
12. E. Agichtein, E. Brill, and S. Dumais. Improving web search ranking by incorporating user behavior information. In *Proceedings of The 29th SIGIR*, pp. 19–26, Seattle, WA, USA, 2006.
13. R.J.K. Jacob and K.S. Karn. Eye tracking in human-computer interaction and usability research: Ready to deliver the promises. In J. Hyona, R. Radach, and H. Deubel, editors, *The*

Mind's Eye: Cognitive and Applied Aspects of Eye Movement Research, pp. 573–605, Elsevier Science, Amsterdam, 2003.

14. L.A. Granka, T. Joachims, and G. Gay. Eye-tracking analysis of user behavior in www search. In *Proceedings of The 27th SIGIR*, pp. 478–479, Sheffield, United Kingdom, 2004.

15. Y-H. Park and P.S. Fader. Modeling browsing behavior at multiple websites. *Marketing Science*, 23:280–303, 2004.

16. R.E. Bucklin and C. Sismeiro. A model of web site browsing behavior estimated on click-stream data. *Journal of Marketing Research*, 40:249–267, 2003.

17. M. Ahuya, B. Gupta, and P. Raman. An empirical investigation of online consumer purchasing behavior. *Communications of the ACM*, 46:145–151, 2003.

18. W.W. Moe. Buying, searching, or browsing: Differentiating between online shoppers using in-store navigational clickstream. *Journal of Consumer Psychology*, 13:29–39, 2003.

19. M. Deshpande and G. Karypis. Selective markov models for predicting web page accesses. *ACM Transactions on Internet Technology*, 4:163–184, 2004.

20. H. Wu, M. Gordon, K. DeMaagd, and W. Fan. Mining web navigaitons for intelligence. *Decision Support Systems*, 41:574–591, 2006.

21. J.D. Martín-Guerrero, P.J.G. Lisboa, E. Soria-Olivas, A. Palomares, and E. Balaguer. An approach based on the adaptive resonance theory for analysing the viability of recommender systems in a citizen web portal. *Expert Systems with Applications*, 33(3):743–753, 2007.

22. I. Zukerman and D.W. Albrecht. Predictive statistical models for user modeling. *User Modeling and User-Adapted Interaction*, 11:5–18, 2001.

23. L.M. de Campos, J.M. Fernández-Luna, and J.F. Huete. A collaborative recommender system based on probabilistic inference from fuzzy observations. *Fuzzy Sets and Systems*, 159(12):1554–1576, 2008.

24. M. Zanker and S. Gordea. Recommendation-based browsing assistance for corporate knowledge portals. In *Proceedings of the 2006 ACM Symposium on Applied Computing*, pp. 1116–1117, New York, NY, USA, 2006. ACM.

25. J. Jozefowska, A. Lawrynowicz, and T. Lukaszewski. Faster frequent pattern mining from the semantic web. *Intelligent Information Processing and Web Mining, Advances in Soft Computing*, pp. 121–130, 2006.

26. M. Shyu, C. Haruechaiyasak, and S. Chen. Mining user access patterns with traversal constraint for predicting web page requests. *Knowledge and Information Systems*, 10(4): 515–528, 2006.

27. P. Symeonidis, A. Nanopoulos, A.N. Papadopoulos, and Y. Manolopoulos. Collaborative recommender systems: Combining effectiveness and efficiency. *Expert Systems with Applications*, 34(4):2995–3013, 2008.

28. Z. Dezso, E. Almaas, A. Lukacs, B. Racz, I. Szakadat, and A.-L. Barabasi. Dynamics of information access on the web. *Physical Review*, E73:066132(6), 2006.

29. A.-L. Barabasi. The origin of bursts and heavy tails in human dynamics. *Nature*, 435: 207–211, 2005.

30. L. Catledge and J. Pitkow. Characterizing browsing strategies in the world wide web. *Computer Networks and ISDN Systems*, 27:1065–1073, 1995.

31. M.V. Thakor, W. Borsuk, and M. Kalamas. Hotlists and web browsing behavior – An empirical investigation. *Journal of Business Research*, 57:776–786, 2004.

32. P. Géczy, S. Akaho, N. Izumi, and K. Hasida. Long tail attributes of knowledge worker intranet interactions. (P. Perner, Ed.), *Machine Learning and Data Mining in Pattern Recognition*, pp. 419–433, Springer-Verlag, Heidelberg, 2007.

33. A.B. Downey. Lognormal and pareto distributions in the internet. *Computer Communications*, 28:790–801, 2005.

Chapter 15
Using Web Text Mining to Predict Future Events: A Test of the Wisdom of Crowds Hypothesis

Scott Ryan and Lutz Hamel

Abstract This chapter describes an algorithm that predicts events by mining Internet data. A number of specialized Internet search engine queries were designed to summarize results from relevant web pages. At the core of these queries was a set of algorithms that embody the wisdom of crowds hypothesis. This hypothesis states that under the proper conditions the aggregated opinion of a number of nonexperts is more accurate than the opinion of a set of experts. Natural language processing techniques were used to summarize the opinions expressed from all relevant web pages. The specialized queries predicted event results at a statistically significant level. It was hypothesized that predictions from the entire Internet would outperform the predictions of a smaller number of highly ranked web pages. This hypothesis was not confirmed. This data replicated results from an earlier study and indicated that the Internet can make accurate predictions of future events. Evidence that the Internet can function as a wise crowd as predicted by the wisdom of crowds hypothesis was mixed.

15.1 Introduction

This chapter describes an extension of a system that predicts future events by mining Internet data [14]. Mining Internet data is difficult because of the large amount of data available. It is also difficult because there is no simple way to convert text into a form that computers can easily process. In the current chapter a number of search engine queries were crafted and the results were counted in order to summarize the text of all of the web pages that are indexed by the Yahoo! search engine. At first glance it may seem unwise to include the opinions of all writers, as opposed to the opinions of experts only. The Internet is very open and anyone can write anything

Scott Ryan · Lutz Hamel
University of Rhode Island, Kingston, RI, USA,
e-mail: scott.ryan@uconn.edu;hamel@cs.uri.edu

R. Stahlbock et al. (eds.), *Data Mining,* Annals of Information Systems 8,
DOI 10.1007/978-1-4419-1280-0_15, © Springer Science+Business Media, LLC 2010

without having credentials. It may seem better to rely on a smaller number of web pages that are well respected. A recent book entitled *The Wisdom of Crowds: Why the Many Are Smarter Than the Few and How Collective Wisdom Shapes Business, Economies, Societies and Nations* [16] draws on decades of research in psychology and behavioral economics to suggest that, given certain assumptions, experts give inferior answers when compared to the averaged answers of a large crowd.

An excellent example of the accuracy of a large group occurs when one is trying to guess a quantity, such as an individual's weight or the number of jelly beans in a jar. One striking example from Surowiecki [16] was a contest to guess the weight of an ox. There were approximately 800 guesses, and a scientist computed the average of all of the guesses. The average of the guesses was 1197 pounds, and the actual weight of the ox was 1198 pounds. This aggregate guess was better than any of the 800 individual guesses. This demonstrates the idea behind the wisdom of crowds hypothesis: The group as a whole can be very accurate even if no individual in the group is very accurate. The core idea is that some people will be too high, others too low, but in the end these biases will cancel out and an accurate measure will emerge.

Another obvious example of the wisdom of crowds is an open market. Many economists believe that open markets, such as stock or commodities markets, are so accurate that it is impossible to predict future prices. This is the well-known "efficient market hypothesis" [12]. The efficient market hypothesis states that because each market participant has some information about what the price of an asset should be, when these people all participate in the market they combine their information to discover the correct price. Because the group knows what the current price should be, the asset cannot be overvalued or undervalued, so its future price cannot be determined.

It is important to note that a crowd is not always more accurate than an expert. Specific conditions must be present [16]. If a great deal of expertise is required then the expert may outperform the crowd. For example, if a decision about the result of a complex physics experiment were required, an expert may outperform a large crowd. In a chess match, a world champion would probably beat a random crowd of 1000 people that voted on every move. A crowd tends to be most wise when it is similar to a random sample of a population. In statistics the idea of the random sample is that if one randomly selects people from a population, one should get a diverse, representative group. When a crowd is making a decision, in order to avoid bias, diversity of opinion is very important. Each person should use some private information, even if it is only their personal interpretation of publicly known facts. Another factor that tends to make the crowd wise is independence. If individuals' opinions are determined by people around them, then the crowd may simply represent the opinion of the most persuasive member.

The basic measure used to summarize web pages in the current study is counting results from Internet search engines. Counting Internet search results has received little attention from the computer science community. Most research has involved studying the relationship between merit and the number of results returned by a Google search [3, 15]. Bagrow and his coauthors studied the relationship between the number of publications a scientist has produced and the number of search results that were returned by Google. A total of 449 scientists were randomly chosen from the fields of condensed matter and statistical physics. The searches took the form of "Author's name" AND "condensed matter" OR "statistical physics" OR "statistical mechanics." The relationship between the number of search results and the number of publications in an electronic archive was found to be linear with an R^2 value of approximately 0.53. This result implies that there is a relationship between the number of publications and the number of results retrieved from an Internet search engine.

The study discussed in this chapter is an extension of an earlier study that used the wisdom of crowds hypothesis to mine Internet data [14]. The study used natural language processing techniques to summarize the opinions expressed on all relevant web pages. The core of this summary was a count of how many web pages made a given prediction. This system attempted to predict economic indicators, sporting events, and elections. The hypothesis was that results from Internet search queries would correlate with the results of the events studied. Algorithms based on computational linguistics were used to produce counts summarizing predictions. The counts were then correlated with the actual results. For example, if most web pages expressed the opinion that the New York Yankees would win the World Series, then the New York Yankees should win. For the election and sporting event data, the web search results correlated significantly with the results of the events. The economic data did not correlate significantly with the web counts, possibly because the economic data was too dynamic to be predicted by web pages that did not change as quickly as the economic data. The current study extends the predictions to other areas and replicates previous results.

15.2 Method

15.2.1 Hypotheses and Goals

The goal of this project is to apply the wisdom of crowds hypothesis to the Internet. The hypothesis is that results from Internet search queries will correlate with the predictions of an open market and with the results of the events at a level significantly greater than zero. A previous study [14] attempted to predict sporting events, economic data, and the US elections. The current study will replicate the sporting events data. There were no events comparable to the 2006 congressional and gubernatorial elections, so these results were not replicated. The economic web results were not correlated significantly with the actual results or market data, so there was no reason to attempt to replicate the economic results. A great deal has been written recently concerning the Internet and popular culture. With many people able to edit the Internet directly using sites such as myspace.com, many individuals are able to express their opinions. Popular culture, by definition, will be written about a great deal. Much has been written about the fact that more votes are cast for reality show contestants than presidential candidates. With such a great deal of information available, we will be attempting to predict popular culture events. The popular culture events we are attempting to predict are reality television program winners. These events were chosen because they are popular culture contests that have a clear winner.

The general methodology used in this chapter is to try to predict the outcome of events by counting and integrating results from a series of Internet search queries. In all cases these search counts will be compared to the actual results of the event being predicted. In the case of the results from reality television programs and sporting events, the results will be compared to Internet betting markets. The betting market prediction is often expressed in probabilities. For example, the counts could be compared to the sports betting market, which will assign a certain team a higher probability of winning an event such as the Super Bowl. The sports betting market, like most open markets, is assumed by many to be efficient [11]. Therefore the web count prediction is unlikely to outperform or even perform equally to any market, but may be expected to make similar predictions.

The wisdom of crowds hypothesis makes a specific prediction. The prediction is not simply that the crowd will be "accurate" because that is very difficult to define operationally. The more specific hypothesis is that under certain conditions the crowd will be wiser than a smaller number of experts. To test this hypothesis, the first 20 search results were examined in order to determine the opinion of the experts. This group of experts is referred to as the "web top 20." In Internet search, the results that are returned first are supposed to have a higher "page rank," indicating more expertise [7]. Therefore, these results may be representative of a small group of experts. The web top 20 were compared to the results from the search of the entire Internet. If a large crowd is wiser than a smaller number of experts, then the counts for the entire Internet should be more predictive of an event than the counts for only the top 20 websites. This hypothesis may be suspect because the top Internet search results themselves are determined by all available websites. Page rank is mostly determined by how many web pages link to a given site [7]. Because of this the web top 20 may already incorporate the wisdom of the entire Internet. If that is the case then we would expect a statistically significant correlation between the web counts measure and the web top 20 measure. If the web top 20 is a measure of the wisdom of crowds rather than the experts, then this will not be an adequate test of experts vs. crowd.

Because the algorithms used in this chapter have a great deal of noise associated with them, the hypothesis is that the web count predictions will outperform a chance level prediction at a statistically significant level. Any predictions should be more accurate than a chance prediction but certainly not close to 100% accuracy. In summary, the main hypothesis is that the correlations between the Internet counts and the market data, and the correlations between the Internet counts and the actual results, will be significantly greater than zero at the $p < 0.05$ level. A secondary hypothesis is that the counts from the entire Internet should outperform the counts from only the top 20 results.

15.2.2 General Methodology

Web searches were performed with the Yahoo! search engine [24]. The Yahoo! search web services API was used along with the Java programming language in order to automate the search process [25]. One of the problems with counting Internet search results is that the dates of creation for most web pages are not available [18]. To deal with this problem, searches were also performed on the Yahoo! News website. Yahoo! News searches provide the exact date and time of the publication of each result [28]. The news searches gave results no more than 1 month old. It may be suggested that if the news dates are so accurate, then only the news results should be used. Unfortunately, the number of results from news searches is very low, so the general web search was used in order to be assured that the number of results would not often be zero.

Other details of the methodology used are specific to the area that is being predicted.

15.2.3 The 2006 Congressional and Gubernatorial Elections

We attempted to predict the results of all of the US Senate races, all of the gubernatorial races, and all of the House of Representatives races considered "key races" by CNN [10]. We also attempted to predict all of the House of Representatives races in the states with the seven largest number of House seats: California, Texas, New York, Florida, Ohio, Pennsylvania, and Illinois. If CNN reported a candidate as running unopposed then the race was not included in the study. The data was taken from CNN websites [8, 9]. Two candidates were selected to be studied for each race. The two candidates chosen were the ones most likely to win according to prediction market data [17].

The first part of making election predictions was determining which phrases to use in order to determine that someone was expressing the idea that a candidate would win. For example, in the case of Hillary Clinton, possible phrases could be "Clinton will win," "Clinton will win the seat," or "Hillary Clinton will win the Senate seat." More details of this process can be found in [14]. The two final phrases that were selected were simply "*name* will win" and "*name* will beat." These phrases allow for a large number of false positives, but the hope was that there would be enough of a signal to be detected above the noise. Most of the work involved in determining the phrases to be used was done manually because it could not be completely automated. In some cases entire paragraphs needed to be read and understood in context in order to determine if the proposition that a candidate would win was being expressed.

The nature of this election counting system does not allow us to measure the web top 20 for elections, because we are simply searching for counts of phrases such as "Clinton will win," rather than actually asking the question "Who will win the New York Senate race?"

15.2.4 Sporting Events and Reality Television Programs

The methodology for predicting sporting events and reality television programs was the same because in both cases one is trying to predict who will win a particular event. Automating the data gathering for these events relied heavily on examples from the "question answering" literature [13]. The basic algorithm for predicting sporting and reality television contests is given in Fig. 15.1 and described further in [14]. Sample data is provided in Table 15.1. The first column of Table 15.1 is the baseball team, followed by the probability of winning the World Series according to the sports betting market, then by the web and news counts, and finally by the actual finishing position of the team.

Algorithm:

searchQuery = *"will win"* + targetEvent
for counter = *1 to 200*
priorWords = *three words prior to* searchQuery
newPhrase = priorWords + searchQuery
parse newPhrase
properNounArray[counter]= *firstProperNoun*(newPhrase)
end for
get all unique properNouns
for each uniqueProperNoun + searchQuery
nounCountArray = *count of web search results*
end for
nounCountMax = *maximum*(nounCountArray)
for each nounCount
if(nounCount < *1000 and* nounCount < *0.01* * nounCountMax)
    delete nounCount *from* nounCountArray
end if
end for
result = nounCountArray

Fig. 15.1 Basic algorithm for predicting sporting and reality television contests

Table 15.1 Sample data for the World Series winner

Team	Market probability	Web	News	Finishing position
NY YANKEES	0.50	2580	2	5
NY METS	0.20	328	1	3
MIN TWINS	0.10	0	1	5
SD PADRES	0.09	0	0	5

The basic algorithm for determining the counts of the top 20 ranked websites is given in Fig. 15.2 and described in the following paragraphs. The algorithm is similar to the one described previously for determining the web counts.

Algorithm:

searchQuery = *"will win"* + targetEvent
counter = *0*
while counter < *20*
priorWords = *three words prior to* searchQuery
newPhrase = priorWords + searchQuery
parse newPhrase
if (newPhrase *contains a proper noun*)
   counter++
   if(nounCountArray *contains* properNoun)
      increment nounCountArray[properNoun *position*]
   else
      add properNoun *to* nounCountArray
   end if
end if
end while
result = nounCountArray

Fig. 15.2 Basic algorithm for determining the counts of the top 20 ranked websites

This algorithm starts with a search phrase such as "will win the Super Bowl." It then searches through all of the results until 20 proper nouns have been found and counts each instance of a proper noun. The results will not simply be the top 20 search results, but the top 20 that specifically mention a proper noun. Theoretically it could take hundreds of results to get 20 proper nouns. The output is similar to the output for the sports algorithm for the entire Internet.

The sporting events that were predicted are described in [14]. The reality television programs that were predicted were "The Bachelor," "America's Next Top Model," "The Amazing Race," "The Biggest Loser," "Dancing With the Stars," "Survivor: Cook Island," and "Project Runway." All of these shows aired between August and December of 2006. Results were taken from Wikipedia [23, 22] and ABC.com [1]. Results were based on when individuals were eliminated from the contests. Along with attempting to predict the results of the programs, there was an attempt to predict the probabilities of winning based on the betting markets. The probabilities of winning were taken from Bodog.com [5].

15.2.5 Movie Box Office Receipts and Music Sales

The methodology for predicting music sales and movie box office receipts is very similar and therefore the two processes are described together. This methodology is the most simple and most subject to noise. The test is simply whether the mere mention of a movie or music album will make it more likely to be successful. By its nature this data does not have any consensus or market prediction to use as a comparison, and it also is not amenable to the format of gauging the top 20 results. Therefore the only comparison will be to the actual album and movie sales. The hypothesis is that the movie and album web result counts will be correlated with their sales.

For movies, the Yahoo! Movies website [26] was searched to determine which movies that were opening in "wide release." These searches were done on Monday in order to predict the movies that were starting on the following Wednesday or Friday. Unfortunately the Yahoo! search API is limiting in that it cannot combine phrases in quotes with other words, such as "Casino Royale" + movie. Therefore the search queries used were simply the movie name in quotes. The names of the movies were searched and the results were counted for the web in general and the news for the month. Table 15.2 displays sample movie data.

Table 15.2 Sample movie count data

Movie	Web	News
Casino Royale	7,240,000	1427
Happy Feet	4,390,000	517
Let's Go To Prison	3,750,000	66

The relationship between the web and news counts and the amount of money generated by the movies in the opening weekend was studied. The box office money intake was taken from the Yahoo! Movies website [27].

For music albums, the "Amazon.com: New and Future Releases: Music" website [2] was used to determine which albums were being released. The albums were converted into the form *album artist*, such as "There Is A Season The Byrds." These queries were then searched and the numbers of results were counted for the web in general and the news results. The relationship between the web and the news counts and the appearance on the Billboard 200 [4] chart ranking the week after the release was studied. Only the finishers ranking in the top 10 of the Billboard 200 were noted. All data were collected weekly from September 23, 2006, to January 21, 2007.

15.2.6 Replication

In order to further test the techniques used in the prior sections, more data was gathered after all of the preceding data had been analyzed. The new data was analyzed in the same way as the previous data had been analyzed. This paradigm is similar to that used in data mining. In data mining one often trains a model on a certain data set and then tests the model on another data set. Replication is also important because it is an integral part of the scientific process. For the sports and reality television contests, only the market data, as opposed to the actual results, were predicted. Using the market data allowed us to analyze the data immediately rather than waiting for all of the events to actually occur.

The sports results analyzed were the NBA finals of professional basketball, the Stanley Cup championship of professional hockey, and the national championship of college basketball. All of these events were from 2007. The queries used were "will win the NBA finals," "will win the Stanley Cup," and "will win the NCAA tournament." There was a greater challenge with this sports data than with the earlier sports data because there is no popular name for the NBA finals or the NCAA tournament. This is in contrast to the earlier events predicted: the World Series, the Super Bowl, and the BCS. All of these queries were searched on March 14, 2007. The betting odds were taken from VegasInsider.com [19, 21, 20].

The reality television programs that were predicted in this replication test were the versions of "American Idol" and "The Apprentice" that were in progress during March, 2007. The data was sampled on March 14, 2007. The betting odds were taken from Bodog.com [6].

The replicated movie and music data were taken weekly between January 29, 2007 and March 12, 2007.

15.3 Results and Discussion

15.3.1 The 2006 Congressional and Gubernatorial Elections

Table 15.3 displays the correlations between the news and web results and the outcomes of the congressional and gubernatorial elections. The "corr." column indicates the correlation between the results and measure named in the first column. The "N" column is the number of observations, and the last two columns are the 95% upper and lower limits of the confidence interval for the correlations.

Table 15.3 Predicting event results and market probabilities

	Corr.	N	95% c.i. lower	95% c.i. upper
Election results				
Web	0.27	478	0.19	0.35
News	0.18	158	0.02	0.32
Market	0.89	162	0.85	0.92
Election market				
Web	0.49	162	0.36	0.60
News	0.33	80	0.12	0.51

The idea behind these correlations is that if a great deal of web pages made a prediction, then the event should occur, and the market should assign a high probability to the event occurring. All of the results confirmed the primary hypothesis.

The correlations are significantly different from zero because the confidence intervals do not include zero. As seen in the row labeled "market," the correlation was highest between the market probabilities and the actual events.

Earlier it was mentioned that there was a great deal of noise in the queries that were used to test whether a candidate would win. In order to lessen this noise, the top 50 search results were examined manually to determine which ones referred to winning the election and which ones did not. There were 495 searches done. Examining 50 results from each would result in 24,750 manual examinations. Rather than doing all of these examinations, the results were broken down by the total number of search results into deciles. The results were broken down by the total number of search results because it was expected that the candidates with the highest number of search results would contain the most noise. For each of these deciles, the three candidates whose number of search results was closest to the averages of each of the deciles were examined. Table 15.4 displays the name of the candidate, the total number of web search results, the number of results that correctly expressed

Table 15.4 Percent correctly identifying candidate

Name	Web count	Number correct	Sample	Prob. correct
Johnson	1815	0	50	0
White	1856	0	50	0
Clinton	1858	2	50	0.04
Roberts	455	0	50	0
Menendez	454	47	50	0.94
Rounds	491	0	50	0
Ehrlich	199	38	50	0.76
Massa	197	9	50	0.18
Cardin	195	50	50	1
Courtney	101	9	40	0.23
Palin	100	31	35	0.89
Shannon	100	0	38	0
Bean	56	16	30	0.53
Sweeney	54	22	30	0.73
Roth	54	0	20	0
Akaka	39	20	20	1
Giffords	40	27	27	1
Snowe	40	33	33	1
Kuhl	27	15	17	0.88
Roskam	26	14	16	0.88
Pombo	26	35	35	1
Lantos	10	4	4	1
Farr	10	0	4	0
Melancon	10	16	16	1
Bilirakis	4	3	3	1
Tubbs-Jones	4	3	3	1
Lipinski	4	2	3	0.67
LaTourette	1	2	2	1
Regula	1	2	2	1
Altmire	1	4	4	1

the opinion that the candidate would win, the number of results examined, and the percentage that correctly expressed the opinion that the candidate would win.

The correlation between the total results and the percentage correct was –0.56, which was statistically significant. This confirmed the hypothesis that those with higher counts had more false positives. For example, "Johnson will win" often referred to a boxer winning a fight or a driver winning a race, "White will win" often referred to the "white" color in chess winning the match, and "Clinton will win" referred to Hillary Clinton winning the 2008 presidential nomination. An examination of the data indicated that there was a large increase in accuracy at the count of 26. The accuracy of the decile representing a result count of 26 or lower was 0.87. The accuracy of the deciles above 26 was 0.35. The value at the midpoint of this decile and the one above, which was 41, was also tested. The average value of the percentage correct for result counts below 41 was 0.89. The average value of the percentage correct for result counts 41 or above was 0.34. Therefore the noise for the results below 41 was less than the noise of the results 41 or above. With less noise present, we expected to have more accurate predictions when examining only the candidates with result counts below 41.

Table 15.5 displays the correlations between the web measures and the election results and the election prediction market including races in which both candidates had result counts less than or greater than 41.

Table 15.5 Predicting event results and market probabilities

	Corr.	N	95% c.i. lower	95% c.i. upper
Election results				
≤ 41 web	0.46	124	0.30	0.58
> 41 web	0.19	354	0.08	0.30
Market data				
≤ 41 web	0.79	20	0.53	0.91
> 41 web	0.44	142	0.30	0.56

For the market data and the actual results, there was a marginally statistically significant difference between those with a count above 41 and those with a count below or equal to 41. As predicted, eliminating some of the noise in the election data led to an improvement in accuracy.

15.3.2 Sporting Events and Reality Television Programs

Table 15.6 displays the correlations between the news and web results and the outcomes of sporting and reality television contests.

The correlations for the sports results are negative because those with the highest counts should have the lowest position; for example, first place is considered

Table 15.6 Predicting event results and market probabilities

	Corr.	N	95% c.i. lower	95% c.i. upper
Sports results				
Web	−0.38	119	−0.52	−0.21
News	−0.29	119	−0.45	−0.12
Web Top 20	−0.48	119	−0.61	−0.33
Market	−0.62	119	−0.72	−0.50
Sports market				
Web	0.55	119	0.41	0.66
News	0.47	119	0.32	0.60
Web Top 20	0.44	119	0.29	0.58
Replicated sports market				
Web	0.26	64	0.01	0.47
News	0.41	64	0.18	0.60
Web Top 20	0.41	64	0.18	0.60
Reality television results				
Web	−0.45	13	−0.80	0.13
Web Top 20	−0.59	13	−0.86	−0.06
Market	−0.84	13	−0.95	−0.55
Reality television market				
Web	0.56	13	0.01	0.85
Web Top 20	0.75	13	0.34	0.92
Replicated reality TV market				
Web	0.88	12	0.62	0.97

position number one. For the reality television programs, all of the news counts were zero, so the news correlations could not be computed.

All of the results confirmed the primary hypothesis, which was that the various web count measures would predict the event results. The correlations for the sports and reality television contests were significantly different from zero. As seen in the rows labeled "market," the correlations were highest between the market probabilities and the actual events. The results all replicated successfully, with all of the replicated web and news counts significantly greater than zero.

Contrary to our secondary hypothesis, the web top 20 count correlation was slightly higher than the web count in four out of five cases although not significantly higher. Because the confidence intervals overlapped, there is no direct evidence that the web top 20 outperformed the entire web. It was mentioned in the Methodology section that the web top 20 and the web counts may actually be measuring similar phenomena because the web top 20 already incorporated information from the entire web. There was evidence that this was the case. The correlations between the web counts and the web top 20 were 0.68 (sports), 0.77 (sports replication), and 0.94 (reality). This is evidence that the web top 20 already incorporates some of the information available on the rest of the web.

15.3.3 *Movie and Music Album Results*

In order to eliminate some of the noise in the movie and music data, if the web count was over 5 million for the movies or 50,000 for the music albums then the top 50 results were inspected manually in order to determine how many of the results actually referred to the movie. This sample was used to determine the signal-to-noise ratio. If most of the observations from this sample did not refer to the movie, then many of the total results may not have referred to the movie. If the number of correct movie mentions was below 40 out of 50, then the data was excluded. This same technique was used in the replication phase to test whether this technique was valid and not simply a post hoc overfitting.

Table 15.7 displays the correlation between the web and news month search result counts and the amount of money generated in the first weekend of a movie's release.

Table 15.7 Correlation between movie web counts and money earned (first weekend of a movie's release)

	Corr.	N	95% c.i. lower	95% c.i. upper
Original				
Web	0.40	36	0.08	0.65
News	0.26	36	−0.07	0.54
Replication				
Web	0.69	12	0.19	0.91
News	0.66	12	0.14	0.90

The web count data was a statistically significant predictor of box office success because the confidence interval for the correlation does not include zero. The correlation for the news was positive but not significantly greater than zero. The correlations for the replication were even higher than the original correlations and further indicated that the counts were significant predictors of movie success.

Table 15.8 displays the correlation between the web and news month search counts and the position of the album on the Billboard 200 chart.

Table 15.8 Correlation between music album web counts and chart position (Billboard 200)

	Corr.	N	95% c.i. lower	95% c.i. upper
Original				
Web	−0.45	93	−0.60	−0.27
News	−0.54	93	−0.67	−0.38
Replication				
Web	−0.50	56	−0.67	−0.27
News	−0.51	56	−0.68	−0.29

These correlations are negative because a lower position is more indicative of success. For example, chart position number one is the best seller. These results indicate that the web count and news month data are statistically significant predictors of the position of an album on the Billboard 200 charts because the confidence intervals for the correlations do not include zero.

Overall these results are similar to those for the movies. The relationship between the counts and the success of the albums is somewhat strong. The replication correlations were significantly greater than zero.

15.4 Conclusion

The evidence collected for this project indicates that the Internet can be used to make predictions that are more accurate than chance levels. The web search results and the news search results correlated significantly with the actual results and the market data. The highest correlations were between the market predictions and the actual events, which is a confirmation of the wisdom of crowds hypothesis and the efficient market hypothesis.

The hypothesis that the predictions of the entire web would outperform the predictions of the top 20 websites was not supported, and there was mild evidence that the web top 20 outperformed the entire web. The general prediction of the wisdom of crowds is that a "large" crowd will outperform a "small" number of experts. It could be the case in the current study that 20 experts represented a large enough crowd to contain the aggregating advantages of a crowd. It is possible that a slightly large number of experts are better than an even larger number of nonexperts, which is in contradiction to the wisdom of crowds. The current study does not support all of the facets of the wisdom of crowds hypothesis. However, Section 15.2 of this chapter reviewed evidence that the wisdom of crowds is actually used to choose the top websites. Therefore, the wisdom of crowds may already be represented by the web top 20. Given the results of this chapter, the evidence that the Internet conforms well to the wisdom of crowds hypothesis is mixed. This research indicates that a new hypothesis may be tested. The hypothesis would be that the top 50% of websites on a given topic are the most accurate, followed by all of the websites on a given topic, and then by the single top website on a given topic.

There is a great deal of other future work that could be done in this research area. Future research could further automate and generalize the techniques used in this chapter. More specific queries could be used, and more advanced computational linguistics techniques could eliminate some of the false positives and false negatives that were encountered in the searches. The techniques that were used could also be used in a more general manner, predicting the outcomes of a large number of different events.

Although this data has told us a great deal about how the Internet can be mined to make predictions, it tells us even more about the Internet's reliability. Because the Internet, or at least a small subset of the Internet, appears to be able to operate as an efficient market and a wise crowd, it tells us that the Internet shares some of the traits of a wise crowd. First, it tells us that the opinions on the Internet are diverse. Second, it tells us that the opinions on the Internet are independent of other opinions. Finally, and most importantly, it tells us that the Internet as a whole appears to contain accurate information that can be used to predict future events.

References

1. ABC (2007) Dancing with the Stars. Available:
   `http://abc.go.com/indexCited01Feb2007`
2. Amazon.com (2007) New and Future Releases: Music. Available:
   `http://www.amazon.com/New-Future-Releases-Music/b/ref=sv_m_2?ie=UTF8`
   `\&node=465672Cited31Jan2007`
3. Bagrow JP, Rozenfeld HD, Bollt EM Ben-Avraham, D (2004) How famous is a scientist? -Famous to those who know us. Europhys Lett 67(4):511-516
4. Billboard.com (2007) Billboard Album Charts - Top 100 Albums - Music Retail Sales. Available:
   `http://www.billboard.com/bbcom/charts/chart_display.jsp?g=Albums\&f=`
   `The+Billboard+200Cited31Jan2007`
5. Bodog.com (2007) Television and Movie Betting at Bodog Sportsbook. Available:
   `http://www.bodog.com/sports-betting/tv-film-movie-props.jspCited01Feb2007`
6. Bodog.com (2007) Television and Movie Betting, American Idol Odds at Bodog Sportsbook. Available:
   `http://www.bodog.com/sports-betting/tv-film-movie-props.jspCited01Feb2007`
7. Brin S, Page L (2007) The Anatomy of a Large-Scale Hypertextual Web Search Engine. Available:
   `http://infolab.stanford.edu/~backrub/google.htmlCited24Jan2007`
8. CNN (2006) CNN.com - Elections 2006. Available:
   `http://www.cnn.com/ELECTION/2006/pages/results/Senate/Cited21Dec2006`
9. CNN (2006) CNN.com - Elections 2006. Available:
   `http://www.cnn.com/ELECTION/2006/pages/results/governor/Cited21Dec2006`
10. CNN (2006) CNN.com - Elections 2006. Available:
    `http://www.cnn.com/ELECTION/2006/pages/results/house/Cited21Dec2006`
11. Debnath S, Pennock DM, Giles CL, Lawrence S (2003) Information incorporation in online in-game sports betting markets. In: Proceedings of the 4th ACM conference on electronic commerce. ACM, New York
12. Fama EF (1965) Random Walks in Stock Market Prices. Financial Anal J September/October
13. Gelbukh A (2006) Computational Linguistics and Intelligent Text Processing. Springer, Berlin
14. Pion S, Hamel L (2007) The Internet Democracy: A Predictive Model Based on Web Text Mining. In: Stahlbock R et al. (eds) Proceedings of the 2007 International Conference on Data Mining. CSREA Press, USA
15. Simkin MV, Roychowdhury VP (2006) Theory of Aces: Fame by chance or merit? Available:
    `http://www.citebase.org/cgibin/fulltext?format=application/pdf\&identifier=`
    `oai:arXiv.org:cond-mat/0310049Cited28Sep2006`
16. Surowiecki J (2004) The Wisdom of Crowds: Why the Many Are Smarter Than the Few and How Collective Wisdom Shapes Business, Economies, Societies and Nations. Doubleday Publishing, Westminster, MD
17. TradeSports.com (2006) Available:
    `http://www.tradesports.com/aav2/trading/tradingCited10Oct2006`

18. Tyburski G (2006) It's Tough to Get a Good Date with a Search Engine. Available:
    http://searchenginewatch.com/showPage.html?page=2160061Cited16Dec2006
19. VegasInsider.com (2007) College Basketball Future Book Odds at VegasInsider.com, the
    leader in Sportsbook and Gaming information - College Basketball Odds, College Basketball
    Futures, College Basketball Future Odds. Available:
    http://www.vegasinsider.com/college-basketball/odds/futures/Cited14Mar2007
20. VegasInsider.com (2007) College Basketball Future Book Odds at VegasInsider.com, the
    leader in Sportsbook and Gaming information - NHL Odds, NHL Futures, Pro Hockey Odds,
    Pro Hockey Futures. Available:
    http://www.vegasinsider.com/nhl/odds/futures/Cited14Mar2007
21. VegasInsider.com (2007) NBA Future Odds at VegasInsider.com, The Leader in Sportsbook
    and Gaming Information - NBA Odds, NBA Futures, NBA Future Odds. Available:
    http://www.vegasinsider.com/nba/odds/futures/index.cfm#1479Cited14Mar2007
22. Wikipedia (2007) Project Runway. Available:
    http://en.wikipedia.org/wiki/Future_tenseCited01Feb2007
23. Wikipedia (2007) Survivor: Cook Islands. Available:
    http://en.wikipedia.org/wiki/Survivor:_Cook_IslandsCited01Feb2007
24. Yahoo! (2006) Available:
    http://www.yahoo.com/Cited16Dec2006
25. Yahoo! (2006) Yahoo! search web services. Available:
    http://developer.yahoo.com/search/Cited16Dec2006
26. Yahoo! Movies (2007) Yahoo! Movies - In Theaters This Weekend. Available:
    http://movies.yahoo.com/feature/thisweekend.htmlCited31Jan2007
27. Yahoo! Movies (2007) Yahoo! Movies - Weekend Box Office and Buzz. Available:
    http://movies.yahoo.com/mv/boxoffice/Cited31Jan2007
28. Yahoo! News (2006) Available:
    http://news.search.yahoo.com/news/search?fr=sfp\&ei=UTF-8\&p=testCited16
    Dec2006

Part VI
Privacy-Preserving Data Mining

Chapter 16
Avoiding Attribute Disclosure with the (Extended) p-Sensitive k-Anonymity Model

Traian Marius Truta and Alina Campan

Abstract Existing privacy regulations together with large amounts of available data created a huge interest in data privacy research. A main research direction is built around the k-anonymity property. Several shortcomings of the k-anonymity model were addressed by new privacy models such as p-sensitive k-anonymity, l-diversity, (α, k)-anonymity, t-closeness. In this chapter we describe two algorithms (*GreedyPKClustering* and *EnhancedPKClustering*) for generating (extended) p-sensitive k-anonymous microdata. In our experiments, we compare the quality of generated microdata obtained with the mentioned algorithms and with another existing anonymization algorithm (*Incognito*). Also, we present two new branches of p-sensitive k-anonymity, the constrained p-sensitive k-anonymity model and the p-sensitive k-anonymity model for social networks.

16.1 Introduction

The increased availability of individual data combined with today's significant computational power and the tools available to analyze this data, have created major privacy concerns not only for researchers but also for the public [16] and legislators [3]. Privacy has become an important aspect of regulatory compliance, and the ability to automate the privacy enforcement procedures would lead to reduced cost for enterprises. Policies must be developed and modeled to describe how data has to be stored, accessed, manipulated, processed, managed, transferred, and eventually deleted in any organization that stores confidential data. Still, many of these aspects of data management have not been rigorously analyzed from a privacy perspective [15].

Traian Marius Truta · Alina Campan
Department of Computer Science, Northern Kentucky University, Highland Heights, KY 41099, USA, e-mail: `trutat1@nku.edu;campana1@nku.edu;ford1@nku.edu`

R. Stahlbock et al. (eds.), *Data Mining,* Annals of Information Systems 8,
DOI 10.1007/978-1-4419-1280-0_16, © Springer Science+Business Media, LLC 2010

Data privacy researchers have presented several techniques that aim to avoid the disclosure of confidential information by processing sensitive data before public release ([1, 23], etc.). Among them, the *k*-anonymity model was recently introduced [18, 19]. This model requires that in the *released* (also referred as *masked*) *microdata* (data sets where each tuple belongs to an individual entity, e.g., a person, a company) every tuple will be undistinguishable from at least *k*–1 other tuples with respect to a subset of attributes called *key* or *quasi-identifier* attributes.

Although the model's properties and the techniques used to enforce it on data have been extensively studied ([2, 5, 10, 18, 20], etc.), recent results have shown that *k*-anonymity fails to protect the privacy of individuals in all situations ([13, 21, 24], etc.). New enhanced privacy models have been proposed in the literature to deal with *k*-anonymity's limitations with respect to *sensitive attributes disclosure* [9]. These models include *p*-sensitive *k*-anonymity [22] with its expansion called extended *p*-sensitive *k*-anonymity [6], *l*-diversity [13], (α, *k*)-anonymity [24], *t*-closeness [12], *m*-confidentiality [25], personalized anonymity [26], etc.

In this chapter we describe two algorithms, called *GreedyPKClustering* [7] and *EnhancedPKClustering* [22], that anonymize a microdata set such that its released version will satisfy *p*-sensitive *k*-anonymity. We tailored both algorithms to also generate extended *p*-sensitive *k*-anonymous microdata. We compare the results obtained by our algorithms with the results produced by the *Incognito* algorithm [10], which was adapted to generate *p*-sensitive *k*-anonymous microdata.

Additionally, new branches developed out of the *p*-sensitivity *k*-anonymity model are presented. The first of these two new extensions, called the constrained *p*-sensitive *k*-anonymity model, allows quasi-identifiers generalization boundaries to be specified and *p*-sensitive *k*-anonymity is achieved within the imposed boundaries. This model has the advantage of protecting against identity and attribute disclosure, while controlling the microdata modifications within allowed boundaries. The other new *p*-sensitive *k*-anonymity extension targets the social networks field. A social network can be anonymized to comply with *p*-sensitive *k*-anonymity model, and this model will provide protection against disclosure of confidential information in social network data.

The chapter is structured as follows. Section 16.2 presents the *p*-sensitive *k*-anonymity model, the extended *p*-sensitive *k*-anonymity model, and the anonymization algorithms. Section 16.3 contains an extensive set of experiments. The new branches of *p*-sensitive *k*-anonymity model are defined in Section 16.4. This chapter ends with conclusions and future work directions (Section 16.5).

16.2 Privacy Models and Algorithms

16.2.1 The p-Sensitive k-Anonymity Model and Its Extension

P-sensitive *k*-anonymity is a natural extension of *k*-anonymity that avoids several shortcomings of this model [21]. Next, we present these two models.

Let IM be the initial data set (called initial microdata). IM is described by a set of attributes that are classified into the following three categories:

- $I_1, I_2, \ldots, I_m$ are identifier attributes such as *Name* and *SSN* that can be used to identify a record.
- $K_1, K_2, \ldots, K_q$ are key or quasi-identifier attributes such as *ZipCode* and *Sex* that may be known by an intruder.
- $S_1, S_2, \ldots, S_r$ are confidential or sensitive attributes such as *Diagnosis* and *Income* that are assumed to be unknown to an intruder.

In the released data set (called *masked microdata* and labeled MM) only the quasi-identifier and confidential attributes are preserved; identifier attributes are removed as a prime measure for ensuring data privacy. In order to rigorously and succinctly express the *k*-anonymity property, we use the following concept.

Definition 16.1 (*QI-Cluster*). Given a microdata, a *QI-cluster* consists of all the tuples with identical combination of quasi-identifier attribute values in that microdata.

We define *k*-anonymity based on the minimum size of all *QI*-clusters.

Definition 16.2 (*k-Anonymity Property*). The *k-anonymity property* for a MM is satisfied if every *QI*-cluster from MM contains *k* or more tuples.

Unfortunately, *k*-anonymity does not provide the amount of confidentiality required for every individual [12, 18, 21]. *k*-anonymity protects against identity disclosure [8] but fails to protect against attribute disclosure [8] when all tuples of a *QI*-cluster share the same value for one sensitive attribute [18].

The *p*-sensitive *k*-anonymity model considers several sensitive attributes that must be protected against attribute disclosure. It has the advantage of simplicity and allows the data owner to customize the desired protection level by setting various values for *p* and *k*.

Definition 16.3 (*p-Sensitive k-Anonymity Property*). A MM satisfies the *p-sensitive k-anonymity property* if it satisfies *k*-anonymity and the number of distinct values for each confidential attribute is at least *p* within every *QI*-cluster from MM.

To illustrate this property, we consider the masked microdata from Table 16.1 where *Age* and *ZipCode* are quasi-identifier attributes, and *Diagnosis* and *Income* are confidential attributes.

This masked microdata satisfies the 3-anonymity property with respect to *Age* and *ZipCode*. The first *QI*-cluster (the first three tuples in Table 16.1) has two different incomes (*60,000* and *40,000*) and only one diagnosis (*AIDS*): therefore, the highest value of *p* for which *p*-sensitive 3-anonymity holds is 1. As a result, an intruder who searches information about a young person in his twenties that lives in zip code area 41,099 will discover that the target entity suffers from *AIDS*, even if he does not know which tuple in the first *QI*-cluster corresponds to that person. This attribute disclosure problem can be avoided if one of the tuples from the first

Table 16.1 Masked microdata example for p-sensitive k-anonymity property

Age	ZipCode	Diagnosis	Income
20	41099	AIDS	60,000
20	41099	AIDS	60,000
20	41099	AIDS	40,000
30	41099	Diabetes	50,000
30	41099	Diabetes	40,000
30	41099	Tuberculosis	50,000
30	41099	Tuberculosis	40,000

QI-cluster would have a value other than *AIDS* for the *Diagnosis* attribute. In this case, both QI-clusters would have two different illnesses and two different incomes, and, as a result, the highest value of p would be 2.

P-sensitive k-anonymity cannot be enforced on any given IM, for any p and k. Two necessary conditions to generate a masked microdata with p-sensitive k-anonymity property are presented in [22].

This privacy model has a shortcoming related to the "closeness" of the sensitive attribute values within a QI-cluster. To present this situation, we consider the value generalization hierarchy for a sensitive attribute as defined by Sweeney [19]. We use such a hierarchy for the sensitive attribute *Illness* in the following example. We consider that the information that a person has *cancer* (not a leaf value in this case) needs to be protected, regardless of the cancer type she has (*colon cancer*, *prostate cancer*, *breast cancer* are leaf nodes in this generalization hierarchy). If the p-sensitive k-anonymity property is enforced for the released microdata, it is possible that for one QI-cluster all of the *Illness* attribute values are to be descendants of the *cancer* node, therefore leading to disclosure. To avoid such situations, the extended p-sensitive k-anonymity model was introduced [6].

We use the notation $\mathcal{H}_S$ to represent the value generalization hierarchy for the sensitive attribute S. We assume that the data owner has the following requirements in order to release a masked microdata:

- All ground (leaf) values in $\mathcal{H}_S$ must be protected against disclosure.
- Some non-ground values in $\mathcal{H}_S$ must be protected against disclosure.
- All the descendants of a protected non-ground value in $\mathcal{H}_S$ must also be protected.

The following definitions allow us to rigorously define the extended p-sensitive k-anonymity property.

Definition 16.4 (*Strong Value*). A protected value in the value generalization hierarchy $\mathcal{H}_S$ of a confidential attribute S is called **strong** if none of its ascendants (including the root) is protected.

Definition 16.5 (*Protected Subtree*). We define a **protected subtree** of a hierarchy $\mathcal{H}_S$ as a subtree in $\mathcal{H}_S$ that has as root a strong protected value.

Definition 16.6 (*Extended p-Sensitive k-Anonymity Property*). The masked microdata $\mathcal{MM}$ satisfies **extended *p*-sensitive *k*-anonymity property** if it satisfies *k*-anonymity and, for each *QI*-cluster from $\mathcal{MM}$, the values of each confidential attribute *S* within that group belong to at least *p* different protected subtrees in $\mathcal{H}_S$.

At a closer look, extended *p*-sensitive *k*-anonymity is equivalent to *p*-sensitive *k*-anonymity where the confidential attribute values are generalized to their first protected ancestor starting from the hierarchy root (their strong ancestor). Consequently, in order to enforce extended *p*-sensitive *k*-anonymity to a data set, the following two-step procedure can be applied:

- Each value of a confidential attribute is generalized (temporarily) to its strong ancestor.
- Any algorithm which can be used for *p*-sensitive *k*-anonymization is applied to the modified data set. In the resulted masked microdata the original values of the confidential attributes are restored.

The microdata obtained following these steps satisfy the extended *p*-sensitive *k*-anonymity property. Due to this procedure, the algorithms from the next section refer only to *p*-sensitive *k*-anonymity. In the experiments related to the extended model, we applied the above-mentioned procedure.

16.2.2 Algorithms for the p-Sensitive k-Anonymity Model

Besides achieving the properties required by the target privacy model (*p*-sensitive *k*-anonymity or its extension), anonymization algorithms must also consider minimizing one or more cost measure. We know that optimal *k*-anonymization is a NP-hard problem [2]. By simple reduction to *k*-anonymity, it can be easily shown that *p*-sensitive *k*-anonymization is also a NP-hard problem. The algorithms we will describe next are good approximations of the optimal solution.

The microdata *p*-sensitive *k*-anonymization problem can be formulated as follows.

Definition 16.7 (*p-Sensitive k-Anonymization Problem*). Given a microdata $\mathcal{IM}$, the *p*-sensitive *k*-anonymization problem for $\mathcal{MM}$ is to find a partition $S = \{cl_1, cl_2, \ldots, cl_v\}$ of $\mathcal{IM}$, where $cl_j \in \mathcal{IM}$, $j = 1..v$, are called clusters and: $\bigcup_{j=1}^{v} cl_j = \mathcal{IM}$; $cl_i \cap cl_j = \emptyset$, $i, j = 1..v, i \neq j$; $|cl_j| \geq k$ and cl_j is *p*-sensitive, $j = 1..v$; and a cost measure is optimized.

Once a solution S to the above problem is found for a microdata $\mathcal{IM}$, a masked microdata $\mathcal{MM}$ that is *p*-sensitive *k*-anonymous is formed by generalizing the quasi-identifier attributes of all tuples inside each cluster of S to the same values. The generalization method consists in replacing the actual value of an attribute with a less specific, more general value that is faithful to the original [19].

We call *generalization information* for a cluster the minimal covering tuple for that cluster, and we define it as follows.

Definition 16.8 (Generalization Information). Let $cl = \{r_1, r_2, \ldots, r_q\} \in S$ be a cluster, $KN = \{N_1, N_2, \ldots, N_s\}$ be the set of numerical quasi-identifier attributes and $KC = \{C_1, C_2, \ldots, C_t\}$ be the set of categorical quasi-identifier attributes. The **generalization information of cl**, w.r.t. quasi-identifier attribute set $\mathcal{K} = KN \cup KC$ is the "tuple" $gen(cl)$, having the scheme $\mathcal{K}$, where

- For each categorical attribute $C_j \in \mathcal{K}$, $gen(cl)[C_j]$ = the lowest common ancestor in $\mathcal{H}_{C_j}$ of $\{r_1[C_j], \ldots, r_q[C_j]\}$, where $\mathcal{H}_C$ denotes the hierarchies (domain and value) associated to the categorical quasi-identifier attribute C.
- For each numerical attribute $N_j \in \mathcal{K}$, $gen(cl)[N_j]$ = the interval $[min\{r_1[N_j], \ldots, r_q[N_j]\}, max\{r_1[N_j], \ldots, r_q[N_j]\}]$.

For a cluster cl, its generalization information $gen(cl)$ is the tuple having as value for each quasi-identifier attribute the most specific common generalized value for all the attribute values from cl. In $\mathcal{MM}$, each tuple (its quasi-identifier part) from the cluster cl will be replaced by $gen(cl)$, and thus forming a QI-cluster.

There are several possible cost measures that can be used as optimization criterion for the p-sensitive k-anonymization problem ([4, 5], etc.). A simple cost measure is based on the size of each cluster from S. This measure, called *discernibility metric (DM)* [4], assigns to each record x from $\mathcal{IM}$ a penalty that is determined by the size of the cluster containing x:

$$DM(S) = \sum_{j=1}^{v} |cl_j|^2 \qquad (16.1)$$

LeFevre introduced the alternative measure called *normalized average cluster size metric (AVG)* [11]:

$$AVG(S) = \frac{n}{v \cdot k} \qquad (16.2)$$

where n is the size of the $\mathcal{IM}$, v is the number of clusters, and k is as in k-anonymity. It is easy to notice that the AVG cost measure is inversely proportional to the number of clusters, and minimizing AVG is equivalent to maximizing the total number of clusters.

Another cost measure described in the literature is the *information loss (IL)* caused by generalizing each cluster to a common tuple [5].

While k-anonymity is satisfied for each individual cluster when its size is k or more, the p-sensitive property is not so obvious to achieve. For this, two diversity measures that quantify, with respect to sensitive attributes, the *diversity between a tuple and a cluster* and the *homogeneity of a cluster* were introduced [22].

The *GreedyPKClustering* algorithm is briefly described below. A complete presentation including a pseudocode-like algorithm can be found in [7].

The QI-clusters are formed one at a time. For forming one QI-cluster, a tuple in $\mathcal{IM}$ not yet allocated to any cluster is selected as a seed for the new cluster. Then the algorithm gathers tuples to this currently processed cluster until it satisfies both requirements of the p-sensitive k-anonymity model. At each step, the current cluster grows with one tuple. This tuple is selected, of course, from the tuples not yet allocated to any cluster. If the p-sensitive part is not yet satisfied for the current cluster, then the chosen tuple is the one most probable to enrich the diversity of the current cluster with regard to the confidential attribute values. This selection is made by the diversity measure between a tuple and a cluster. If the p-sensitive part is already satisfied for every confidential attribute, then the least different or diverse tuple (w.r.t. the confidential attributes) of the current cluster is chosen. This selection is justified by the need to spare other different confidential values, not present in the current cluster, in order to be able to form as many as possible new p-sensitive clusters. When a tie happens, i.e., multiple candidate tuples exist conforming to the previous selection criteria, then the tuple that minimizes the cluster's IL growth will be preferred.

It is possible that the last constructed cluster will contain less than k tuples or it will not satisfy the p-sensitivity requirement. In that case, this cluster needs to be dispersed between the previously constructed groups. Each of its tuples will be added to the cluster whose IL will minimally increase by that tuple addition. At the end, a solution for p-sensitive k-anonymity problem is found.

The *EnhancedPKClustering* algorithm is an alternative solution for the p-sensitive k-anonymization problem. It considers AVG (or the partition cardinality) that has to be maximized as the cost measure. Its complete presentation can be found in [22].

This algorithm starts by enforcing the p-sensitive part using the properties proved for the p-sensitive k-anonymity model [22]. The tuples from IM are distributed to form p-sensitive clusters with respect to the sensitive attributes. After p-sensitivity is achieved, the clusters are further processed to satisfy k-anonymity requirement as well. A more detailed description of how the algorithm proceeds follows.

In the beginning, the algorithm determines the p-sensitive equivalence classes [22], orders the attributes based on the harder to make sensitive relation [22], and computes the value *iValue* that divides the p-sensitive equivalence classes into two categories: one with less frequent values for the hardest to anonymize attribute and one with more frequent values. Now, the QI-clusters are created using the following steps:

- First, the tuples in the least frequent category of p-sensitive equivalence classes are divided into *maxClusters* clusters (maximum possible number of clusters can be computed in advance based on frequency distributions of sensitive attributes [21]) such that each cluster will have *iValue* tuples with unique values within each cluster for the harder to make sensitive attribute [22].
- Second, the remaining p-sensitive equivalence classes are used to fill the clusters such that each of them will have exactly p tuples with p distinct values for S_1.
- Third, the tuples not yet assigned to any cluster are used to add diversity for all remaining sensitive attributes until all clusters are p-sensitive. If no tuples are available, some of the less diverse (more homogenous) clusters are removed and their tuples are reused for the remaining clusters. At the end of this step all clusters are p-sensitive.

- Fourth, the tuples not yet assigned to any cluster are used to increase the size of each cluster to k. If no tuples are available, some of the less populated clusters are removed and their tuples are reused for the remaining clusters. At the end of this step all clusters are p-sensitive k-anonymous.

Along all the steps, when a choice is to be made, one or more optimization criteria are used (diversity between a tuple and a cluster, and increase in information loss).

While both of these algorithms achieve the p-sensitive k-anonymous data sets, their approach is different. *GreedyPKClustering* is an extension of the *greedy_k_member_clustering* [6], a clustering algorithm used for k-anonymity, and *Enhanced-PKClustering* is a novel algorithm that takes advantage of the p-sensitive k-anonymity model properties and does not have an equivalent for k-anonymity only.

16.3 Experimental Results

16.3.1 Experiments for p-Sensitive k-Anonymity

In this section we compare the performance of *EnhancedPKClustering*, *Greedy-PKClustering*, and an adapted version of *Incognito* [10].

The first two algorithms are explained in the previous section, and *Incognito* is the first efficient algorithm that generates a k-anonymous data set. This algorithm finds a full-domain generalization that is k-anonymous by creating a multi-domain generalization lattice for the domains of the quasi-identifiers attributes. Starting with the least general domain at the root of the lattice, the algorithm performs a breadth-first search, checking whether each generalization encountered satisfies k-anonymity. This algorithm can be used to find a single (weighted) minimal generalization, or it can be used to find the set of all k-anonymous minimal domain generalizations [10]. We easily adapted this algorithm by testing for p-sensitive k-anonymity (instead of k-anonymity) at every node in the generalization lattice.

All three algorithms have been implemented in Java, and tests were executed on a dual CPU machine running Windows 2003 Server with 3.00 GHz and 1 GB of RAM.

A set of experiments has been conducted for an IM consisting of 10,000 tuples randomly selected from the *Adult* data set [17]. In all the experiments, we considered *age*, *work-class*, *marital-status*, *race*, *sex*, and *native-country* as the set of quasi-identifier attributes; and *education-num*, *education*, and *occupation* as the set of confidential attributes. Among the quasi-identifier attributes, *age* was numerical, and the other five attributes were categorical. The value generalization hierarchies of the quasi-identifier categorical attributes were as follows: for *work-class*, *race*, and *sex* two-level hierarchies (i.e., ground level and root level); for *marital-status* a three-level hierarchy; and for *native-country* a four-level hierarchy. The value hierarchy for the *native-country* quasi-identifier attribute, the most complex among the hierarchies for all our quasi-identifiers, is depicted in Fig. 16.1.

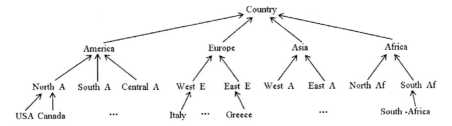

Fig. 16.1 The value hierarchy for the quasi-identifier categorical attribute *Country*

P-sensitive *k*-anonymity was enforced with respect to all six quasi-identifier attributes and all three confidential attributes. Figures 16.2 and 16.3 show comparatively the *AVG* and *DM* values of the three algorithms, *EnhancedPKClustering*, *GreedyPKClustering*, and *Incognito*, produced for *p* = 3, respectively, *p* = 10, and different *k* values. As expected, the results for the first two algorithms clearly outperform *Incognito* results in all cases. We also notice that *EnhancedPKClustering*

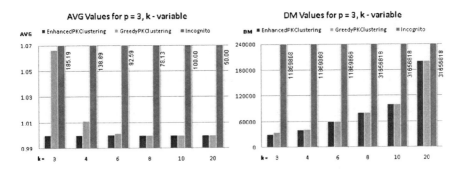

Fig. 16.2 *AVG* and *DM* for *EnhancedPKClustering*, *GreedyPKClustering*, and *Incognito*, *p*=3 and *k* variable

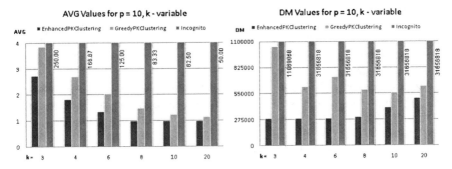

Fig. 16.3 *AVG* and *DM* for *EnhancedPKClustering*, *GreedyPKClustering*, and *Incognito*, *p*=10 and *k* variable

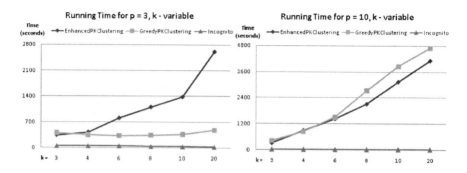

Fig. 16.4 Running time for *EnhancedPKClustering*, *GreedyPKClustering*, and *Incognito* algorithms

is able to improve the performances of the *GreedyPKClustering* algorithm in cases where solving the *p*-sensitivity part takes prevalence over creating clusters of size *k*.

Figure 16.4 shows the time required to generate the masked microdata by all three algorithms, for *p* = 3, respectively, *p* = 10, and different *k* values. Since *Incognito* uses global recording and our domain generalization hierarchies for this data set have a low height, its running time is very fast. The *GreedyPKClustering* is faster than the new algorithm for small values of *p*, but when it is more difficult to create *p*-sensitivity within each cluster the *EnhancedPKClustering* has a slight advantage.

Based on these results, it is worth noting that a combination of *GreedyPKClustering* (for low values of *p*, in our case 3) and *EnhancedPKClustering* (for high values of *p*, in our experiment 10) would be desirable in order to improve both running time and the selected cost measure (*AVG* or *DM*).

16.3.2 Experiments for Extended p-Sensitive k-Anonymity

The *EnhancedPKClustering* and *GreedyPKClustering* algorithms can easily be adapted to generate extended *p*-sensitive *k*-anonymous microdata. In order to do so, the algorithms are applied to a modified IM in which the sensitive attributes are replaced with their strong ancestors. In the resulting MM the sensitive attributes are restored to their original values.

In this section we compare the performance of *EnhancedPKClustering* and *GreedyPKClustering* algorithms for the extended *p*-sensitive *k*-anonymity model. A set of experiments were conducted for the same IM as in the previous section. We also reused the generalization hierarchies of all six quasi-identifier categorical attributes. All three confidential attributes were considered categorical, and their value hierarchies and strong values are depicted in Fig. 16.5 – strong values are bolded and delimited by * characters. We make an observation with regard to the *education* sensitive attribute's hierarchy. This hierarchy is not balanced, but this has

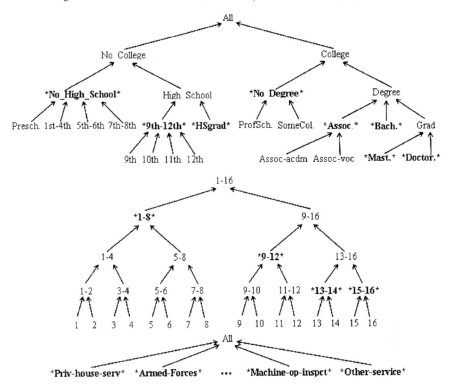

Fig. 16.5 The value hierarchies and strong values for the sensitive categorical attributes *education*, *education-num*, and *occupation*

no influence on the algorithm's performance or results' quality, as long as no cost measures are computed w.r.t. generalization performed according to this hierarchy; its only role is to give guidance about the sensitivity of the values of the confidential attribute *education*.

Another set of experiments used synthetic data sets, where the quasi-identifier and the sensitive attribute values were generated to follow some predefined distributions. For our experiments, we generated four microdata sets using normal and uniform distributions. All four data sets have identical schema *(QI_N; QI_C1; QI_C2; QI_C3; S_C1; S_C2)* where the first attribute *(QI_N)* is a numerical quasi-identifier (*age* like), the next three *(QI_C1; QI_C2; QI_C3)* are categorical quasi-identifiers and the last two *(S_C1* and *S_C2)* are categorical sensitive attributes. The distributions followed by each attribute for the four data sets are illustrated in Table 16.2.

Figure 16.6 depicts the common value generalization hierarchy for the categorical quasi-identifiers of the synthetic data sets. Figure 16.7 shows the value generalization hierarchies and the strong values for the sensitive attributes of the synthetic data sets.

For the numerical attribute we use *age*-like values 0, 1, ..., 99. To generate a uniform distribution for this range we use the mean 99/2 and standard deviation 99/6.

Table 16.2 Data distribution in the synthetic data sets

	All *QI* attributes	All sensitive attributes
Dataset_UU	Uniform	Uniform
Dataset_UN	Uniform	Normal
Dataset_NU	Normal	Uniform
Dataset_NN	Normal	Normal

Fig. 16.6 The value generalization hierarchy of the categorical attributes of the synthetic data sets

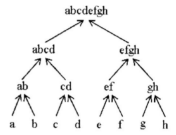

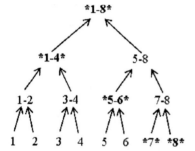

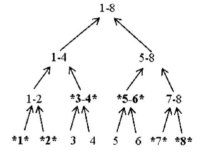

Fig. 16.7 The value generalization hierarchies and strong values for the sensitive attributes of the synthetic data sets: *S_C*1 and *S_C*2

For each categorical attribute we use eight values that are grouped in a hierarchy as shown in Fig. 16.6. To generate a uniform-like distribution for the categorical attributes we use the range 0–8 with mean 8/2 and standard deviation 8/6 and the mapping shown in Table 16.3 (val is the value computed by the generator).

Table 16.3 Mapping between 0 and 8 range and discrete values

val < 1	1 ≤ *val* < 2	2 ≤ *val* < 3	...	6 ≤ *val* < 7	*val* ≤ 8
a	b	c	...	g	h

Next, for each of the five experimental data sets used, we present the *AVG*, *DM*, and some of the execution time cost measure values for each of the two algorithms, *EnhancedPKClustering* and *GreedyPKClustering* for $p = 3$ and different k values (Figs. 16.8, 16.9, 16.10, 16.11, 16.12, and 16.13).

The *AVG* and *DM* results are very similar. We notice that when the *p*-sensitive part is difficult to achieve, the *EnhancedPKClustering* algorithm performs better. These results are similar with the ones obtained for *p*-sensitive *k*-anonymity property.

The following observations are true for both *p*-sensitive *k*-anonymity and its extension. The *GreedyPKClustering* is faster than the *EnhancedPKClustering* algorithm for large values of k, but when it is more difficult to create *p*-sensitivity within each cluster the *EnhancedPKClustering* has a slight advantage. We also notice that the running time of *GreedyPKClustering* algorithm is influenced by the sensitive attributes' distribution.

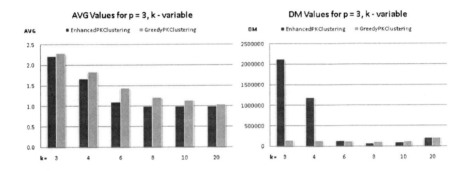

Fig. 16.8 *AVG*, *DM* for *EnhancedPKClustering* and *GreedyPKClustering*, Adult Dataset

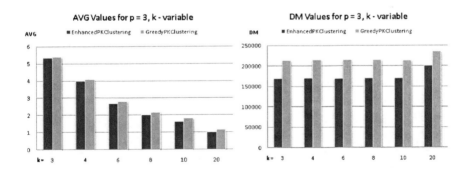

Fig. 16.9 *AVG*, *DM* for *EnhancedPKClustering* and *GreedyPKClustering*, Dataset_NN

16.4 New Enhanced Models Based on *p*-Sensitive *k*-Anonymity

16.4.1 Constrained p-Sensitive k-Anonymity

In general, the existing anonymization algorithms use different quasi-identifier's generalization/tuple suppression strategies in order to obtain a masked microdata that is *k*-anonymous (or satisfies an extension of *k*-anonymity) and conserves as much information intrinsic to the initial microdata as possible. To our knowledge, none of these models limits the amount of generalization that is permitted to be performed for specific quasi-identifier attributes. The ability to limit the amount of allowed generalization could be valuable, and, in fact, indispensable for real-life data sets and applications. For example, for some specific data analysis tasks, available masked microdata with the address information generalized beyond the US state level could be useless. Our approach consists of specifying quasi-identifiers generalization boundaries and achieving *p*-sensitive *k*-anonymity within the imposed boundaries. Using this approach we recently introduced a similar model for *k*-anonymity only, entitled constrained *k*-anonymity. In this subsection we present

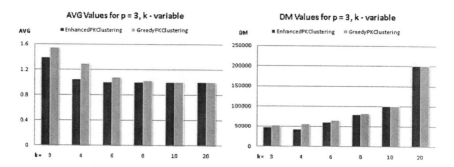

Fig. 16.10 *AVG, DM* for *EnhancedPKClustering* and *GreedyPKClustering*, Dataset_NU

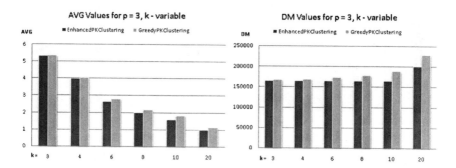

Fig. 16.11 *AVG, DM* for *EnhancedPKClustering* and *GreedyPKClustering*, Dataset_UN

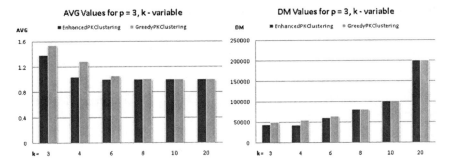

Fig. 16.12 *AVG, DM* for *EnhancedPKClustering* and *GreedyPKClustering*, Dataset_UU

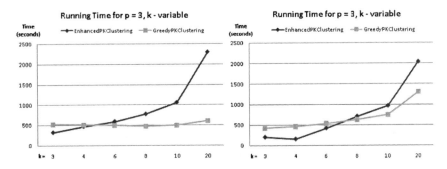

Fig. 16.13 Running time for *EnhancedPKClustering* and *GreedyPKClustering*, Adult, and Dataset_NU

the constrained *p*-sensitive *k*-anonymity privacy model. A complete presentation of constrained *k*-anonymity can be found in [14].

In order to specify a generalization boundary, we introduced the concept of a maximal allowed generalization value that is associated with each quasi-identifier attribute value. This concept is used to express how far the owner of the data thinks that the quasi-identifier's values could be generalized, such that the resulted masked microdata would still be useful. Limiting the amount of generalization for quasi-identifier attribute values is a necessity for various uses of the data. The data owner is often aware of the way various researchers are using the data and, as a consequence, he/she is able to identify maximal allowed generalization values. For instance, when the released microdata is used to compute various statistical measures related to the US states, the data owner will select the states as maximal allowed generalization values.

Definition 16.9 (*Maximal Allowed Generalization Value*). Let Q be a quasi-identifier attribute (categorical or numerical), and $\mathcal{H}_Q$ its predefined value generalization hierarchy. For every leaf value $v \in \mathcal{H}_Q$, the **maximal allowed generalization value** of v, $MAGVal(v)$, is the value (leaf or not-leaf) in $\mathcal{H}_Q$ situated on the path from v to the root, such that

- for any released microdata, the value v is permitted to be generalized only up to $MAGVal(v)$ and
- when several $MAGVals$ exist on the path between v and the hierarchy root, then the $MAGVal(v)$ is the first $MAGVal$ that is reached when following the path from v to the root node.

Figure 16.14 contains an example of defining $MAGVals$ for a subset of values for the *Location* attribute.

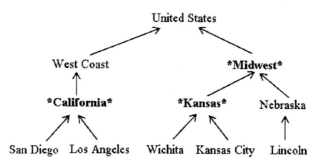

Fig. 16.14 Examples of maximal allowed generalization values

The $MAGVals$ for the leaf values "San Diego" and "Lincoln" are "California," and, respectively, "Midwest" (the maximal allowed generalization values are bolded and marked by * characters that delimit them). This means that the quasi-identifier *Location*'s value "San Diego" may be generalized to itself or "California," but not to "West Coast" or the "United States." Also, "Lincoln" may be generalized to itself, "Nebraska," or "Midwest," but it may not be generalized to the "United States."

Usually, the data owner has generalization restrictions for most of the quasi-identifiers. If for a particular quasi-identifier attribute Q there are not any restrictions with respect to its generalization, then the $\mathcal{H}_Q$'s root value will be considered the maximal allowed generalization value for all the leaf values.

Definition 16.10 (*Constraint Violation*). We say that the masked microdata $\mathcal{MM}$ has a constraint violation if one quasi-identifier value, v, in $\mathcal{IM}$, is generalized in one tuple in $\mathcal{MM}$ beyond its specific maximal generalization value, $MAGVal(v)$.

Definition 16.11 (*Constrained p-Sensitive k-Anonymity*). The masked microdata $\mathcal{MM}$ satisfies the constrained p-sensitive k-anonymity property if it satisfies p-sensitive k-anonymity and it does not have any constraint violation.

We illustrate the above concept with the following example. The initial microdata set $\mathcal{IM}$ in Table 16.4 is characterized by the following attributes: *Name* and *SSN* are identifier attributes (removed from the $\mathcal{MM}$), *Age* and *Location* are the quasi-identifier attributes, and *Diagnosis* is the sensitive attribute. The attribute *Location* values and their $MAGVals$ are described in Fig. 16.14. *Age* does not have any generalization boundary requirements. This microdata set has to be masked such that the

Table 16.4 An initial microdata set IM

Record	Name	SSN	Age	Location	Diagnosis
r_1	Alice	123456789	20	San Diego	AIDS
r_2	Bob	323232323	40	Los Angeles	Asthma
r_3	Charley	232345656	20	Wichita	Asthma
r_4	Dave	333333333	40	Kansas City	Tuberculosis
r_5	Eva	666666666	40	Wichita	Asthma
r_6	John	214365879	20	Kansas City	Asthma

corresponding masked microdata will satisfy constrained *p*-sensitivity *k*-anonymity, where the user wants that the *Location* attribute values not to be generalized in the masked microdata further than the specified maximal allowed generalization values shown in Fig. 16.14.

Tables 16.5 and 16.6 illustrate two possible masked microdata MM_1 and MM_2 for the initial microdata IM. The first one, MM_1, satisfies 2-sensitive 2-anonymity (it is actually 2-sensitive 3-anonymous), but contradicts constrained 2-sensitive 2-anonymity w.r.t. *Location* attribute's maximal allowed generalization. On the other hand, the second microdata set, MM_2, satisfies constrained 2-sensitive 2-anonymity: every *QI*-cluster consists of at least two tuples, there are two distinct values for the sensitive attribute in each cluster, and none of the *Location* initial attribute's values are generalized beyond its *MAGVal*.

Table 16.5 A masked microdata set MM_1 for the initial microdata IM

Record	Age	Location	Diagnosis
r_1	20	United States	AIDS
r_3	20	United States	Asthma
r_6	20	United States	Asthma
r_2	40	United States	Asthma
r_4	40	United States	Tuberculosis
r_5	40	United States	Asthma

Table 16.6 A masked microdata set MM_2 for the initial microdata IM

Record	Age	Location	Diagnosis
r_1	20–40	California	AIDS
r_2	20–40	California	Asthma
r_3	20–40	Kansas	Asthma
r_4	20–40	Kansas	Tuberculosis
r_5	20–40	Kansas	Asthma
r_6	20–40	Kansas	Asthma

16.4.2 p-Sensitive k-Anonymity in Social Networks

The advent of social networks in the last few years has accelerated the research in this field. Online social interaction has become very popular around the globe and most sociologists agree that this trend will not fade away. Privacy in social networks is still in its infancy, and practical approaches are yet to be developed. K-anonymity model has been recently extended to social networks [8, 27] by requiring that every node (individual) in the social network to be undistinguishable from other $(k-1)$ nodes. While this seems similar with the microdata case, the requirement of indistinguishability includes the similar network (graph) structure.

We consider the social network modeled as a simple undirected graph $G = (\mathcal{N}, \mathcal{E})$, where $\mathcal{N}$ is the set of nodes and $\mathcal{E} \subseteq \mathcal{N} \times \mathcal{N}$ is the set of edges. Each node represents an individual entity. Each edge represents a relationship between two entities.

The set of nodes, $\mathcal{N}$, is described by a set of attributes that are classified into identifier, quasi-identifier, and confidential categories. If we exclude the relationship between nodes, the social network data resembles a microdata set.

We allow only binary relationships in our model. Moreover, we consider all relationships as being of the same type and, as a result, we represent them via unlabeled undirected edges. We consider also this type of relationships to be of the same nature as all the other "traditional" quasi-identifier attributes. We will refer to this type of relationship as the *quasi-identifier relationship*. In other words, the graph structure may be known to an intruder and used by matching it with known external structural information, therefore serving in privacy attacks that might lead to identity and/or attribute disclosure.

To create a p-sensitive k-anonymous social network we reuse the generalization technique for quasi-identifier attributes. For the quasi-identifier relationship we use the generalization approach employed in [26] which consists of collapsing clusters together with their component nodes' structure.

Given a partition of nodes for a social network G, we are able to create an anonymized graph by using generalization information and quasi-identifier relationship generalization (for more details about this generalization see [8].

Definition 16.12 (Masked Social Network). Given an initial social network, modeled as a graph $G = (\mathcal{N}, \mathcal{E})$, and a partition $S = \{cl_1, cl_2, \ldots, cl_v\}$ of the nodes set $\mathcal{N}$, $\bigcup_{j=1}^{v} cl_j = \mathcal{N}$; $cl_i \cap cl_j = \emptyset$, $i, j = 1..v$, $i \neq j$; the corresponding **masked social network** $\mathcal{M}G$ is defined as $\mathcal{M}G = (\mathcal{M}\mathcal{N}, \mathcal{M}\mathcal{E})$, where

- $\mathcal{M}\mathcal{N} = \{Cl_1, Cl_2, \ldots, Cl_v\}$, Cl_i is a node corresponding to the cluster $cl_j \in S$ and is described by the "tuple" $gen(cl_j)$ (the generalization information of cl_j, w.r.t. quasi-identifier attribute set) and the intra-cluster generalization pair $(|cl_j|, |E_{cl_j}|)$ ($|cl|$ – the number of nodes in the cluster cl; $|E_{cl}|$ – the number of edges between nodes from cl);

- $\mathcal{ME} \subseteq \mathcal{MN} \times \mathcal{MN}$; $(Cl_i, Cl_j) \in \mathcal{ME}$ iif $Cl_i, Cl_j \in \mathcal{MN}$ and $\exists X \in cl_j, Y \in cl_j$, such that $(X, Y) \in \mathcal{E}$. Each generalized edge $(Cl_i, Cl_j) \in \mathcal{ME}$ is labeled with the inter-cluster generalization value $|\mathcal{E}_{cl_i, cl_j}|$ (the number of edges between nodes from cl_i and cl_j).

By construction, all nodes from a cluster *cl* collapsed into the generalized (masked) node *Cl* are undistinguishable from each other.

In order to have *p*-sensitive *k*-anonymity property for a masked social network, we need to add two extra conditions to Definition 16.12, first that each cluster from the initial partition is of size at least *k* and second that each cluster has at least *p* distinct values for each sensitive attribute. The formal definition of a masked social network that is *p*-sensitive *k*-anonymous is presented below.

Definition 16.13 (*p*-Sensitive *k*-Anonymous Masked Social Network). A masked social network $\mathcal{MG} = (\mathcal{MN}, \mathcal{ME})$, where $\mathcal{MN} = \{Cl_1, Cl_2, \ldots, Cl_v\}$, and $Cl_j = [gen(cl_j), (|cl_j|, |E_{cl_j}|)]$, $j = 1, \ldots, v$ is *p*-sensitive *k*-anonymous if and only if $|cl_j| \geq k$ for all $j = 1, \ldots, v$ (the *k*-anonymity requirement) and for each sensitive attribute *S* and for each $Cl \in \mathcal{MN}$, the number of distinct values for *S* in *Cl* is greater than or equal to *p* (the *p*-sensitive requirement).

We illustrate the above concept with the following example. We consider a social network $\mathcal{G}$ as depicted in Fig. 16.15. The quasi-identifier attributes are *Zip-Code* and *Gender*, and the sensitive attribute is *Disease*. This social network can be anonymized to comply with 2-sensitive 3-anonymity, and one possible masked social network that corresponds to $\mathcal{G}$ is depicted in Fig. 16.16.

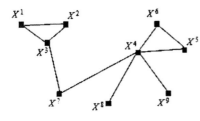

Node	ZipCode	Gender	Disease
X^1	41076	male	Diabetes
X^2	41075	male	Asthma
X^3	41076	male	Asthma
X^4	41099	male	Diabetes
X^5	48201	female	Diabetes
X^6	41075	female	Diabetes
X^7	41099	male	Asthma
X^8	41099	male	Asthma
X^9	41075	female	Asthma

Fig. 16.15 A social network G, its structural information, and its node's attribute values

$cl_2 = \{X^1, X^2, X^3\}$

(3, 3)

$cl_3 = \{X^5, X^6, X^9\}$

(3, 1)

1

3

(3, 2)

$cl_1 = \{X^4, X^7, X^8\}$

Node	ZipCode	Gender	Disease
X^1	4107*	male	Diabetes
X^2	4107*	male	Asthma
X^3	4107*	male	Asthma
X^4	41099	male	Diabetes
X^7	41099	male	Asthma
X^8	41099	male	Asthma
X^5	4****	female	Diabetes
X^6	4****	female	Diabetes
X^9	4****	female	Asthma

Fig. 16.16 A masked social network MG, its structural information, and its node's attribute values

16.5 Conclusions and Future Work

Our extensive experiments showed that both *GreedyPKClustering* and *Enhanced-PKClustering* produce quality masked microdata that satisfy (extended) *p*-sensitive *k*-anonymity and outperform anonymization algorithms based on global recoding.

The new privacy models are a promising avenue for future research; we currently work on developing efficient algorithms for constrained *p*-sensitive *k*-anonymity and *p*-sensitive *k*-anonymity for social networks models. We expect both *Enhanced-PKClustering* and *GreedyPKClustering* to be adjustable for achieving data anonymization in agreement with both these new models.

Another research direction is to adapt the *EnhancedPKClustering* and *Greedy-PKClustering* for enforcing similar privacy requirements such as (α, k)-anonymity, *l*-diversity.

References

1. N.R. Adam and J.C. Wortmann, Security Control Methods for Statistical Databases: A Comparative Study, *ACM Computing Surveys* 21(4) (1989), pp. 515–556.
2. G. Aggarwal, T. Feder, K. Kenthapadi, R. Motwani, R. Panigrahy, D. Thomas, and A. Zhu, Anonymizing Tables, in: *Proceedings of the International Conference on Database Theory*, 2005, pp. 246–258.
3. R. Agrawal, J. Kiernan, R. Srikant, R. and Y. Xu. Hippocratic Databases, in: *Proceedings of the Very Large Data Base Conference*, 2002, pp. 143–154.
4. R.J. Bayardo and R. Agrawal, Data Privacy through Optimal k-Anonymization, in: *Proceedings of the IEEE International Conference on Data Engineering*, 2005, pp. 217–228.
5. J.W. Byun, A. Kamra, E. Bertino and N. Li, Efficient k-Anonymity using Clustering Techniques, in: *Proceedings of Database Systems for Advanced Applications*, 2006, pp. 188–200.
6. A. Campan and T.M. Truta, Extended P-Sensitive K-Anonymity, *Studia Universitatis Babes-Bolyai Informatica* 51(2) (2006), pp. 19–30.
7. A. Campan, T.M. Truta, J. Miller and R.A. Sinca, Clustering Approach for Achieving Data Privacy, in: *Proceedings of the International Data Mining Conference*, 2007, pp. 321–327.
8. A. Campan and T.M. Truta, A Clustering Approach for Data and Structural Anonymity in Social Networks, in: *Proceedings of the Privacy, Security, and Trust in KDD Workshop*, 2008.
9. D. Lambert, Measures of Disclosure Risk and Harm, *Journal of Official Statistics* 9 (1993), pp. 313–331.
10. K. LeFevre, D. DeWitt and R. Ramakrishnan, Incognito: Efficient Full-Domain K-Anonymity, in: *Proceedings of the ACM SIGMOD*, 2005, pp. 49–60.
11. K. LeFevre, D. DeWitt and R. Ramakrishnan, Mondrian Multidimensional K-Anonymity, in: *Proceedings of the IEEE International Conference on Data Engineering*, 2006, 25.
12. N. Li, T. Li and S. Venkatasubramanian, T-Closeness: Privacy Beyond k-Anonymity and l-Diversity, in: *Proceedings of the IEEE International Conference on Data Engineering*, 2007, pp. 106–115.
13. A. Machanavajjhala, J. Gehrke and D. Kifer, L-Diversity: Privacy beyond K-Anonymity, in: *Proceedings of the IEEE International Conference on Data Engineering*, 2006, 24.
14. J. Miller, A. Campan and T.M. Truta, Constrained K-Anonymity: Privacy with Generalization Boundaries, in: *Proceedings of the Practical Preserving Data Mining Workshop*, 2008.
15. M.C. Mont, S. Pearson and R. Thyne, A Systematic Approach to Privacy Enforcement and Policy Compliance Checking in Enterprises, in: *Proceedings of the Trust and Privacy in Digital Business Conference*, 2006, pp. 91–102.

16. MSNBC, Privacy Lost, 2006, *Available online at http://www.msnbc.msn.com/id/15157222*.
17. D.J. Newman, S. Hettich, C.L. Blake and C.J. Merz, UCI Repository of Machine Learning Databases, UC Irvine, 1998, *Available online at www.ics.uci.edu/ mlearn/MLRepository.html*.
18. P. Samarati, Protecting Respondents Identities in Microdata Release, *IEEE Transactions on Knowledge and Data Engineering* 13(6) (2001), pp. 1010–1027.
19. L. Sweeney, k-Anonymity: A Model for Protecting Privacy, *International Journal on Uncertainty, Fuzziness, and Knowledge-based Systems* 10(5) (2002), pp. 557–570.
20. L. Sweeney, Achieving k-Anonymity Privacy Protection Using Generalization and Suppression, *International Journal on Uncertainty, Fuzziness, and Knowledge-based Systems* 10(5) (2002), pp. 571–588.
21. T.M. Truta and V. Bindu, Privacy Protection: P-Sensitive K-Anonymity Property, in: *Proceedings of the ICDE Workshop on Privacy Data Management*, 2006, 94.
22. T.M. Truta, A. Campan and P. Meyer, Generating Microdata with P-Sensitive K-Anonymity Property, in: *Proceedings of the VLDB Workshop on Secure data Management*, 2007, pp. 124–141.
23. L. Willemborg and T. Waal (ed), Elements of Statistical Disclosure Control, Springer Verlag, New York, 2001.
24. R.C.W. Wong, J. Li, A.W.C. Fu and K. Wang, (α, k)-Anonymity: An Enhanced k-Anonymity Model for Privacy-Preserving Data Publishing, in: *Proceedings of the ACM International Conference on Knowledge Discovery and Data Mining*, 2006, pp. 754–759.
25. R.C.W. Wong, J. Li, A.W.C. Fu and J. Pei, Minimality Attack in Privacy-Preserving Data Publishing, in: *Proceedings of the Very Large Data Base Conference*, 2007, pp. 543–554.
26. X. Xiao and Y. Tao, Personalized Privacy Preservation, in: *Proceedings of the ACM SIGMOD*, 2006, pp. 229–240.
27. B. Zhou and J. Pei, Preserving Privacy in Social Networks against Neighborhood Attacks, in: *Proceedings of the IEEE International Conference on Data Engineering*, 2008, pp. 506–515.

Chapter 17
Privacy-Preserving Random Kernel Classification of Checkerboard Partitioned Data

Olvi L. Mangasarian and Edward W. Wild

Abstract We propose a privacy-preserving support vector machine (SVM) classifier for a data matrix A whose input feature columns as well as individual data point rows are divided into groups belonging to different entities. Each entity is unwilling to make public its group of columns and rows. Our classifier utilizes the entire data matrix A while maintaining the privacy of each block. This classifier is based on the concept of a random kernel $K(A, B')$ where B' is the transpose of a random matrix B, as well as the reduction of a possibly complex pattern of data held by each entity into a checkerboard pattern. The proposed nonlinear SVM classifier, which is public but does not reveal any of the privately held data, has accuracy comparable to that of an ordinary SVM classifier based on the entire set of input features and data points all made public.

17.1 Introduction

Recently there has been wide interest in privacy-preserving support vector machine (SVM) classification. Basically the problem revolves around generating a classifier based on data, parts of which are held by private entities who, for various reasons, are unwilling to make it public.

Ordinarily, the data used to generate a classifier is considered to be either owned by a single entity or available publicly. In *privacy-preserving classification*, the data is broken up between different entities, which are unwilling or unable to disclose

Olvi L. Mangasarian
Computer Sciences Department, University of Wisconsin, Madison, WI 53706, USA; Department of Mathematics, University of California at San Diego, La Jolla, CA 92093, USA, e-mail: olvi@cs.wisc.edu

Edward W. Wild
Computer Sciences Department, University of Wisconsin, Madison, WI 53706, USA, e-mail: wildt@cs.wisc.edu

R. Stahlbock et al. (eds.), *Data Mining*, Annals of Information Systems 8, DOI 10.1007/978-1-4419-1280-0_17, © Springer Science+Business Media, LLC 2010

their data to the other entities. We present a method by which entities may collaborate to generate a classifier *without* revealing their data to the other entities. This method allows entities to obtain a more accurate classifier while protecting their private data, which may include personal or confidential information. For example, hospitals might collaborate to generate a classifier that diagnoses a disease more accurately but without revealing personal information about their patients. In another example, lending organizations may jointly generate a classifier which more accurately detects whether a customer is a good credit risk, without revealing their customers' data. As such, privacy-preserving classification plays a significant role in data mining in information systems, and the very general and novel approach proposed here serves both a theoretical and a practical purpose for such systems.

When each entity holds its own group of input feature values for all individuals while other entities hold other groups of feature values for the same individuals, the data is referred to as *vertically partitioned*. This is so because feature values are represented by columns of a data matrix while individuals are represented by rows of the data matrix. In [22], privacy-preserving SVM classifiers were obtained for vertically partitioned data by adding random perturbations to the data. In [20, 21], *horizontally partitioned* privacy-preserving SVMs and induction tree classifiers were obtained for data where different entities hold the same input features for different groups of individuals. Other privacy preserving classifying techniques include cryptographically private SVMs [8], wavelet-based distortion [11], and rotation perturbation [2]. More recently [15, 14] a random kernel $K(A, B')$, where B' is the transpose of a random matrix B, was used to handle vertically partitioned data [15] as well as horizontally partitioned data [14].

In this work we propose a highly efficient privacy-preserving SVM (PPSVM) classifier for vertically and horizontally partitioned data that employs a random kernel $K(A, B')$. Thus, the $m \times n$ data matrix A with n features and m data points, each of which in R^n, is partitioned in a possibly complex way among p entities as depicted, for example, among $p = 4$ entities as shown in Fig. 17.1. Our task is to construct an SVM classifier based on the entire data matrix A without requiring the contents of each entity's matrix block be made public.

Our approach will be to first subdivide a given data matrix A that is owned by p entities into a checkerboard pattern of q cells, with $q \geq p$, as depicted, for example in Fig. 17.2. Second, each cell block A_{ij} of the checkerboard will be utilized to generate the random kernel block $K(A_{ij}, B_{.j}')$, where $B_{.j}$ is a random matrix of appropriate dimension. It will be shown in Section 17.2 that under mild assumptions, the random kernel $K(A_{ij}, B_{.j}')$ will safely protect the data block A_{ij} from discovery by entities that do not own it, while allowing the computation of a classifier based on the entire data matrix A.

We now briefly describe the contents of the chapter. In Section 17.2 we present our method for a privacy-protecting linear SVM classifier for checkerboard partitioned data, and in Section 17.3 we do the same for a nonlinear SVM classifier. In Section 17.4 we give computational results that show the effectiveness of our approach, including correctness that is comparable to ordinary SVMs that use the entire data set. Section 17.5 concludes the chapter with a summary and some ideas for future work.

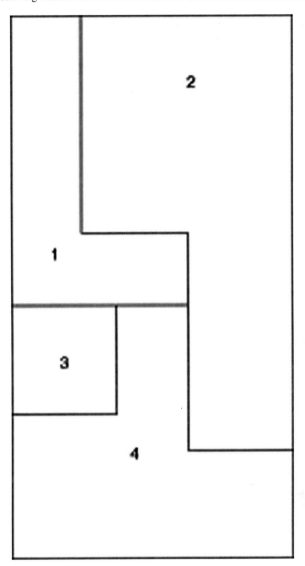

Fig. 17.1 A data matrix A partitioned into $p = 4$ blocks with each block owned by a distinct entity

We describe our notation now. All vectors will be column vectors unless transposed to a row vector by a prime $'$. For a vector $x \in R^n$ the notation x_j will signify either the j th component or the j th block of components. The scalar (inner) product of two vectors x and y in the n-dimensional real space R^n will be denoted by $x'y$. For $x \in R^n$, $\|x\|_1$ denotes the 1-norm: $(\sum_{i=1}^{n}|x_i|)$. The notation $A \in R^{m \times n}$ will

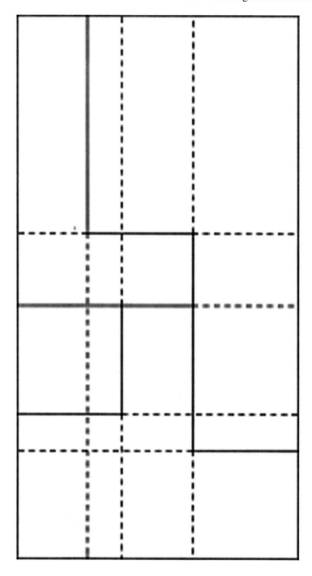

Fig. 17.2 The checkerboard pattern containing $q = 20$ cell blocks generated from the data matrix A of Fig. 17.1

signify a real $m \times n$ matrix. For such a matrix, A' will denote the transpose of A, A_i will denote the i th row or i th block of rows of A and $A_{.j}$ the j th column or the j th block of columns of A. A vector of ones in a real space of arbitrary dimension will be denoted by e. Thus, for $e \in R^m$ and $y \in R^m$ the notation $e'y$ will denote the sum of the components of y. A vector of zeros in a real space of arbitrary dimension will be denoted by 0. For $A \in R^{m \times n}$ and $B \in R^{k \times n}$, a *kernel* $K(A, B')$

maps $R^{m \times n} \times R^{n \times k}$ into $R^{m \times k}$. In particular, if x and y are column vectors in R^n then, $K(x', y)$ is a real number, $K(x', B')$ is a row vector in R^k, and $K(A, B')$ is an $m \times k$ matrix. The base of the natural logarithm will be denoted by ε. A frequently used kernel in nonlinear classification is the Gaussian kernel [18, 17, 12] whose ij th element, $i = 1, \ldots \ldots, m$, $j = 1, \ldots \ldots, k$, is given by $(K(A, B'))_{ij} = \varepsilon^{-\mu \| A_i - B._j' \|^2}$, where $A \in R^{m \times n}$, $B \in R^{k \times n}$, and μ is a positive constant. We shall not assume that our kernels satisfy Mercer's positive definiteness condition [18, 17, 3]; however, we shall assume that they are separable in the following sense:

$$K([E\ F], [G\ H]') = K(E, G') + K(F, H') \text{ or } K([E\ F], [G\ H]') = K(E, G') \odot K(F, H'),$$
$$(17.1)$$

where the symbol $\odot$ denotes the Hadamard component-wise product of two matrices of the same dimensions [5], $E \in R^{m \times n_1}$, $F \in R^{m \times n_2}$, $G \in R^{k \times n_1}$, and $H \in R^{k \times n_2}$. It is straightforward to show that a linear kernel $K(A, B') = AB'$ satisfies (17.1) with the $+$ sign and a Gaussian kernel satisfies (17.1) with the $\odot$ sign. The abbreviation "s.t." stands for "subject to."

17.2 Privacy-Preserving Linear Classifier for Checkerboard Partitioned Data

The data set that we wish to obtain for a classifier consists of m points in R^n represented by the m rows of the matrix $A \in R^{m \times n}$. The matrix columns of A are partitioned into s vertical blocks of $n_1, n_2, \ldots$ and n_s columns in each block such that $n_1 + n_2 + \ldots + n_s = n$. Furthermore, all of the column blocks are identically partitioned into r horizontal blocks of $m_1, m_2, \ldots \ldots$ and m_r rows in each block such that $m_1 + m_2 + \ldots + m_r = m$. This checkerboard pattern of data similar to that of Fig. 17.2 may result from a more complex data pattern similar to that of Fig. 17.1. We note that each cell block of the checkerboard is owned by a separate entity but with the possibility of a single entity owning more than one checkerboard cell. No entity is willing to make its cell block(s) public. Furthermore, each individual row of A is labeled as belonging to the class $+1$ or -1 by a corresponding diagonal matrix $D \in R^{m \times m}$ of ± 1's. The linear kernel classifier to be generated based on all the data will be a separating plane in R^n:

$$x'w - \gamma = x'B'u - \gamma = 0, \tag{17.2}$$

which classifies a given point x according to the sign of $x'w - \gamma$. Here, $w = B'u$, $w \in R^n$ is the normal to the plane $x'w - \gamma = 0$, $\gamma \in R$, determines the distance of the plane from the origin in R^n, and B is a random matrix in $R^{k \times n}$. The change of variables $w = B'u$ is employed in order to kernelize the data and is motivated by the fact that when $B = A$ and hence $w = A'u$, the variable u is the dual variable for a 2-norm SVM [12]. The variables $u \in R^k$ and $\gamma \in R$ are to be determined by an optimization problem such that the labeled data A satisfy, to the extent possible, the separation condition:

$$D(AB'u - e\gamma) \geq 0. \tag{17.3}$$

This condition (17.3) places the $+1$ and -1 points represented by A on opposite sides of the separating plane (17.2). In general, the matrix B which determines a transformation of variables $w = B'u$ is set equal to A. However, in reduced support vector machines [10, 7] $B = \bar{A}$, where $\bar{A}$ is a submatrix of A whose rows are a small subset of the rows of A. However, B can be a random matrix in $R^{\bar{m} \times n}$ with $n \leq \bar{m} \leq m$ if $m \geq n$ and $\bar{m} = m$ if $m \leq n$. This random choice of B holds the key to our privacy-preserving classifier and has been used effectively in SVM classification problems [13]. Our computational results of Section 17.4 will show that there is no substantial difference between using a random B or a random submatrix of $\bar{A}$ of the rows of A as in reduced SVMs [10, 9]. One justification for these similar results can be given for the case when $\bar{m} \geq n$ and the rank of the $\bar{m} \times n$ matrix B is n. For such a case, when B is replaced by A in (17.3), this results in a regular linear SVM formulation with a solution, say $v \in R^m$. In this case, the reduced SVM formulation (17.3) can match the regular SVM term $AA'v$ by the term $AB'u$, since $B'u = A'v$ has a solution u for any v because B' has rank n.

We shall now partition the n columns of the random matrix $B \in R^{\bar{m} \times n}$ into s column blocks with column block $B_{.j}$ containing n_j columns for $j = 1, \ldots, s$. Furthermore, each column block $B_{.j}$ will be generated by entities owning the $m \times n_j$ column block of $A_{.j}$ and is never made public. Thus, we have

$$B = [B_{.1} \; B_{.2} \ldots \ldots B_{.s}]. \tag{17.4}$$

We will show that under the assumption that

$$n_j > \bar{m}, \quad j = 1, \ldots, s, \tag{17.5}$$

the privacy of each checkerboard block privacy is protected.

We are ready to state our algorithm which will provide a linear classifier for the data without revealing privately held checkerboard cell blocks A_{ij}, $i = 1, \ldots, r$, $j = 1, \ldots, s$. The accuracy of this algorithm will, in general, be comparable to that of a linear SVM using a publicly available A instead of merely $A_{.1}B_{.1}', A_{.2}B_{.2}', \ldots \ldots, A_{.s} B_{.s}'$, as will be the case in the following algorithm.

Algorithm 17.2.1 Linear PPSVM Algorithm

(I) *All entities agree on the same labels for each data point, that is $D_{ii} = \pm 1$, $i = 1, \ldots \ldots, m$, and on the magnitude of $\bar{m}$, the number of rows of the random matrix $B \in R^{\bar{m} \times n}$ which must satisfy (17.5).*

(II) *All entities $i = 1, \ldots, r$ sharing the same column block j, $1 \leq j \leq s$, with n_j features must agree on using the same $\bar{m} \times n_j$ random matrix $B_{.j}$ which is privately held by themselves.*

(III) *Each entity $i = 1, \ldots, r$ owning cell block A_{ij} makes public its linear kernel $A_{ij}B_{.j}'$, but not A_{ij}. This allows the public computation of the full linear kernel:*

$$(AB')_i = A_{i1}B_{.1}' + \ldots \ldots + A_{is}B_{.s}', \; i = 1, \ldots, r. \tag{17.6}$$

(IV) A publicly calculated linear classifier $x'Bu - \gamma = 0$ is computed by some standard method such as 1-norm SVM [12, 1] for some positive parameter ν:

$$\min_{(u,\gamma,y)} \quad \nu\|y\|_1 + \|u\|_1$$
$$\text{s.t. } D(AB'u - e\gamma) + y \geq e, \qquad (17.7)$$
$$y \geq 0.$$

(V) For each new $x \in R^n$, the component blocks $x_j'B._j'$, $j = 1,\ldots,s$, are made public from which a public linear classifier is computed as follows:

$$x'B'u - \gamma = (x_1'B._1' + x_2'B._2' + \ldots\ldots + x_s'B._s')u - \gamma = 0, \qquad (17.8)$$

which classifies the given x according to the sign of $x'Bu - \gamma$.

Remark 17.2.2 *Note that in the above algorithm no entity ij which owns cell block A_{ij} reveals its data set nor its components of a new data point x_j. This is so because it is impossible to compute the $m_i n_j$ numbers constituting $A_{ij} \in R^{m_i \times n_j}$ given only the $m_i \bar{m}$ numbers constituting $(A_{ij}B._j') \in R^{m_i \times \bar{m}}$, because $m_i n_j > m_i \bar{m}$. Similarly it is impossible to compute the n_j numbers constituting $x_j \in R^{n_j}$ from the $\bar{m}$ constituting $x_j'B._j' \in R^{\bar{m}}$ because $n_j > \bar{m}$. Hence, all entities share the publicly computed linear classifier (17.8) using AB' and $x'B'$ without revealing either the individual data sets or the new point components.*

We turn now to nonlinear classification.

17.3 Privacy-Preserving Nonlinear Classifier for Checkerboard Partitioned Data

The approach to nonlinear classification is similar to that for the linear one, except that we make use of the Hadamard separability of a nonlinear kernel (17.1) which is satisfied by a Gaussian kernel. Otherwise, the approach is very similar to that of a linear kernel. We state that approach explicitly now.

Algorithm 17.3.1 Nonlinear PPSVM Algorithm

(I) All s entities agree on the same labels for each data point, that is $D_{ii} = \pm 1$, $i = 1,\ldots\ldots,m$, and on the magnitude of $\bar{m}$, the number of rows of the random matrix $B \in R^{\bar{m} \times n}$ which must satisfy (17.5).

(II) All entities $i = 1,\ldots,r$ sharing the same column block j, $1 \leq j \leq s$, with n_j features must agree on using the same $\bar{m} \times n_j$ random matrix $B._j$ which is privately held by themselves.

(III) Each entity $i = 1,\ldots,r$ owning cell block A_{ij} makes public its nonlinear kernel $K(A_{ij}, B._j')$, but not A_{ij}. This allows the public computation of the full nonlinear kernel:

$$K(A,B')_i = K(A_{i1},B_{.1}') \odot \ldots\ldots \odot K(A_{is},B_{.s}'), \ i = 1,\ldots,r. \qquad (17.9)$$

(IV) A publicly calculated linear classifier $K(x',B')u - \gamma = 0$ is computed by some standard method such as 1-norm SVM [12, 1] for some positive parameter ν:

$$\min_{(u,\gamma,y)} \quad \nu\|y\|_1 + \|u\|_1$$
$$s.t. \ D(K(A,B')u - e\gamma) + y \geq e, \qquad (17.10)$$
$$y \geq 0.$$

(V) For each new $x \in R^n$, the component blocks $K(x^{j'},B_{.j}')$, $j = 1,\ldots,s$, are made public from which a public nonlinear classifier is computed as follows:

$$K(x',B')u - \gamma = (K(x_1',B_{.1}') \odot K(x_2',B_{.2}') \odot \ldots\ldots \odot K(x_s',B_{.s}'))u - \gamma = 0, \ (17.11)$$

which classifies the given x according to the sign of $K(x',B')u - \gamma$.

Remark 17.3.2 *Note that in the above algorithm no entity ij which owns cell block A_{ij} reveals its data set nor its components of a new data point x_j. This is so because it is impossible to compute the $m_i n_j$ numbers constituting $A_{ij} \in R^{m_i \times n_j}$ given only the $m_i \bar{m}$ numbers constituting $K(A_{ij},B_{.j}') \in R^{m_i \times \bar{m}}$ because $m_i n_j > m_i \bar{m}$. Similarly it is impossible to compute the n_j numbers constituting $x_j \in R^{n_j}$ from the $\bar{m}$ constituting $K(x_j',B_{.j}') \in R^{\bar{m}}$ because $n_j > \bar{m}$. Hence, all entities share the publicly computed nonlinear classifier (17.11) using $K(A,B')$ and $K(x',B')$ without revealing either the individual data sets or the new point components.*

Before turning to our computational results, it is useful to note that Algorithms 17.2.1 and 17.3.1 can be used easily with other kernel classification algorithms instead of the 1-norm SVM, including the ordinary 2-norm SVM [17], the proximal SVM [4], and the logistic regression [19].

We turn now to our computational results.

17.4 Computational Results

To illustrate the effectiveness of our proposed privacy preserving SVM (PPSVM), we used seven data sets from the UCI Repository [16] to simulate a situation in which data are distributed among several different entities. We formed a checkerboard partition which divided the data into blocks, with each entity owning exactly one block. Each block had data for approximately 25 examples, and we carried out experiments in which there were one, two, four, and eight vertical partitions (for example, the checkerboard pattern in Fig. 17.2 has four vertical partitions). Thus, the blocks in each experiment contained all, one-half, one-fourth, or one-eighth of the total number of features. With one vertical partition, our approach is the same as the technique for horizontally partitioned data described in [14], and these results provide a baseline for the experiments with more partitions. We note that the errors with no sharing represent a worst-case scenario in that a different entity owns each block of data. If entities owned multiple blocks, their errors without sharing might decrease. Nevertheless, it is unlikely that such entities would generally do better than our PPSVM approach, especially in cases in which the PPSVM is close to the ordinary 1-norm SVM.

We compare our PPSVM approach to a situation in which each entity forms a classifier only using its own data, with no sharing, and to a situation in which all entities share the reduced kernel $K(A, \bar{A}')$ without privacy, where $\bar{A}$ is a matrix whose rows are a random subset of the rows of A [10]. Results for one, two, four, and eight vertical partitions are reported in Table 17.1. All experiments were run using the commonly used Gaussian kernel described in Section 17.1. In every result, $\bar{A}$ consisted of 10% of the rows of A randomly selected, while B was a completely random matrix with the same number of columns as A. The number of rows of B was set to the minimum of $n - 1$ and the number of rows of $\bar{A}$, where n is the number of features in the vertical partition. Thus, we ensure that the condition (17.5) discussed in the previous sections holds in order to guarantee that the private data A_{ij} cannot be recovered from $K(A_{ij}, B')$. Each entry of B was selected independently from a uniform distribution on the interval $[0, 1]$. All data sets were normalized so that each feature was between 0 and 1. This normalization can be carried out if the entities disclose only the maximum and minimum of each feature in their data sets. When computing tenfold cross validation, we first divided the data into folds and set up the training and testing sets in the usual way. Then each entity's dataset was formed from the training set of each fold. The accuracies of all classifiers were computed on the testing set of each fold.

To save time, we used the tuning strategy described in [6] to choose the parameters ν of (17.10) and μ of the Gaussian kernel. In this Nested Uniform Design approach, rather than evaluating a classifier at each point of a grid in the parameter space, the classifier is evaluated only at a set of points which is designed to "cover" the original grid to the extent possible. The point from this smaller set on which the classifier does best is then made the center of a grid which covers a smaller range of parameter space, and the process is repeated. Huang et al. [6] demonstrate empirically that this approach finds classifiers with similar misclassification error as a brute-force search through the entire grid. We set the initial range of $\log_{10} \nu$ to $[-7, 7]$ and the initial range of $\log_{10} \mu$ as described in [6]. Note that we set the initial range of $\log_{10} \mu$ independently for each entity using only that entity's examples and features. We used a Uniform Design with 30 runs from http://www.math.hkbu.edu.hk/UniformDesign for both nestings and used leave-one-out cross validation on the training set to evaluate each (ν, μ) pair when the entities did not share and fivefold cross validation on the training set when they did. We used leave-one-out cross validation when not sharing because only about 25 examples were available to each entity in that situation.

To illustrate the improvement in error rate of PPSVM compared to an ordinary 1-norm SVM based only on the data for each entity with no sharing, we provide a graphical presentation of some results in Table 17.1. Figure 17.3 shows a scatterplot comparing the error rates of our data-sharing PPSVM vs. the 1-norm

Table 17.1 Comparison of error rates for entities sharing entire data without privacy through the reduced kernel $K(A, \bar{A}')$, sharing data using our PPSVM approach, and not sharing data. When there are enough features, results are given for situations with one, two, four, and eight vertical partitions using a Gaussian kernel

Dataset Examples × Features	No. of vertical partitions	Rows of B	Ideal error using entire data without privacy $K(A, \bar{A}')$	PPSVM error sharing protected data $K(A, B')$	Error using individual data without sharing $K(A_{is}, A_{is}')$
Cleveland heart (CH)	1	12	0.17	0.15	0.24
297 × 13	2	5	0.19	0.19	0.28
	4	2	0.17	0.24	0.30
Ionosphere (IO)	1	33	0.07	0.09	0.19
351 × 34	2	16	0.06	0.11	0.20
	4	7	0.05	0.17	0.21
	8	3	0.06	0.26	0.24
WDBC (WD)	1	29	0.03	0.03	0.11
569 × 30	2	14	0.02	0.04	0.10
	4	6	0.03	0.06	0.12
	8	2	0.03	0.11	0.16
Arrhythmia (AR)	1	45	0.21	0.27	0.38
452 × 279	2	45	0.22	0.28	0.36
	4	45	0.23	0.27	0.40
	8	33	0.24	0.29	0.40
Pima Indians (PI)	1	7	0.23	0.25	0.36
768 × 8	2	3	0.23	0.31	0.35
	4	1	0.23	0.34	0.38
Bupa liver (BL)	1	5	0.30	0.40	0.42
345 × 6	2	2	0.30	0.42	0.42
German credit (GC)	1	23	0.24	0.24	0.34
1000 × 24	2	11	0.24	0.29	0.34
	4	5	0.24	0.30	0.34
	8	2	0.24	0.30	0.33

no-sharing-reduced SVM using Gaussian kernels. The diagonal line in both figures marks equal error rates. Note that points below the diagonal line represent data sets for which PPSVM has a lower error rate than the average error of the entities using only their own data. Figure 17.3 shows a situation in which there are two vertical partitions of the data set, while Fig. 17.4 shows a situation in which there are four vertical partitions. Note that in Fig. 17.3, our PPSVM approach has a lower error rate for six of the seven data sets, while in Fig. 17.4, PPSVM has a lower error rate on all six data sets.

17.5 Conclusion and Outlook

We have proposed a linear and a nonlinear privacy-preserving SVM classifier for a data matrix, arbitrary blocks of which are held by various entities that are unwilling to make their blocks public. Our approach divides the data matrix into a

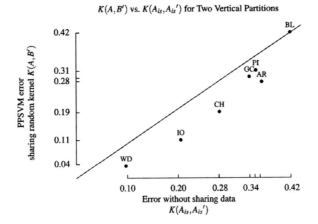

Fig. 17.3 Error rate comparison of our PPSVM with a random kernel $K(A,B')$ vs. 1-norm nonlinear SVMs sharing no data for checkerboard data with two vertical partitions. For points below the diagonal, PPSVM has a better error rate. The diagonal line in each plot marks equal error rates. Each point represents the result for the data set in Table 17.1 corresponding to the letters attached to the point

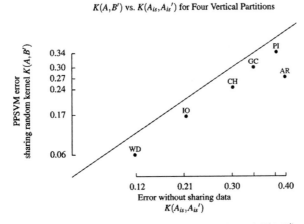

Fig. 17.4 Error rate comparison of our PPSVM with a random kernel $K(A,B')$ vs. 1-norm nonlinear SVMs sharing no data for checkerboard data with four vertical partitions. For points below the diagonal, PPSVM has a better error rate. The diagonal line in each plot marks equal error rates. Each point represents the result for the dataset in Table 17.1 corresponding to the letters attached to the point. Note that there are not enough features in the Bupa Liver dataset for four vertical partitions

checkerboard pattern and then creates a linear or a nonlinear kernel matrix from each cell block of the checkerboard together with a suitable random matrix that preserves the privacy of the cell block data. Computational comparisons indicate that the accuracy of our proposed approach is comparable to full and reduced data classifiers. Furthermore, a marked improvement of accuracy is obtained by the privacy-preserving SVM compared to classifiers generated by each entity using its own data alone. Hence, by making use of a random kernel for each cell block, the proposed approach succeeds in generating an accurate classifier based on privately held data without revealing any of that data.

Future work will entail combining our approach with other ones such as those of rotation perturbation [2], cryptographic approach [8], and data distortion [11].

Acknowledgments The research described in this Data Mining Institute Report 08-02, September 2008, was supported by National Science Foundation Grant IIS-0511905.

References

1. P. S. Bradley and O. L. Mangasarian. Feature selection via concave minimization and support vector machines. In J. Shavlik, editor, *Proceedings 15th International Conference on Machine Learning*, pages 82–90, San Francisco, California, 1998. Morgan Kaufmann. ftp://ftp.cs.wisc.edu/math-prog/tech-reports/98-03.ps.

2. K. Chen and L. Liu. Privacy preserving data classification with rotation perturbation. In *Proceedings of the Fifth International Conference of Data Mining (ICDM'05)*, pages 589–592. IEEE, 2005.

3. N. Cristianini and J. Shawe-Taylor. *An Introduction to Support Vector Machines*. Cambridge University Press, Cambridge, 2000.

4. G. Fung and O. L. Mangasarian. Proximal support vector machine classifiers. In F. Provost and R. Srikant, editors, *Proceedings KDD-2001: Knowledge Discovery and Data Mining*, August 26–29, 2001, San Francisco, CA, pages 77–86, New York, 2001. Association for Computing Machinery. ftp://ftp.cs.wisc.edu/pub/dmi/tech-reports/01-02.ps.

5. R. A. Horn and C. R. Johnson. *Matrix Analysis*. Cambridge University Press, Cambridge, England, 1985.

6. C.-H. Huang, Y.-J. Lee, D.K.J. Lin, and S.-Y. Huang. Model selection for support vector machines via uniform design. In *Machine Learning and Robust Data Mining of Computational Statistics and Data Analysis*, Amsterdam, 2007. Elsevier Publishing Company. http://dmlab1.csie.ntust.edu.tw/downloads/papers/UD4SVM013006.pdf.

7. S.Y. Huang and Y.-J. Lee. Theoretical study on reduced support vector machines. Technical report, National Taiwan University of Science and Technology, Taipei, Taiwan, 2004. yuh-jye@mail.ntust.edu.tw.

8. S. Laur, H. Lipmaa, and T. Mielikäinen. Cryptographically private support vector machines. In *KDD '06: Proceedings of the 12th ACM SIGKDD international conference on Knowledge discovery and data mining*, pages 618–624, New York, NY, USA, 2006. ACM.

9. Y.-J. Lee and S.Y. Huang. Reduced support vector machines: A statistical theory. *IEEE Transactions on Neural Networks*, 18:1–13, 2007.

10. Y.-J. Lee and O. L. Mangasarian. RSVM: Reduced support vector machines. In *Proceedings First SIAM International Conference on Data Mining*, Chicago, April 5–7, 2001, CD-ROM, 2001. ftp://ftp.cs.wisc.edu/pub/dmi/tech-reports/00-07.pdf.

11. L. Liu, J. Wang, Z. Lin, and J. Zhang. Wavelet-based data distortion for privacy-preserving collaborative analysis. Technical Report 482-07, Department of Computer Science, University of Kentucky, Lexington, KY 40506, 2007. http://www.cs.uky.edu/jzhang/pub/MINING/lianliu1.pdf.

12. O. L. Mangasarian. Generalized support vector machines. In A. Smola, P. Bartlett, B. Schölkopf, and D. Schuurmans, editors, *Advances in Large Margin Classifiers*, pages

135–146, Cambridge, MA, 2000. MIT Press. ftp://ftp.cs.wisc.edu/math-prog/tech-reports/98-14.ps.

13. O. L. Mangasarian and M. E. Thompson. Massive data classification via unconstrained support vector machines. *Journal of Optimization Theory and Applications*, 131:315–325, 2006. ftp://ftp.cs.wisc.edu/pub/dmi/tech-reports/06-01.pdf.

14. O. L. Mangasarian and E. W. Wild. Privacy-preserving classification of horizontally partitioned data via random kernels. Technical Report 07-03, Data Mining Institute, Computer Sciences Department, University of Wisconsin, Madison, Wisconsin, November 2007. Proceedings of the 2008 International Conference on Data Mining, DMIN08, Las Vegas July 2008, Volume II, 473–479, R. Stahlbock, S.V. Crone and S. Lessman, Editors.

15. O. L. Mangasarian, E. W. Wild, and G. M. Fung. Privacy-preserving classification of vertically partitioned data via random kernels. Technical Report 07-02, Data Mining Institute, Computer Sciences Department, University of Wisconsin, Madison, Wisconsin, September 2007. ACM Transactions on Knowledge Discovery from Data (TKDD), Volume 2, Issue 3, pages 12.1–12.16, October 2008.

16. P. M. Murphy and D. W. Aha. UCI machine learning repository, 1992. www.ics.uci.edu/~mlearn/MLRepository.html.

17. B. Schölkopf and A. Smola. *Learning with Kernels*. MIT Press, Cambridge, MA, 2002.

18. V. N. Vapnik. *The Nature of Statistical Learning Theory*. Springer, New York, second edition, 2000.

19. G. Wahba. Support vector machines, reproducing kernel Hilbert spaces and the randomized GACV. In B. Schölkopf, C. J. C. Burges, and A. J. Smola, editors, *Advances in Kernel Methods – Support Vector Learning*, pages 69–88, Cambridge, MA, 1999. MIT Press. ftp://ftp.stat.wisc.edu/pub/wahba/index.html.

20. M.-J. Xiao, L.-S. Huang, H. Shen, and Y.-L. Luo. Privacy preserving id3 algorithm over horizontally partitioned data. In *Sixth International Conference on Parallel and Distributed Computing Applications and Technologies (PDCAT'05)*, pages 239–243. IEEE Computer Society, 2005.

21. H. Yu, X. Jiang, and J. Vaidya. Privacy-preserving SVM using nonlinear kernels on horizontally partitioned data. In *SAC '06: Proceedings of the 2006 ACM symposium on Applied computing*, pages 603–610, New York, NY, USA, 2006. ACM Press.

22. H. Yu, J. Vaidya, and X. Jiang. Privacy-preserving svm classification on vertically partitioned data. In *Proceedings of PAKDD '06*, volume 3918 of *LNCS: Lecture Notes in Computer Science*, pages 647–656. Springer-Verlag, January 2006.